# Extra Class

## FCC License Preparation for Element 4

BY
GORDON WEST
WB6NOA

FOURTH EDITION

Master Publishing, Inc.

*This book was developed and published by:*
Master Publishing, Inc.
Lincolnwood, Illinois

*Editing by:*
Pete Trotter, KB9SMG
Gerald Luecke, KB5TZY

*Thanks to Fred Maia, W5YI Group, for his assistance with this book. Thanks also to: KE5XV, WB6HOO, Jeanie and Fay, KM4AS, KC2FMH, N9ESM, KD6VLF, KE4KNS, KG4INA, KE5XV, KD5CLN, Alan Carney, John Konen, P.E., KC8JP, Ken Kalbfell, N6HQV, and Jim Ford, N6JF.*

*Printing by:*
Arby Graphic Service
Lincolnwood, Illinois

*Photograph Credit:*
All photographs that do not have a source identification are either courtesy of RadioShack, the author, or Master Publishing, Inc. originals.
Cover photo by Julian Frost, N3JF


Master Publishing, Inc.
7101 N. Ridgeway Avenue
Lincolnwood, Illinois 60712-2621
847/763-0916 (voice)
847/763-0918 (fax)
E-Mail: MasterPubl@aol.com


Visit Master Publishing
on the Internet at:
www.masterpublishing.com



Fourth Edition
9 8 7 6 5 4 3 2 1

# Table of Contents



## QUESTION POOL NOMENCLATURE

The latest nomenclature changes recommended by the Volunteer Examiner Coordinator's Question Pool Committee (QPC) for the Element 4 question pool effective July 1, 2002 and valid to June 30, 2006 have been incorporated in this book.

Our interpretation and implementation may be different from that of other publishers. We are sure that any slight differences will not contribute to improper understanding of the question or its answer.

## FCC RULES, REGULATIONS AND POLICIES

The NCVEC QPC releases revised question pools on a regular cycle, and deletions as necessary. The FCC releases changes to FCC rules, regulations and policies as they are implemented. This book includes the most recent information released by the FCC when this copy was printed.

# Preface

The Amateur Extra Class is the highest-level Federal Communications Commission license you can hold in the amateur radio service. Achieving the Extra Class license immediately opens up additional sub-bands for voice operation on worldwide frequencies, along with additional, exclusive sub-bands for Morse code. Your Extra Class license gives you full operating privileges across the complete spectrum of every ham radio band!

The old 20-wpm Morse code test requirement prevented thousands of very-technically-capable hams from ever reaching Extra Class status. In fact, it was such an obstacle that only 11 percent of all hams ever achieved an Extra Class license. But all of that changed on April 15, 2000, when the FCC Report & Order of December 30, 1999 restructuring ham licensing went into effect. Morse code proficiency for the General and Extra class licenses was lowered to just 5-wpm. The simple, 5-wpm code test is easy to pass, and if you currently hold a General Class or Advanced Class license, you will not need to take the 5-wpm code test again. All you will need to do is pass a 50-question, Element 4 multiple-choice exam.

Achieving the Extra Class license allows you to administer all amateur radio exams and the 5-wpm code test. As an Extra Class licensee, you may be accredited as an official Volunteer Examiner to prepare and administer ham tests as a member of a 3-person VE team. You also could qualify to become a contact volunteer examiner to help coordinate new test teams in your area.

The new Element 4 Extra Class question pool included in this book is valid for written examinations from July 1, 2002, through June 30, 2006. Although the revised pool now includes 801 total possible questions – up 20 percent from the previous pool – the overall question pool and 50-question exam have not increased in complexity or math difficulty. This book explains EVERYTHING about the Extra Class:

- ♦ Chapter 1 reviews the new Extra Class license requirements and privileges.
- ♦ Chapter 2 contains all 801 Element 4 questions and answers, along with my fun description of the correct answer.
- ♦ Chapter 3 tells you what to expect on examination day, and reviews latest FCC e-filing requirements.

Welcome Advanced and General Class operators to Extra Class test preparation! Get set for new privileges and increased VE opportunities. I hope to hear you on our top bands very soon!

Gordon West

73!

*Gordon West, WB6NOA*

# 1

# The Extra Class License & Priveleges

## INTRODUCTION

The Extra Class amateur operator/primary station license is the highest-level license in the United States amateur radio service. The Extra Class license is obtained by successive passing of written examinations for Element 2 Technician, Element 3 General, and Element 4 Extra, and the 5-wpm, Element 1 Morse code test required for General Class.

If you are going from your present General Class to Extra Class, you already have passed the required code test, and this book will help you breeze through the new 50-question Extra Class written exam.

If you are presently a grandfathered Advanced Class operator, portions of your upcoming 50-question Extra Class exam should look mighty familiar – more than 20 questions will probably be repeats of what you have already passed on that original Advanced Class examination.

The Extra Class license allows you to give something back to the amateur service as a volunteer conducting examinations. As an Extra Class amateur operator, you may become a lead Volunteer Examiner. You are qualified to prepare and administer *all* levels of amateur radio testing as a member of a 3-person accredited VE team. You also can help recruit and organize other hams to serve on your VE team. This includes grandfathered Advanced Class operators who are permitted to administer Element 2 (Technician) and Element 3 (General) written exams, and the Element 1 code test. It also includes General Class operators, who are permitted to administer the Element 2 (Technician) written exam, and the Element 1 code test. This is a great opportunity to help our hobby grow and prosper!

When you pass your Extra Class examination, you will gain a 25-kHz "window" of *exclusive* CW privileges at the bottom end of each of the 80-, 40-, 20-, and 15-meter bands. You also gain 25-kHz "windows" of *exclusive* Extra Class phone privileges on 75-, 20-, and 15-meters. If you are a General Class operator upgrading to Extra, you will gain a total of 350 kHz of operating privileges on the Advanced and Extra Class bands. Remember, grandfathered Advanced Class operators continue to enjoy their own sub-band privileges indefinitely.

And, as an Extra Class operator, you will be eligible for the AA-ALx2 or 2x1 short-block call signs that are reserved exclusively for those amateurs who have made it to the very highest level of license in the amateur service. You may wish to take advantage of the Vanity Call Sign program, too, to select specific, available call letters of your choice (see Chapter 3 for details on this).

And probably best of all, when you reach Extra Class you do not have to prepare for anymore license upgrades. You're the top!

Before we launch into the privileges of the Extra Class license, we'd like to take a brief look at the history of ham radio licensing in the United States. After all, as an Extra Class operator administering exams and code tests, you should be able to answer those "rookie" questions about the history of our hobby.

## A BRIEF HISTORY OF AMATEUR RADIO LICENSING

**Marconi in 1896**
(Courtesy Marconi Co. Archives)

Amateur radio really got started around the turn of the *last* century when Italian inventor Guglielmo Marconi flashed his first wireless signal across the English Channel in 1899. Two years later, he telegraphed the letter "s" from England to Newfoundland. This was the first successful trans-Atlantic radio transmission. *Marconi considered himself an amateur* and he inspired hundreds of others to experiment with radio communications. He won the 1909 Nobel Prize in physics for his work in wireless telegraphy.

Before government licensing of amateur operators and stations was instituted in 1912, radio amateurs could operate on any wavelength they chose and could even select their own call letters. The Radio Act of 1912 mandated the first Federal licensing of radio stations and banished amateurs to the short wavelengths of less than 200 meters. But these restrictions didn't stop them! Within a few short years, there were thousands of licensed ham operators in the United States.

Since electromagnetic signals do not respect national boundaries, radio is international in scope. National governments enact and enforce radio laws within a framework of international agreements which are overseen by the International Telecommunications Union. The ITU is a worldwide United Nations agency headquartered in Geneva, Switzerland. The ITU divides the radio spectrum into a number of segments or frequency bands. Each band is reserved for a particular use. Amateur radio is fortunate to have many bands allocated to it all across the radio spectrum.

**In the 1920s, Secretary of Commerce Herbert Hoover understood the need to regulate the use of airwaves for radio and used his authority to issue operator licenses.**
Courtesy Herbert Hoover Presidential Library Museum

In the United States, the Federal Communications Commission is the Federal agency responsible for the regulation of wire and radio communications. The FCC further allocates frequency bands to the various services in accordance with the ITU plan – including the Amateur Service – and regulates stations and operators.

By international agreement, in 1927 the alphabet was apportioned among various nations for basic call sign use. The prefix letters K, N and W were assigned to the United States, which also shares the letter A with some other countries.

## The Early Years of Amateur Radio Licensing

In the early years of amateur radio licensing in the U.S., the classes of licenses were designated by the letters "A," "B," and "C." The highest license class with the most privileges was "A."

In 1951, the FCC dropped the letter designations and gave the license classes names. They also added a new Novice Class – a one-year, non-renewable license for beginners that required a 5-wpm Morse code speed proficiency and a 20-question written examination on elementary theory and regulations, with both tests taken before one licensed ham.

In 1967, the Advanced Class was added to the Novice, Technician, General and Extra classes. The General Class required 13-wpm code speed, and the Extra Class required 20-wpm. Each of the five written examinations were progressively more comprehensive and formed what came to be known as the *Incentive Licensing System*. The idea was to get General Class amateurs to upgrade their license. Advanced and Extra Class amateurs were awarded tiny slivers of voice and CW spectrum – the "incentive" to upgrade – in exchange for increased telegraphy skill and electronic knowledge. Many General Class amateurs were furious that they had lost some spectrum privileges and had to pass more examinations to get them back. Some are still angry to this day!

In the '70s, the Technician Class license became very popular because of the number of repeater stations appearing on the air that extended the range of VHF and UHF mobile stations and some very-large, hand-held equipment. It also was very fashionable to be able to patch your mobile radio into the telephone system, a practice which allowed hams to make telephone calls from their automobiles.

**The advent of the Tech No-Code license in 1991 opened the airwaves above 30 MHz to a broader group of amateurs.**

In 1979, the international Amateur Service regulations were changed to permit all countries to waive the manual Morse code proficiency requirement for "...stations making use exclusively of frequencies above 30 MHz." This set the stage for the Technician "no-code" license.

## Self-Testing In The Amateur Service

Among the things that happened in the '80s that encouraged the growth of ham radio was the advent of the volunteer examination program that occurred when the FCC turned over responsibility for the actual test giving to the amateur community. Until 1983, all amateur radio operator examinations were administered by FCC personnel at various FCC Field Offices around the country. The following year, the FCC adopted a two-tier system beneath it called the VEC System to handle license examinations in the amateur service. It also increased the length of the term of amateur radio licenses from five to ten years.

The VEC (Volunteer Examiner Coordinator) System was formed after Congress passed laws that allowed the FCC to accept the services of Volunteer Examiners (or VEs) to prepare and administer amateur service license examinations. The testing activity of VEs is managed by Volunteer Examiner Coordinators (or VECs). A VEC acts as the administrative liaison between the VEs who administer the various ham examinations and the FCC, which grants the license.

A team of three VEs, who must be approved by a VEC, is needed to conduct amateur radio examinations. General Class amateurs may serve as examiners for the Technician Class and the 5-wpm code test. Advanced Class amateurs may administer exam Elements 1, 2 and 3; that is, all except Element 4. The Extra Class written Element 4 may only be administered by a VE who holds an Extra Class license.

For the first couple of years of the VEC System, the FCC handled the development and revision of the examination questions. The questions supposedly were "secret," but word eventually got around as to their content. To combat this situation, the FCC changed to a question pool system that had been successfully used by the Federal Aviation Administration for its written examinations: henceforth, all exam questions would be selected from lists of publicly-available question pools.

In 1986, the FCC turned over responsibility for maintenance of the exam questions to the National Conference of VECs, which appointed a Question Pool Committee (QPC) to develop and revise the various question pools according to a schedule. The QPC is required by the FCC to have at least ten times as many questions in each of the pools as may appear on an examination. As a rule, each of the three question pools is changed once every four years on a rotating schedule. The QPC and the question pool system continues in operation today.

Since the inception of the VE system in 1984, more than one million examinations have been administered to applicants for amateur radio operator licenses under the VEC System at essentially no cost to the government or taxpayer.

## Novice Enhancement

In 1987, there was an attempt to generate additional enthusiasm among entry-level amateurs. The "Novice Enhancement" proceeding allowed beginners to operate on the 220-MHz and 1270-MHz bands at reduced power, and the sub-band on 10 meters for Novices and Technicians was enlarged to 28.1 to 28.5 MHz (CW) and 28.3 to 28.5 MHz (CW and SSB). This provided an incentive for Novices to pass the 13-wpm code test and other written theory exams so they could obtain access to still more worldwide frequencies.

The Novice written Element 2 was increased from 20 to 30 questions, and written Element 3 was split into two parts with the Technician (VHF-oriented) questions being placed into the Element 3-A pool and the General (HF-oriented) questions into the Element 3-B pool. These so-called "old Techs" would ultimately be granted complete credit for the new General Class license since they had been examined on code and HF operation. The new RF safety standards introduced in 1998 further increased the Novice exam questions from 30 to 35, and the Technician and General from 25 to 30 questions each.

By this time, the number of Amateur Service examinations had been increased to eight. Five different license classes could now be obtained by passing combinations of five written and three telegraphy examinations.

A major boost to the hobby occurred in 1991, when the 5-wpm Morse code requirement for the Technician Class was eliminated. New licensees were now permitted to operate on all amateur bands above 30 MHz. Applicants for the no-code Technician license had to pass the 35-question Novice and 30-question Technician class written examinations but, for the first time, not a Morse code test.

Technician Class amateurs who also passed a 5-wpm code test were awarded a Technician-Plus license, now creating a total of six Amateur Service license classes. Besides their 30 MHz and higher no-code frequency privileges, they gained the Novice CW privileges and a sliver of the 10 meter voice spectrum. It was during this time of upgrading in the late '80s and '90s that there were all sorts of different Technician Class categories – "Old Techs," "Classic" Technicians, Technician "No-Coders," "Tech Plus" code, and Technicians with code credit that didn't show up on their older Tech licenses. Much of this confusion continues to this day as Volunteer Examiners try to sort out who passed the original 5-wpm Morse code test, and when!

## The Amateur Service Is Restructured

As we approached the end of the 1990s, the majority of licensed amateur operators realized there was a lot more to HF operation than just passing a Morse code test. In 1998, the FCC began a review of the amateur radio service with the objective of streamlining the licensing process, eliminating unnecessary and duplicated rules, and to possibly reduce the emphasis on the Morse code tests. The result of this review was a complete restructuring of the U.S. amateur service that became effective April 15, 2000. Henceforth, applicants for amateur service licenses will only be able to be examined for three license classes:

- Technician Class – the VHF/UHF entry level license;
- General Class – the HF entry level license, and
- Amateur Extra Class – a technically-oriented senior license.

In addition, there now is only one Morse code test speed at 5 words-per-minute (now called Element 1), which is required for the General Class and Extra Class licenses.

Individuals with licenses issued before April 15, 2000, have been "grandfathered" under the new rules. This means that current Novice and Advanced Class amateurs will be able to modify and renew their licenses indefinitely. Technician-Plus amateur licenses will be renewed as Technician Class, but these licensees will retain their HF operating privileges indefinitely. The FCC elected *not* to change the operating privileges of any class. The previously-mandated ten written exam topics were eliminated and the VECs' Question Pool Committee (QPC) will now decide the content of each the three written examinations. Both the Technician Class Element 2 and General Class Element 3 written examinations contain 35 multiple-choice questions. The Extra Class Element 4 written examination has 50 questions.

Finally, there were no automatic upgrades. Currently-licensed Technicians (pre-1987) with 5-wpm code and Element 3B credit qualify immediately for the General Class license, but must apply at VE session showing evidence of having held a Technician license prior to 1987.

*Table 1-1* details the new amateur service license structure and required examinations with three license classes and one Morse code test. *Table 1-2* details the *former* license structure, which had six classes of license and three Morse code tests. Individuals who hold these former licenses have been "grandfathered" under the new rules and may continue to enjoy their present radio frequency privileges, holding that specific license indefinitely as long as they renew it every 10 years. Comparing the two charts will give you an excellent idea of how much easier the FCC has made it to get to the top ham radio license.

**Table 1-1: Current Amateur License Classes and Exam Requirements (Effective April 15, 2000)**

| License Class | Exam Element | Type of Examination |
|---|---|---|
| Technician Class | 2 | 35-question, multiple-choice written examination. Minimum passing score is 26 questions answered correctly (74%). |
| General Class | 3 | 35-question, multiple-choice written examination. Minimum passing score is 26 questions answered correctly (74%). Also requires passing Element 1 Morse code test. |
| Extra Class | 4 | 50-question, multiple-choice written examination. Minimum passing score is 37 questions answered correctly (74%). |
| Morse Code | 1 | Demonstrate ability to receive Morse code at a 5-word-per-minute rate. |

**Table 1-2. Previous Amateur License Classes and Exam Requirements**
*(Prior to April 15, 2000)*

| License Class | Exam/Test Elements | Type of Examination |
|---|---|---|
| Novice | Element 2 and Element 1A | 35-question written examination and 5-wpm code test |
| Technician | Element 2 and 3A | 65-question written examination in 2 parts (35 Element 2 plus 30 Element 3A questions) (No Morse code requirement) |
| Technician-Plus | Element 2 and Element 3A and Element 1A | 35-question written examination, and 30-question written examination, and 5-wpm code test |
| General | Element 3B and Element 1B | 25-question written examination, and 13-wpm code test |
| Advanced | Element 4A | 50-question written examination (No additional Morse code requirement) |
| Extra | Element 4B and Element 1C | 40-question written examination, and 20-wpm code test |

What is the focus of each of the new written examinations, and how does it relate to gaining expanded amateur radio privileges as you move up the ladder toward your Extra Class license? *Table 1-3* summarizes the subjects covered in each written examination element. *Table 1-4* shows the examination subelement topics, total number of questions within each pool, and the number of questions taken from the various subelements for each of the three written examinations.

***Table 1-3. Question Element Subjects***

| Exam Element | License Class | Subjects |
|---|---|---|
| Element 2 | Technician | Elementary operating procedures, radio regulations, and a smattering of beginning electronics. Emphasis on VHF and UHF operating. |
| Element 3 | General | HF (high-frequency) operating privileges, amateur practices, radio regulations, and a little more electronics. Emphasis is on HF bands. |
| Element 4 | Extra | Basically a technical examination. Covers specialized operating procedures, more radio regulations, formulas and heavy math. Also covers the specifics on amateur testing procedures. |

***Table 1-4. Examination Topic Distribution Over License Classes***

| Written Examination Subelement | Element 2 – Technician Class | | Element 3 – General Class | | Element 4 – Extra Class | |
|---|---|---|---|---|---|---|
| | Pool | Exam | Pool | Exam | Pool | Exam |
| 1 Commission's Rules | 112 | 9 | 66 | 6 | 117 | 7 |
| 2 Operating Procedures | 55 | 5 | 66 | 6 | 65 | 5 |
| 3 Radio Wave Propagation | 33 | 3 | 33 | 3 | 37 | 3 |
| 4 Amateur Radio Practices | 44 | 4 | 55 | 5 | 79 | 5 |
| 5 Electrical Principles | 33 | 3 | 22 | 2 | 140 | 9 |
| 6 Circuit Components | 22 | 2 | 11 | 1 | 87 | 5 |
| 7 Practical Circuits | 22 | 2 | 11 | 1 | 109 | 7 |
| 8 Signals and Emissions | 20 | 2 | 22 | 2 | 73 | 4 |
| 9 Antennas and Feed Lines | 22 | 2 | 44 | 4 | 94 | 5 |
| 0 RF Safety | 31 | 3 | 55 | 5 | 0 | 0 |
| Total Questions | 394 | 35 | 385 | 35 | 801 | 50 |

Well, that completes your history lesson and a look at the new, improved licensing structure that became effective April 15, 2000. Now let's turn our attention to the privileges you'll earn as new Extra Class operator.

**Field Day exercises demonstrate ham radio capabilities**

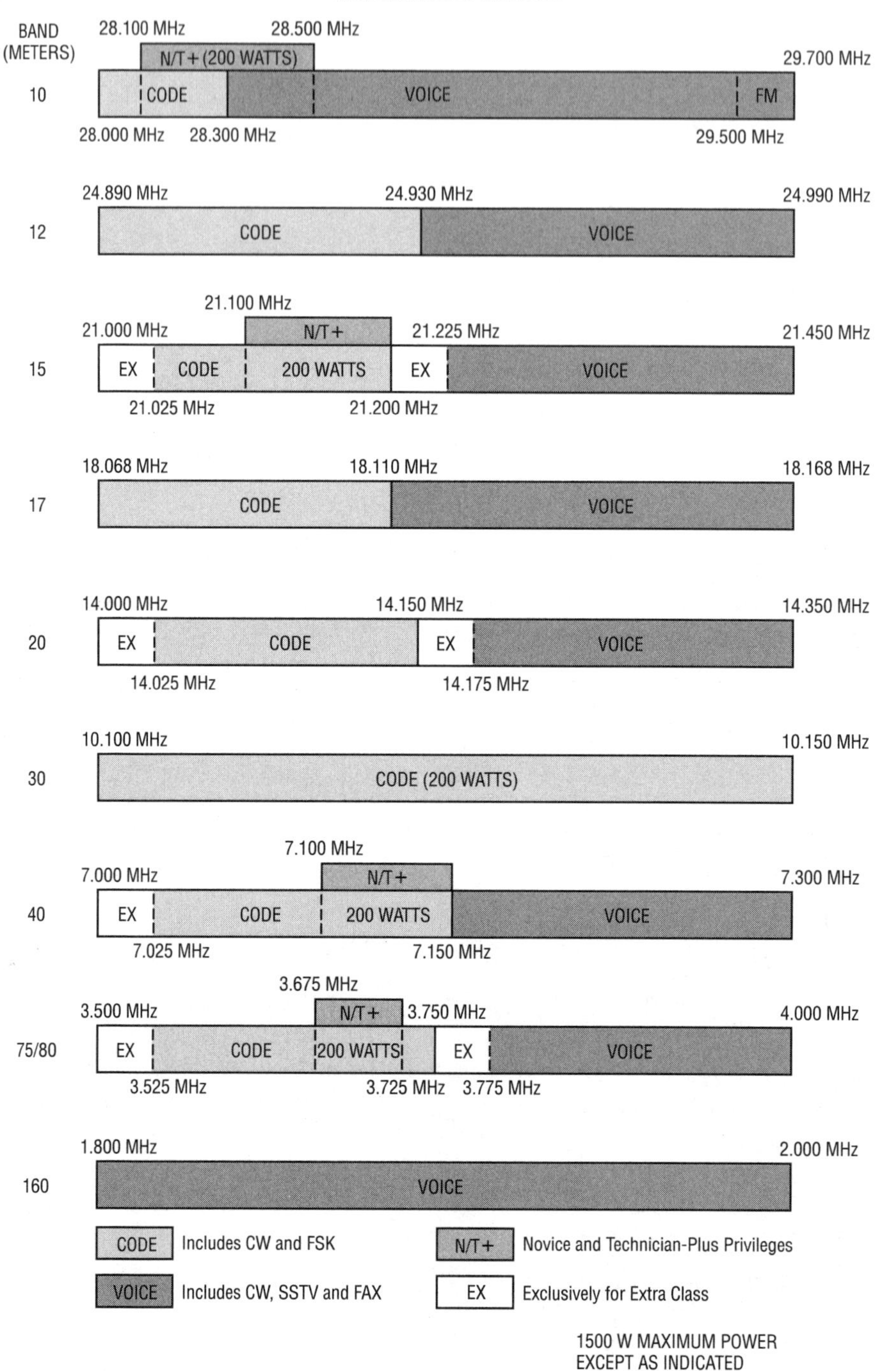
WORLDWIDE SPECTRUM
BAND (METERS)
10
28.100 MHz
28.500 MHz
N/T+(200 WATTS)
29.700 MHz
CODE
VOICE
FM
28.000 MHz
28.300 MHz
29.500 MHz
12
24.890 MHz
24.930 MHz
24.990 MHz
CODE
VOICE
15
21.100 MHz
21.000 MHz
N/T+
21.225 MHz
21.450 MHz
EX
CODE
200 WATTS
EX
VOICE
21.025 MHz
21.200 MHz
17
18.068 MHz
18.110 MHz
18.168 MHz
CODE
VOICE
20
14.000 MHz
14.150 MHz
14.350 MHz
EX
CODE
EX
VOICE
14.025 MHz
14.175 MHz
30
10.100 MHz
10.150 MHz
CODE (200 WATTS)
40
7.100 MHz
7.000 MHz
N/T+
7.300 MHz
EX
CODE
200 WATTS
VOICE
7.025 MHz
7.150 MHz
75/80
3.675 MHz
3.500 MHz
N/T+
3.750 MHz
4.000 MHz
EX
CODE
200 WATTS
EX
VOICE
3.525 MHz
3.725 MHz
3.775 MHz
160
1.800 MHz
2.000 MHz
VOICE
CODE Includes CW and FSK
N/T+ Novice and Technician-Plus Privileges
VOICE Includes CW, SSTV and FAX
EX Exclusively for Extra Class
1500 W MAXIMUM POWER
EXCEPT AS INDICATED

## THE EXTRA CLASS LICENSE

Let's begin by taking a look at *Figure 1-1*, which graphically illustrates your overall Extra Class privileges on the 160-meter medium-frequency (MF) bands from 1.8 to 2.0 MHz, and on the 80-, 40-, 30-, 20-, 17-, 15-, 12-, and 10-meter high frequency (HF) bands from 3.0 to 30 MHz.

Morse code privileges – where only CW and data are allowed – are shown in the designated area on the left side of each band edge. The areas that include voice privileges are shown in the designated areas on the right. In the 80-, 40-, 20-, and 15-meter bands, we illustrate areas exclusively for Extra class that are *not* shared with grandfathered Advanced Class operators. If you are upgrading from Advanced Class, passing the Extra Class exam immediately gives you access to these new Extra Class sub-bands.

If you're upgrading from General Class, you not only gain all Extra Class sub-bands, but you also gain all grandfathered Advanced Class sub-band privileges, too. In other words, as an Extra Class operator, *you get everything!*

Take a look at *Figure 1-1* for Extra Class HF license privileges and see the "EX" – this is where you have *exclusive* voice operating privileges and CW privileges shared with no one other than Extra Class operators.

## EXCLUSIVE EXTRA CLASS LICENSE PRIVILEGES

### 160 Meters: 1.8 MHz – 2.0 MHz.

Your Extra Class privileges are the same as grandfathered Advanced and General on this band. You may operate voice and code from one end of the band to the other. The 160-meter band is great for long-distance nighttime communications. It is located just above the AM broadcast radio frequencies. At night, 160 meters lets you work the world.

### 80 Meters: 3.5 MHz – 4.0 MHz.

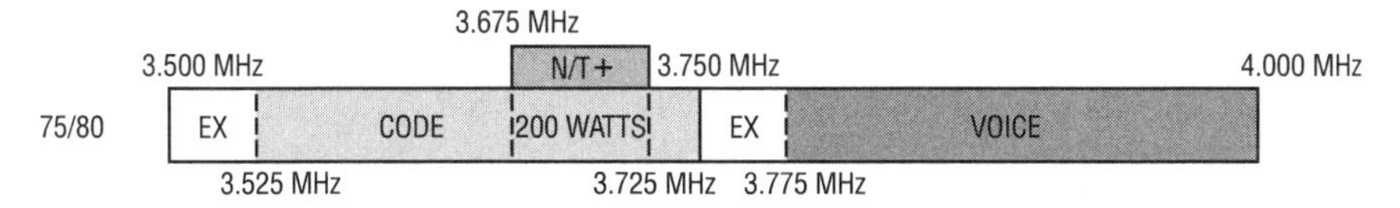


As an Extra Class operator, you get this entire band from bottom to top. Your *exclusive* Extra Class CW privileges, shared only with other Extra Class operators, are from 3500 to 3525 kHz. You also have 25 kHz of exclusive voice privileges from 3750 to 3775 kHz. Only Extra Class operators may use this voice sub-band. During the day, range is limited to about 400 miles. However, at nighttime, you can work well over 6,000 miles away. The 75-meter/80-meter band is a fun one for Extra Class operators with two sections of *exclusive* frequency privileges.

**40 Meters: 7 MHz – 7.3 MHz.**

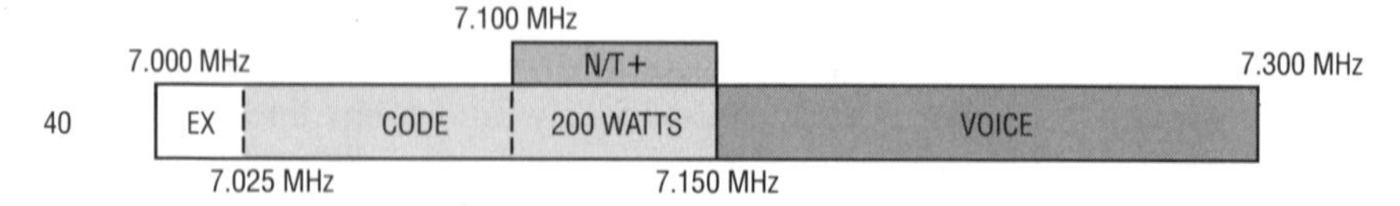


Extra Class operators have an *exclusive* 25 kHz of CW operating at the bottom of the band from 7.0 to 7.025 MHz. Voice privileges begin at 7.150 MHz, and are shared with grandfathered Advanced Class operators, and at and above 7.225 MHz, are also shared with General Class operators. This means the Extra Class operator only gains exclusive privileges for CW on the 40-meter band. There is plenty of CW DX on 40 meters. At nighttime, it's possible to operate both code as well as voice throughout the world on 40 meters. However, 40 meters is shared with worldwide AM shortwave broadcast stations, so at night be prepared to dodge megawatt carriers playing everything from rock and roll to political broadcasts. During daylight hours, the range on 40 meters is limited to approximately 500 miles.

**30 Meters: 10.1 MHz – 10.15 MHz.**

This is a shared band for code only between General, grandfathered Advanced, and Extra Class operators. 30 meters is located just above the 10-MHz WWV time broadcasts on shortwave radio. Voice is not allowed on this band by any class of amateur operator. Maximum PEP power level is limited to 200 watts.

**20 Meters: 14 MHz – 14.350 MHz.**

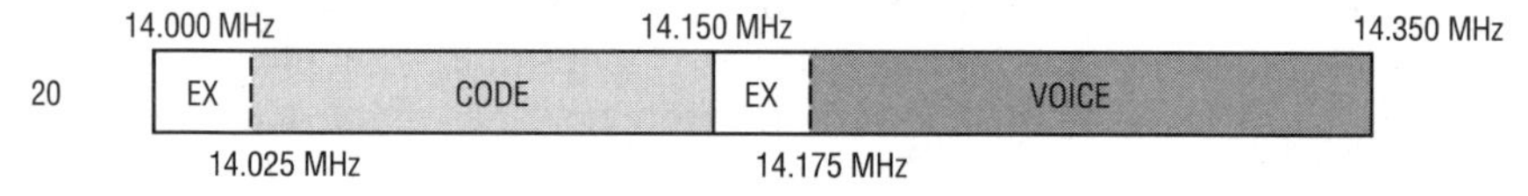


Extra Class operators have two band segments of *exclusive* frequency privileges. For CW, the Extra Class operator has *exclusive* use from 14.0 MHz to 14.025 MHz; for voice, Extra Class operators have *exclusive* use from 14.150 MHz to 14.175 MHz. Extras share the voice portion of the band with Advanced operators up to 14.225 MHz, and from there share with both General and grandfathered Advanced operators to 14.350 MHz. Extra Class operators have access to the entire 20-meter band, from top to bottom.

20 meters is where the *real* DX activity takes place. Almost 24 hours a day, you should be able to work stations in excess of 5,000 miles with a modest antenna set up. If you are a mariner, most maritime mobile operations take place on the 20-meter band. If you are into recreational vehicles (RVs), there are nets all over the country especially for you. And if you enjoy DXing the world and collecting rare QSL cards, 20 meters is where the action is!

## 17 Meters: 18.068 MHz – 18.168 MHz.

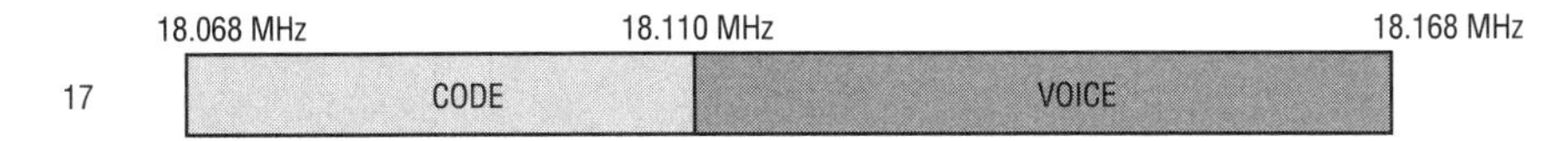


There is plenty of elbow room here with a lot of foreign DX coming in during the day and evenings. While the Extra Class operator does not have any exclusive portions on 17 meters, Extra Class licensees have the entire band from 18.068 MHz all the way up to 18.168 MHz. Voice begins at 18.110 MHz, and is shared with General and grandfathered Advanced operators.

## 15 Meters: 21.0 MHz – 21.450 MHz.

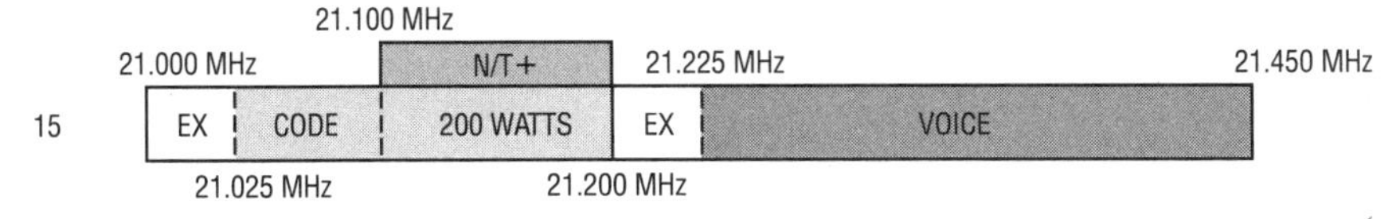


The Extra Class operator has two *exclusive* operating areas shared by no other class of license – *exclusive* CW from 21.0 MHz to 21.025 MHz, and *exclusive* voice from 21.200 MHz to 21.225 MHz. From 21.225 MHz to 21.300 MHz, the voice band is shared with grandfathered Advanced operators, and from 21.3 to 21.450 MHz, General, grandfathered Advanced, and Extra Class operators have voice privileges. Thanks to the two exclusive areas only for Extra Class operators, Extra Class operators love both code as well as voice on 15 meters.

In addition, the 15-meter band is loved by hams throughout the world because it has extremely low noise, and strong signal levels. There is little power line noise on 15 meters, and almost no atmospheric static. Band conditions on 15 meters usually favor daytime and evening contacts in the direction of the sun. Late at night, 15 meters begins to fade away, and you won't get skywave coverage until the next morning. 15 meters is a popular band for mobile operators because antennas are smaller.

## 12 Meters: 24.890 MHz – 24.990 MHz.

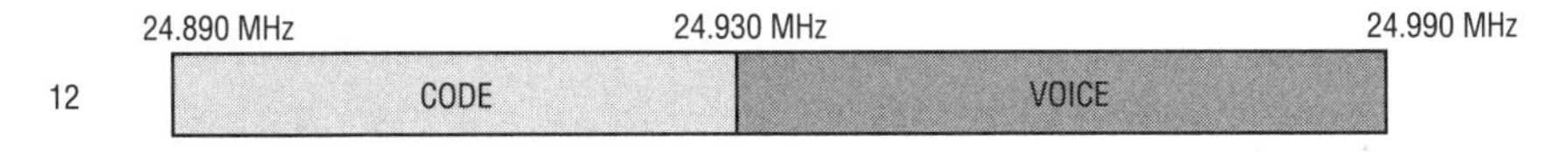


The 12-meter band is a daytime band for the best long-range DX, and the Extra Class operator shares this entire band with General Class and grandfathered Advanced Class operators. CW starts at the bottom of the band at 24.890 MHz, and voice begins in the middle of the band at 24.930 MHz. There is plenty of excitement on the 12-meter band, and antenna requirements are smaller.

**10 Meters: 28.0 MHz – 29.7 MHz.**

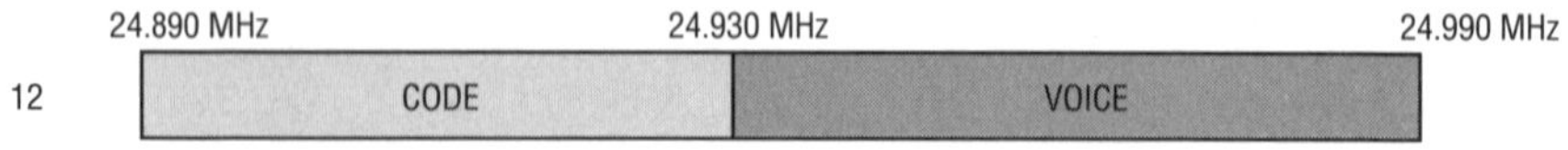

24.890 MHz 24.930 MHz 24.990 MHz

12 CODE VOICE

There is plenty of excitement shared by all classes of amateur operators (except no-code Technician) on the 10-meter band. Worldwide and continental band openings may pop up anytime there are extra amounts of solar activity. And during the summertime, Sporadic-E ionospheric conditions make 1,500-mile band openings exciting for all operators on "10." The Extra Class operator has the entire 10-meter band from top to bottom, with plenty of CW activity at 28.1 MHz, propagation beacons at 28.2 MHz, phone excitement between 28.3 MHz to 29.0 MHz, and all sorts of elbow room, including FM operation, way up there between 29.5 and 29.7 MHz.

## Advanced Class Voice Sub-Bands

If you are upgrading from General to Extra, you also gain full privileges on all Advanced Class sub-bands in addition to your exclusive voice and CW Extra Class sub-bands.

**On 160 meters** you may now operate over the entire 1800 to 2000 kHz band, sharing your operation with General, grandfathered Advanced, and Extra class licensees.

**On 75 meters,** going from General to Extra will give you privileges from 3775 to 3850 kHz, with plenty of room to spread out for voice from 3750 kHz all the way up to 4000 kHz.

**On 40 meters** upgrading to Extra will gain a 100% increase in voice operation elbow room – with total voice operation from 7150 kHz all the way up to 7300 kHz.

**On 20 meters** going from General to Extra Class permits operation on grandfathered Advanced frequencies from 14.175 to 14.225 MHz, with full voice operation from 14.150 to 14.350 MHz.

**On 17 meters** you share the band as an Extra Class operator with Advanced and General class operators with voice privileges from 18.110 to 18.168 MHz.

**On 15 meters** going from General to Extra Class allows you to use Advanced sub-bands from 21.225 to 21.300 MHz, with full Extra Class voice operation from 21.200 to 21.450 MHz.

**On 12 meters** you share voice operation with grandfathered Advanced and General class operators from 24.930 to 24.990 MHz.

**On 10 meters** you share full Extra Class voice privileges with Advanced and General class operators from 28.300 to 29.700 MHz, and also may share voice communications with grandfathered Novice and grandfathered Technician-Plus operators from 28.300 to 28.500 MHz.

What all of this means for the General Class operator going to Extra Class is you gain privileges on both Advanced Class sub-bands as well as *exclusive* Extra Class bands. For the grandfathered Advanced Class operator going to Extra Class, your gain is 150 more kilohertz of *exclusive* Extra Class voice operation beyond your existing grandfathered Advanced Class privileges. Plus 100 more kHz of CW *exclusive* operation!

**6 Meters and Up.** Your Extra Class license allows you unlimited band privileges and unlimited emission privileges shared with all classes of license from Technician Class on up on the following bands:

| *Meters* | *Frequency* |
|---|---|
| 6 meters | 50-54 MHz |
| 2 meters | 144-148 MHz |
| 1.25 meters | 222-225 MHz (219-220 MHz Digital) |
| 70 centimeters | 420-450 MHz |
| 35 centimeters | 902-928 MHz |
| 23 centimeters | 1240-1300 MHz |

**Microwave Bands.** Many Extra Class operators are employed by companies that specialize in microwave electronics. The Extra Class license allows for unlimited privileges and unlimited emission privileges on the following microwave bands:

| *Frequency* | *Frequency* |
|---|---|
| 2300-2310 MHz | 47.0-47.2 GHz |
| 2390-2450 MHz | 75.5-81.0 GHz |
| 3.3-3.5 GHz | 119.98-120.02 GHz |
| 5.65-5.925 GHz | 142-149 GHz |
| 10.0-10.5 GHz (popular X band) | 241-250 GHz |
| 24.0-24.25 GHz | All above 300 GHz |

Look for plenty of Extra Class excitement on the 10-GHz band. This may soon become the new frontier for those amateur operators establishing new records on microwave frequencies.

## BECOME A VOLUNTEER EXAMINER

Here's some additional information to further encourage you to become a Volunteer Examiner once you upgrade to Extra Class.

The Telecommunications Act of 1996 removed many of the burdensome conflict-of-interest, record keeping, and financial-certification requirements previously imposed on VEs and VECs (Volunteer Examiner Coordinators). As a result of these rule changes, there no longer are any regulatory prohibitions preventing VEs or VECs from distributing license preparation (study) materials (with or without profit) to anyone, or from being employed by a company that manufactures or sells amateur radio station equipment or license study material. Amateurs who serve as ham class instructors may now distribute license preparation materials to their students and also serve as Volunteer Examiners for the course examinations. Serving as a VE is a great way to help our hobby grow!

## BE AN EXTRA CLASS ELMER

When you pass your upcoming Extra Class exam, you will have arrived at the TOP! You can share your enthusiasm for our amateur radio service by becoming a Ham Ambassador. The Ham Ambassador program is designed to help spread the word – and training – to hobby radio enthusiasts to encourage them to earn their entry-level ham radio license.

Ham Ambassadors may find kids and scouting as a great place to introduce ham radio and develop amateur radio training classes. Boaters, flyers, and RVers are *all* interested in what amateur radio can do to make their traveling safer and more enjoyable. You might become an Elmer and help them locate club or local weekend classes to earn their entry-level Technician Class license.

**Become a Ham Ambassador and help our hobby grow!**

The new Technician Class license is the ideal starting point for everyone wanting to learn about ham radio. No Morse code test involved! The Technician exam trains would-be hams on techniques of going on the air for the first time, and learning about all the excitement ham radio has to offer, including repeaters and satellites.

As a Ham Ambassador, the industry will support you with free log books, band plan charts, frequency guides, and other training materials. Extra Class Ham Ambassadors will help our service continually grow, and to learn more about how to become a Ham Ambassador and Elmer, call 800-669-9594, and join the outreach program to bring more kids and radio enthusiasts into our hobby and public service.

## SUMMARY

Whether you are upgrading to Extra Class from the grandfathered Advance Class license or the General Class license, you will gain some exciting new privileges on shared and Extra-only voice and CW portions of the bands for worldwide communications. This is where all the rare DX happens, and you will be operating in an area of the lower portion of the sub-bands where General Class operators can only listen in with envy!

But your biggest privilege as an Extra Class operator is the ability for you to work closely with a Volunteer Examination Coordinator and serve as a VE team leader. Now that General Class operators, along with grandfathered Advanced Class operators, may also assist Extra Class operators in administering license exams, your roll as a contact Volunteer Examiner will be an important one for the growth of the testing teams – and the growth of our hobby – throughout the country.

# 2

# Getting Ready for the Written Examination

## ABOUT THIS CHAPTER

This chapter includes the entire, new Element 4 Question Pool questions and answers that can be used to make-up your 50-question Extra Class Element 4 written theory examination *exactly as they will appear on your exam.* Your exam will have 50 questions, and a score of 37 or more correctly-answered questions (74%) will earn you a passing grade – which means you cannot have more than 13 wrong. Answer at least 37 questions correctly, and you pass the Extra Class examination!

> THIS EXTRA CLASS QUESTION POOL IS VALID FROM JULY 1, 2002, THROUGH JUNE 30, 2006.

The Element 4 question pool contains a total of 801 active* questions and multiple-choice answers and distracters (the false answers are called *distracters.*) 50 of these questions will appear on your upcoming Element 4 written examination. All of these examination questions, plus the precise multiple-choice answers – one of which is the correct answer – *are identical to those included in this book.*

The Volunteer Examiners are not permitted to reword the questions, nor are they permitted to change any of the right or wrong answers (but they can change the A B C D *order* of the answers). *Every question in this book, letter for letter, word for word, number for number, and every right and wrong answer, will be exactly the same on your upcoming 50-question Extra Class Element 4 exam.*

In this book, we emphasize the correct answer – and along with each correct answer is your author's discussion on why the answer is indeed the right one. We also give you some hints on how to identify wrong answers that often look very much like the correct one and how to avoid selecting them during the exam. Of course, these explanations *will not appear* on your test papers when you take the Element 4 exam.

***Questions Deleted from Pool**

The initial release of the 2002 Element 4 Question Pool included a total of 806 questions. Following release of the Pool into the public domain, the QPC identified five questions that were either incorrect or based on obsolete information. These questions have been deleted from the pool, but the question numbers remain in the sequence. In this book, you will see these questions identified with the explanation: "Question deleted from Element 4 pool by the QPC." The deleted questions are: E2E09, E2E10, E6B17, E8B09, and E8B16. These five questions are *not* included in the question count shown in Table 2-1, or elsewhere in this book.

The new Element 4 Extra Class question pool was developed by refining the Extra Class question pool issued in April, 2000, inserting some previous Advanced and Extra class questions, editing out duplicate and obsolete questions, and then adding a handful of brand new questions to keep the pool up-to-date with newer technology and current FCC rules and regulations. This question pool was carefully developed by the National Conference of Volunteer Exam Coordinators' Question Pool Committee, under the leadership of Scotty Neustadter, W4WW. Other members of the QPC are Bart Jahnke, W9JJ, Fred Maia, W5YI, and John Johnson, W3BE. *Congratulations, gentlemen, on another excellent question pool update!*

## WHAT THE EXAM CONTAINS

Again, your exam will be comprised of 50 questions taken from the Element 4 question pool. The new pool is divided into 9 sub-elements. *Table 2-1* shows you the subelement topics, the total number of questions for each sub-element, and the number of questions from each subelement that will appear on your exam. Carefully study *Table 2-1* – it will help you understand how the questions are distributed among the 9 sub-elements and how your exam will be constructed.

***Table 2-1. Question Distribution for the Extra Class Element 4 Exam***

| Subelement Number | Subelement Topic | No. of Questions in Pool | No. of Questions on Exam |
|---|---|---|---|
| E1 | Commission's Rules (FCC rules for the Amateur Radio services) | 117 | 7 |
| E2 | Operating Procedures (Amateur station operating procedures) | 65 | 5 |
| E3 | Radio Wave Propagation (Radio wave propagation characteristics) | 37 | 3 |
| E4 | Amateur Radio Practices (Amateur Radio practices) | 79 | 5 |
| E5 | Electrical Principles (Electrical principles as applied to amateur station equipment) | 140 | 9 |
| E6 | Circuit Components (Amateur station equipment circuit components) | 87 | 5 |
| E7 | Practical Circuits (Practical circuits employed in amateur station equipment) | 109 | 7 |
| E8 | Signals and Emissions (Signals and emissions transmitted by amateur stations) | 73 | 4 |
| E9 | Antennas and Feed Lines (Amateur station antennas and feed lines) | 94 | 5 |
| Total | | 801 | 50 |

(Titles in parentheses are the subelement titles on which you will be examined.)

## QUESTION POOL VALID THROUGH JUNE 30, 2006

The Element 4 question pool in this book is valid from July 1, 2002, until June 30, 2006. During this four-year period, the questions and answers are frozen. The exact numerical values found in each Extra Class question and answer will remain precisely the same. If FCC rules change or advancements in technology cause any question to become obsolete, that question will be eliminated from your test and there will be no surprises in the exam room.

The new 2002 Extra Class question pool has 136 more questions than the previous Extra Class pool. The majority of the additional questions are actually taken from the old Advanced Class question pool that became obsolete when the FCC restructured licensing in December, 1999.

How many *absolutely* new questions in the 2002 Extra Class question pool? There are only about 20, mostly found in Rules & Regs. Here is what is new:

Rules on low-power spurious emissions
More ham satellite questions
More volunteer examiner questions
Spread-spectrum emissions
Fast- and slow-scan TV
Contesting
APRS and digital communications
Phase shift keying bandwidth
Message-forwarding rules
Liquid crystal displays
Active audio filters

## QUESTION CODING

Each question in the Element 4 Extra Class pool is assigned a coded number using a system of numbers and letters. *Figure 2-1* explains the coding for question E5G09. "E" stands for Extra Class Element 4 question pool. "5" indicates subelement 5, Electrical Principles. "G" indicates the sub-element group on Circuit Q; Reactive Power; Power Factor. "09" indicates the ninth question dealing with this topic.

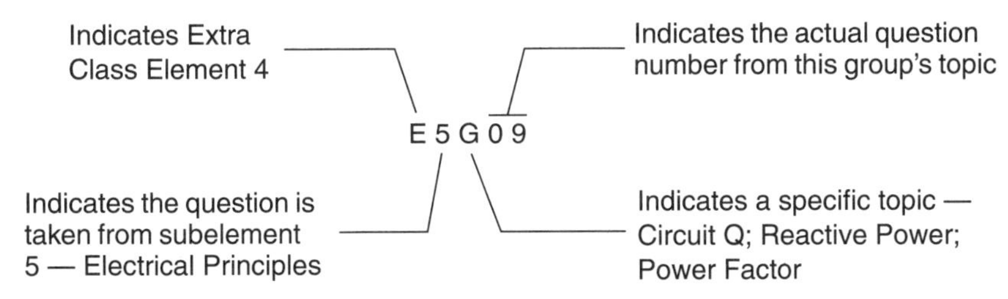


**Figure 2-1. Examination Question Coding**

If you carefully study the numbering scheme, you will find that each subelement is divided into topic groups, and there will be one question on your examination taken from each topic group. If there is a specific group of questions which are giving you a hard time, study them later. Only one question will be taken from any one group.

To facilitate in identifying subelement groups, we describe the subelement contents by individual groups of questions. This is good news for those of you who may be a bit rusty on a specific topic, knowing ahead of time that your upcoming Extra Class examination will only have one question out of each topic group.

## STUDY HINTS

You're probably saying: "801 questions and answers is an overwhelming number of questions and answers to memorize for the new Extra Class exam!" But good news – almost every question is asked at least twice with just slightly different phrasing. Many of the math questions are based on single-formulas; and once you know the formula, you can easily answer the 4 or 5 questions based on that formula.

The mathematical formulas that you will need for the upcoming Extra class test are noted with an ▶ in the question pool that follows, and they are summarized in the Appendix on pages 249 to 253. Memorize the formulas, understand how to use them, and make sure you know the correct units to use. Many times you may have to convert kilohertz (kHz) to megahertz (MHz), or vice versa.

*Work your book!* Put a check mark beside questions you have down cold, and circle those that may need a little bit more study. Highlight those that appear to be brain-busters, and figure out ways to memorize the correct answers. Begin to memorize the formulas, and learn how to apply them to arrive at a correct answer. In many cases, there are handy shortcuts to help you through the longer formulas, and a little humor now and then should keep you on track.

If you study the questions and answers for a half hour in the morning, and a half hour in the evening, it should take you approximately 30 days to prepare for the Extra Class exam. If you are a grandfathered Advanced Class operator, it may take you only 15 days to prepare for your upgrade. One week before the exam, highlight each question's correct answer and spend that week only looking at the question and highlighted correct answer.

## ADDITIONAL STUDY MATERIALS

You are encouraged to use additional textbooks as part of your Extra Class theory preparation to help you better understand everything that goes into the questions and answers. There are many technical books available from your local library, ham radio dealer, and electronics specialty store. Your author has recorded cassettes on Element 4 examination theory available from W5YI Group and at Amateur Radio dealers. They may prove useful to you.

## EXAMINATION QUESTIONS

Now it's time to scan through the examination questions. If you are a grandfathered Advanced Class operator taking the new Extra Class exam, many of the new questions are actually "old questions" you studied when you took your original Advanced exam. If you are going from General Class to Extra Class, you will see that the Question Pool Committee has selected the best of the old Advanced questions and the best of the Extra class questions for your new, 50-question Extra Class Element 4 written exam.

Are you ready to get started preparing for the Extra Class examination? Grandfathered Advanced class operators, get set to see some familiar Q & A's. General Class operators upgrading to Extra, stand by for a whopping increase in operating "elbow room" when you pass that upcoming Element 4, 50-question exam!

## Subelement E1 — Commission's Rules [7 Exam Questions — 7 Groups]

*Note: A Part § 97 reference is enclosed in brackets, i.e. [97], after each correct answer explanation in this subelement.*

### *E1A Operating standards: frequency privileges for Extra class amateurs; emission standards; message forwarding; frequency sharing between ITU Regions; FCC modification of station license; 30-meter band sharing; stations aboard ships or aircraft; telemetry; telecommand of an amateur station; authorized telecommand transmissions*

**E1A01 What exclusive frequency privileges in the 80-meter band are authorized to Amateur Extra Class control operators?**

A. 3525-3775 k Hz
B. 3500-3525 kHz
C. 3700-3750 kHz
D. 3500-3550 kHz

**ANSWER B:** Extra Class operators gain *exclusive* 25-kHz CW operating segments on the 80-meter band, 40-meter band, 20-meter band, and 15-meter band. Use your color marker to highlight this exclusive CW frequency segment on the 80-meter band. Remember, this is the bottom segment of two exclusive Extra Class segments on the 75/80-meter band. [97.301b]

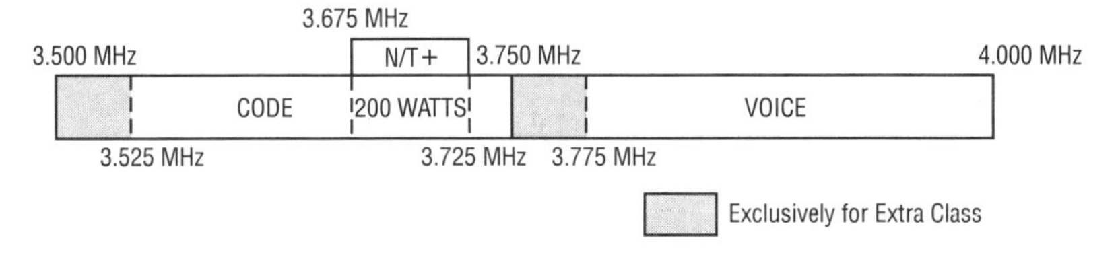


**75/80-Meter Wavelength Band Privileges**

**E1A02 What exclusive frequency privileges in the 75-meter band are authorized to Amateur Extra Class control operators?**

A. 3775-3800 kHz
B. 3800-3850 kHz
C. 3750-3775 kHz
D. 3800-3825 kHz

**ANSWER C:** This is the second Extra Class exclusive 25-kHz segment on the 75/80-meter band. It extends from 3750 to 3775 kHz. Use a color marker to highlight this portion of the band. This top segment includes voice privileges. [97.301b]

**E1A03 What exclusive frequency privileges in the 40-meter band are authorized to Amateur Extra Class control operators?**

A. 7000-7025 kHz
B. 7000-7050 kHz
C. 7025-7050 kHz
D. 7100-7150 kHz

**ANSWER A:** The bottom 25 kHz of the 40-meter band is reserved exclusively for Extra Class operators. [97.301b]

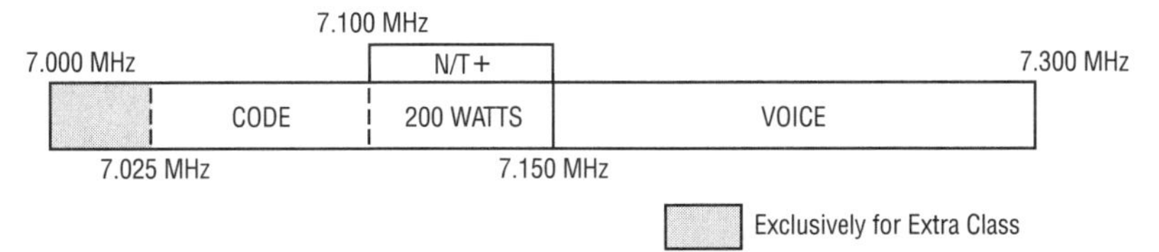


**40-Meter Wavelength Band Privileges**

**E1A04 What exclusive frequency privileges in the 20-meter band are authorized to Amateur Extra Class control operators?**

A. 14.100-14.175 MHz and 14.150-14.175 MHz
B. 14.000-14.125 MHz and 14.250-14.300 MHz
C. 14.025-14.050 MHz and 14.100-14.150 MHz
D. 14.000-14.025 MHz and 14.150-14.175 MHz

**ANSWER D:** The Extra Class operator has two exclusive operating segments on the 20-meter band. For CW, Extras have earned the bottom 25 kHz of the band. Note that the answer is in megahertz, 14.000 MHz to 14.025 MHz. For phone, the Extra exclusive segment is 14.150 MHz to 14.175 MHz. [97.301b]

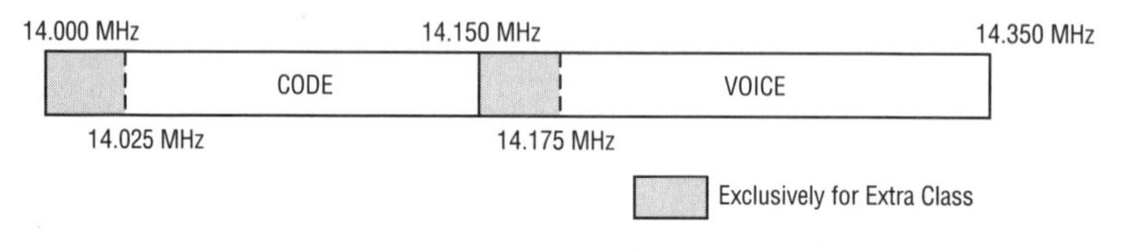


**20-Meter Wavelength Band Privileges**

**E1A05 What exclusive frequency privileges in the 15-meter band are authorized to Amateur Extra Class control operators?**

A. 21.000-21.200 MHz and 21.250-21.270 MHz
B. 21.050-21.100 MHz and 21.150-21.175 MHz
C. 21.000-21.025 MHz and 21.200-21.225 MHz
D. 21.000-21.025 MHz and 21.250-21.275 MHz

**ANSWER C:** The Extra Class operator gains two exclusive operating segments on the 15-meter band. The segments are CW from 21.000 MHz to 21.025 MHz, and voice from 21.200 MHz to 21.225 MHz. [97.301b]

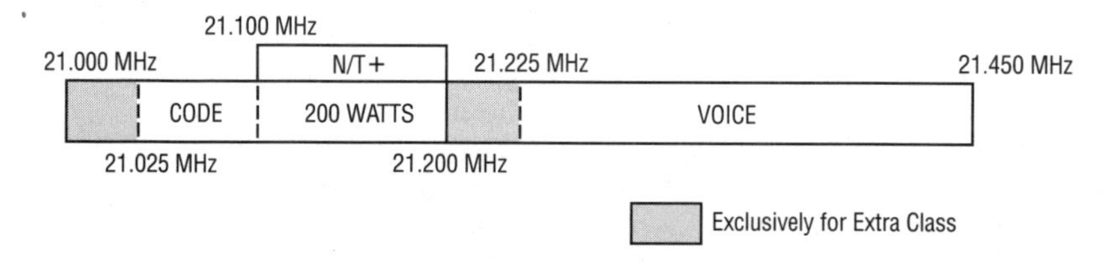


**15-Meter Wavelength Band Privileges**

**E1A06 Which frequency bands contain at least one segment authorized to only control operators holding an Amateur Extra Class operator license?**

A. 80, 75, 40, 20 and 15 meters
B. 80, 40, and 20 meters
C. 75, 40, 30 and 10 meters
D. 160, 80, 40 and 20 meters

**ANSWER A:** On 80 meters, the Extra Class operator has exclusive CW privileges, and on 75 meters, exclusive voice. On 40 meters, the Extra has exclusive CW privileges, and on 20 and 15 meters, the Extra has both CW and voice exclusive band privileges. Won't it be great when you get your Extra Class license! [97.301(b)]

**E1A07 Within the 20-meter band, what is the amount of spectrum authorized to only control operators holding an Amateur Extra Class operator license?**

A. 25 kHz
B. 50 kHz
C. None
D. 25 MHz

**ANSWER B:** This is a tricky answer, because you need to add the 25 kHz of exclusive voice privileges to the 25 kHz of exclusive CW privileges for the Extra operator. If you just consider only CW or voice, you end up with the wrong answer – you must consider both. [97.301(b)]

---

**E1A08 Which frequency bands contain two segments authorized to only control operators holding an Amateur Extra Class operator license, CEPT radio-amateur Class 1 license, or Class 1 IARP?**

A. 80/75, 20 and 15 meters
B. 40, 30 and 20 meters
C. 30, 20 and 17 meters
D. 30, 20 and 12 meters

**ANSWER A:** This answer, again, causes you to remember Extra exclusive CW privileges added to Extra exclusive voice privileges. Only 3 bands have both segments – 15 meters, 20 meters, and the 80/75 meter band. 80 meters is at the bottom of the band, and 75 meters is at the top of that same band; hence, a set of band numbers for just one band. [97.301(b)]

---

**E1A09 What must an amateur station licensee do if a spurious emission from the station causes harmful interference to the reception of another radio station?**

A. Pay a fine each time it happens
B. Submit a written explanation to the FCC
C. Forfeit the station license if it happens more than once
D. Eliminate or reduce the interference

**ANSWER D:** If another ham or the FCC contacts you about spurious emission interference, get your rig over to a specialist to look at your output on a spectrum analyzer. You need to reduce, or preferably eliminate, the interference. [97.307c]

---

**E1A10 What is the maximum mean power permitted for any spurious emission from a transmitter or external RF power amplifier transmitting at a mean power of 5 watts or greater on an amateur service HF band?**

A. The lesser of 50 milliwatts or 40 dB below the mean power of the fundamental emission
B. 60 dB below the mean power of the fundamental emission
C. 10 microwatts
D. The lesser of 25 microwatts or 40 dB below the mean power of the fundamental emission

**ANSWER A:** The following 4 Q & As deal with permissible undesirable spurious emissions from your transmitter or amplifier. On the high-frequency worldwide band, any equipment over 5 watts of power must have less than 50 milliwatts spurious, 40 dB below the mean power of the fundamental emission. Remember, "5 watts = 50 milliwatts." [97.307(d)]

---

**E1A11 What is the maximum mean power permitted for any spurious emission from a transmitter or external RF power amplifier transmitting at a mean power less than 5 watts on an amateur service HF band?**

A. 30 dB below the mean power of the fundamental emission
B. 60 dB below the mean power of the fundamental emission

C. 10 microwatts
D. 25 microwatts

**ANSWER A:** For those of you operating QRP below 5 watts on HF, the spurious emission limitations are slightly less – 30 dB down. [97.307(d)]

---

**E1A12 What is the maximum mean power permitted to any spurious emission from a transmitter or external RF power amplifier transmitting at a mean power greater than 25 watts on an amateur service VHF band?**

A. 60 dB below the mean power of the fundamental emission
B. 40 dB below the mean power of fundamental emission
C. 10 microwatts
D. 25 microwatts

**ANSWER A:** Up on the VHF bands with power output more than 25 watts, "life begins at 60" – 60 dB down on any spurious emissions. [97.307(e)]

---

**E1A13 What is the maximum mean power permitted for any spurious emission from a transmitter having a mean power of 25 W or less on an amateur service VHF band?**

A. The lesser of 25 microwatts or 40 dB below the mean power of the fundamental emission
B. The lesser of 50 microwatts or 40 dB below the mean power of the fundamental emission
C. 20 microwatts
D. 50 microwatts

**ANSWER A:** If you are on VHF and operating with less than 25 watts output, spurious emissions are limited to less than 25 microwatts, or 40 dB below the mean power of the fundamental emission. [97.307(e)] (On these 4 Q & As, you simply need to memorize the correct answer. Don't expect the computer-generated examination to always have the correct answer in the "A" position!)

---

**E1A14 If a packet bulletin board station in a message forwarding system inadvertently forwards a message that is in violation of FCC rules, who is accountable for the rules violation?**

A. The control operator of the packet bulletin board station
B. The control operator of the originating station and conditionally the first forwarding station
C. The control operators of all the stations in the system
D. The control operators of all the stations in the system not authenticating the source from which they accept communications

**ANSWER B:** A few years ago dozens of amateur operators were sent "greetings" from the FCC for taking part in forwarding a packet radio message that contained materials specifically prohibited by the rules and regulations. The hams forwarding the messages claimed everything was in the automatic mode and it was not up to them to censor what was coming down the electronic data line! Hence, new rules now hold the control operator of the station originating the message as primarily accountable for any rule violation, and the control operator of the first forwarding station also accountable for any violation of the rules. More specifically, everyone else "down the line" is no longer responsible for the message content, but must shut down their packet bulletin boards if informed that there is a message being automatically forwarded in violation of the rules. [97.219b/d]

**E1A15 If your packet bulletin board station inadvertently forwards a communication that violates FCC rules, what is the first action you should take?**

A. Discontinue forwarding the communication as soon as you become aware of it
B. Notify the originating station that the communication does not comply with FCC rules
C. Notify the nearest FCC Enforcement Bureau office
D. Discontinue forwarding all messages

**ANSWER A:** If you are advised that your packet station is automatically forwarding communications in violation of the rules, get into the computer and discontinue forwarding this specific communication as soon as you become aware of it. You may continue to forward all other messages. [97.219c]

---

**E1A16 For each ITU Region, how is each frequency band allocated internationally to the amateur service designated?**

A. Primary service or secondary service
B. Primary service
C. Secondary service
D. Co-secondary service

**ANSWER A:** For all ITU regions, each amateur service frequency band allocation is designated either primary or secondary – either one or the other. [97.303]

---

**E1A17 Why might the FCC modify an amateur station license?**

A. To relieve crowding in certain bands
B. To better prepare for a time of national emergency
C. To enforce a radio quiet zone within one mile of an airport
D. To promote the public interest, convenience and necessity

**ANSWER D:** The Federal Communications Commission will sometimes get involved with amateur operators who, in their opinion, are operating outside of the public interest of what ham radio is all about. By necessity the FCC may modify their license to restrict their operation to only certain times of day (quiet hours) or restrict their operation to only certain frequencies or bands. An FCC license modification is rare, but it is one way of insuring things run smoothly on the airwaves. [97.27]

---

**E1A18 What are the sharing requirements for an amateur station transmitting in the 30-meter band?**

A. It must not cause harmful interference to stations in the fixed service authorized by other nations
B. There are no sharing requirements
C. Stations in the fixed service authorized by other nations must not cause harmful interference to amateur stations in the same country
D. Stations in the fixed service authorized by other nations must not cause harmful interference to amateur stations in another country

**ANSWER A:** Our 30-meter band has a limitation of 200 watts maximum power output. Operation on this band for CW and digital modes must not cause harmful interference to any station in the fixed service authorized by other nations. [97.303(d)]

---

**E1A19 If an amateur station is installed on board a ship and is separate from the ship radio installation, what condition must be met before the station may transmit?**

A. Its operation must be approved by the master of the ship
B. Its antenna must be separate from the main ship antennas, transmitting only when the main radios are not in use

C. It must have a power supply that is completely independent of the main ship power supply
D. Its operator must have an FCC Marine endorsement on his or her amateur operator license

**ANSWER A:** If you are planning a long cruise on your pal's boat, make sure you have received approval by the captain (master of the ship) before you begin to transmit. Look at the other 3 incorrect answers – they sound logical, but only the answer we just gave you is correct. [97.11(a)]

**E1A20 What is the definition of the term telemetry?**
A. A one-way transmission of measurements at a distance from the measuring instrument
B. A two-way interactive transmission
C. A two-way single channel transmission of data
D. A one-way transmission to initiate, modify or terminate functions of a device at a distance

**ANSWER A:** We hear the word "telemetry" a lot in satellite service – telemetry is a one-way transmission of MEASUREMENTS at a distance from the measuring instrument. [97.3(a)(45)]

**E1A21 What is the definition of the term telecommand?**
A. A one way transmission of measurements at a distance from the measuring instrument
B. A two-way interactive transmission
C. A two-way single channel transmission of data
D. A one-way transmission to initiate, modify or terminate functions of a device at a distance

**ANSWER D:** Telecommand is a one-way transmission to initiate, modify, or shut down a function on a distant device, such as a satellite or a distant high-frequency remote base. Remember, telemetry is measurement, and telecommand is turning on or off a distant system. [97.3(a)(43)]

**E1A22 When may an amateur station transmit special codes intended to obscure the meaning of messages?**
A. Never under any circumstances
B. Only when a Special Temporary Authority has been obtained from the FCC
C. Only when an Amateur Extra Class operator is the station control operator
D. When sending telecommand messages to a station in space operation

**ANSWER D:** Special uplink codes that are specifically intended to obscure the meaning of messages are perfectly okay when sending telecommand messages to a station in space operation. These special codes, known only to specifically-authorized amateur operators working with satellites in space, allow the designated operator to do many things to that remote space station with complete secrecy to ensure no other operator will "play games" with those satellites. [97.211b]

***E1B Station restrictions: restrictions on station locations; restricted operation; teacher as control operator; station antenna structures; definition and operation of remote control and automatic control; control link***

**E1B01 Which of the following factors might restrict the physical location of an amateur station apparatus or antenna structure?**

A. The land may have environmental importance; or it is significant in American history, architecture or culture
B. The location's political or societal importance
C. The location's geographical or horticultural importance
D. The location's international importance, requiring consultation with one or more foreign governments before installation

**ANSWER A:** The environmental impact of a proposed installation is now a big consideration on anything that may have significant impact on American history, architecture, or culture. For instance, you would not want to place a station on an Indian burial ground. [97.13a]

---

**E1B02 Outside of what distance from an FCC monitoring facility may an amateur station be located without concern for protecting the facility from harmful interference?**

A. 1 mile
B. 3 miles
C. 10 miles
D. 30 miles

**ANSWER A:** The Federal Communications Commission recently automated its monitoring stations. Now there are remote-controlled, unmanned monitoring stations, and all amateur operators within one mile of that station must contact the local FCC engineer in charge to ensure they do not cause harmful interference to that remote-controlled monitoring unit. [97.13b]

---

**E1B03 What must be done before an amateur station is placed within an officially designated wilderness area or wildlife preserve, or an area listed in the National Register of Historical Places?**

A. A proposal must be submitted to the National Park Service
B. A letter of intent must be filed with the National Audubon Society
C. An Environmental Assessment must be submitted to the FCC
D. A form FSD-15 must be submitted to the Department of the Interior

**ANSWER C:** If you plan to install your station on a wildlife preserve, or an area listed as a historical place, you will need to submit an environmental assessment to the FCC for its review and approval. Be prepared for an extremely long wait before your license may be granted! [97.13a]

---

**E1B04 If an amateur station causes interference to the reception of a domestic broadcast station with a receiver of good engineering design, on what frequencies may the operation of the amateur station be restricted?**

A. On the frequency used by the domestic broadcast station
B. On all frequencies below 30 MHz
C. On all frequencies above 30 MHz
D. On the frequency or frequencies used when the interference occurs

**ANSWER D:** Sometimes the FCC may allow you to continue to operate at any time, but may restrict what frequency or frequencies you may operate on, due to interference complaints. [97.121a]

**E1B05 When may an amateur operator accept compensation for serving as the control operator of an amateur station used in a classroom?**

A. Only when the amateur operator does not accept pay during periods of time when the amateur station is used
B. Only when the classroom is in a correctional institution
C. Only when the amateur operator is paid as an incident of a teaching position during periods of time when the station is used by that teacher as a part of classroom instruction at an educational institution
D. Only when the station is restricted to making contacts with similar stations at other educational institutions

**ANSWER C:** If you are a ham instructor in the classroom, you may continue to receive pay as a teacher, and the operating of the ham system is simply "an incident of a teaching position." [97.113(c)]

---

**E1B06 Who may accept compensation for serving as a control operator in a classroom at an educational institution?**

A. Any licensed amateur operator
B. Only an amateur operator accepting such pay as an incident of a teaching position during times when the station is used by that teacher as a part of classroom instruction
C. Only teachers at correctional institutions
D. Only students at educational or correctional institutions

**ANSWER B:** Receiving pay as an instructor is okay if you are operating the equipment as an *"incident of a teaching position."* (Key words) [97.113(c)]

---

**E1B07 If an amateur antenna structure is located in a valley or canyon, what height restrictions apply?**

A. The structure must not extend more than 200 feet above average height of terrain
B. The structure must be no higher than 200 feet above the ground level at its site
C. There are no height restrictions since the structure would not be a hazard to aircraft in a valley or canyon
D. The structure must not extend more than 200 feet above the top of the valley or canyon

**ANSWER B**: An amateur antenna must be no higher than 200 feet. This is measured from the ground level at the base of the tower up to the top of the antenna structure. Even if you live deep in a valley or canyon, you still cannot exceed the base-to-tip 200-feet height restriction. If you need more elevation, reconsider where you are going to mount your antenna tower. [97.15a]

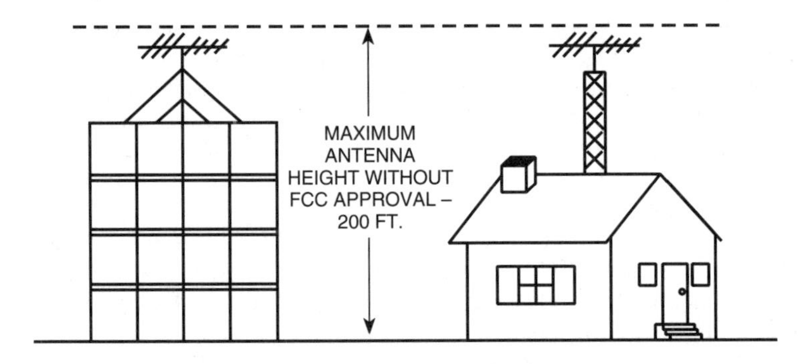


**Maximum Antenna Height**

**E1B08 What limits must local authorities observe when legislating height and dimension restrictions for an amateur station antenna structure?**

A. FAA regulations specify a minimum height for amateur antenna structures located near airports
B. FCC regulations specify a 200 foot minimum height for amateur antenna structures
C. State and local restrictions of amateur antenna structures are not allowed
D. Such regulation must reasonably accommodate amateur service communications and must constitute the minimum practicable regulation to accomplish the state or local authorities legitimate purpose

**ANSWER D:** All amateur operators should be familiar with their local antenna ordinances for the community. Regulations on the books must reasonably accommodate amateur service communications. However, if you sign into a condo complex with rigid antenna restrictions you are probably signing away your right to have any antenna outside at all. Keep this in mind! [97.15b]

---

**E1B09 If you are installing an amateur radio station antenna at a site within 5 miles from a public use airport, what additional rules apply?**

A. You must evaluate the height of your antenna based on the FCC Part 17 regulations
B. No special rules apply if your antenna structure will be less than 200 feet in height
C. You must file an Environmental Impact Statement with the Environmental Protection Agency before construction begins
D. You must obtain a construction permit from the airport zoning authority

**ANSWER A:** If you are located within 5 miles of an airport, you need to take a look at FCC Part 17 regulations that talk about the height of an antenna near an airport. The closer you are to the airport, the lower your antenna system must be. [97.15a]

---

**E1B10 What is meant by a remotely controlled station?**

A. A station operated away from its regular home location
B. Control of a station from a point located other than at the station transmitter
C. A station operating under automatic control
D. A station controlled indirectly through a control link

**ANSWER D:** Popular amateur radio repeaters located high atop mountains or buildings do not have the control operator sitting right in the repeater room. The repeater is controlled through a control link on frequencies above the input channel. Most control link frequencies are found on UHF 430-440 MHz, or up on 1.2 GHz. "Control link" is the key for remote control of an amateur station or repeater. [97.3a38]

---

**E1B11 Which of the following amateur stations may not be operated under automatic control?**

A. Remote control of model aircraft
B. Beacon station
C. Auxiliary station
D. Repeater station

**ANSWER A:** Beacon stations, auxiliary operation, and repeater operation may be automatically controlled without the control operator on duty at the control point. However, a model craft may NOT be under automatic control. This means it is not legal to allow a "third party" to take control of a model airplane or boat on 6-meter ham band privileges without the control operator being physically present at the control point. This means a "third party," unlicensed, may not borrow your 6-meter R/C equipment and use it on the ham bands unless you are physically present at the control point. [97.109d, 97.201d, 97.203d, 97.205d]

**E1B12 What is meant by automatic control of a station?**

A. The use of devices and procedures for control so that the control operator does not have to be present at the control point
B. A station operating with its output power controlled automatically
C. Remotely controlling a station such that a control operator does not have to be present at the control point at all times
D. The use of a control link between a control point and a locally controlled station

**ANSWER A:** Ham operators have come up with *devices and procedures* (key words) so they don't have to stand guard at the repeater block house on a mountain top. [97.3a6, 97.109d]

---

**E1B13 How do the control operator responsibilities of a station under automatic control differ from one under local control?**

A. Under local control there is no control operator
B. Under automatic control the control operator is not required to be present at the control point
C. Under automatic control there is no control operator
D. Under local control a control operator is not required to be present at a control point

**ANSWER B:** Repeaters under automatic control don't have anyone sitting up there at that distant control point. [97.3a6, 97.109d]

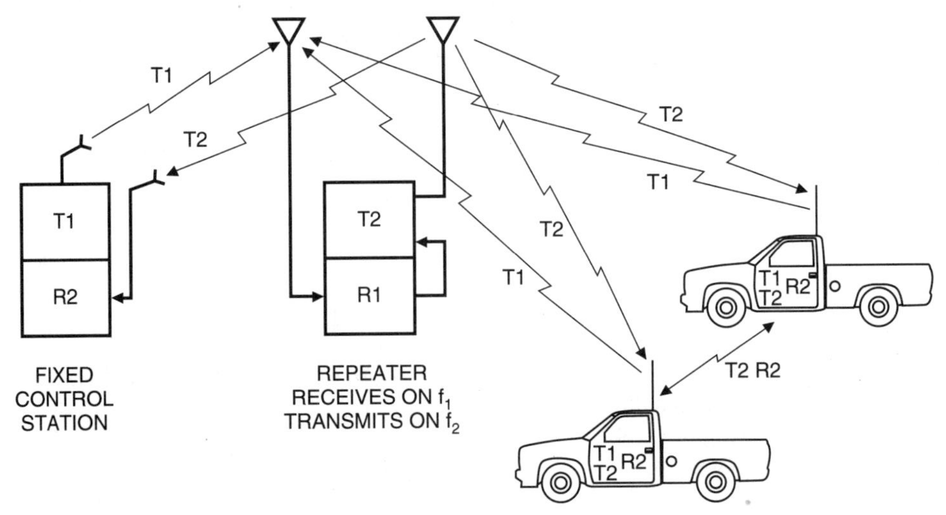


**Repeater**

Source: *Mobile 2-Way Radio Communications,* G. West, Copyright ©1992 Master Publishing, Inc., Lincolnwood, IL

---

**E1B14 What is a control link?**

A. A device that automatically controls an unattended station
B. An automatically operated link between two stations
C. The means of control between a control point and a remotely controlled station
D. A device that limits the time of a station's transmission

**ANSWER C:** When hams talk about their secret "control link," they are talking about that special frequency for remotely controlling a distant station. [97.3a38]

**E1B15 What is the term for apparatus to effect remote control between the control point and a remotely controlled station?**

A. A tone link
B. A wire control
C. A remote control
D. A control link

**ANSWER D:** The path between a control point and a remotely controlled station is called a control link frequency. It may be simplex or duplex. [97.3a38]

***E1C Reciprocal operating: reciprocal operating authority; purpose of reciprocal agreement rules; alien control operator privileges; identification (Note: This includes CEPT and IARP)***

**E1C01 What is an FCC authorization for alien reciprocal operation?**

A. An FCC authorization to the holder of an amateur license issued by certain foreign governments to operate an amateur station in the US
B. An FCC permit to allow a US licensed amateur to operate in a foreign nation except Canada
C. An FCC permit allowing a foreign licensed amateur to handle third-party traffic between the US and the amateur's own nation
D. An FCC agreement with another country allowing the passing of third-party traffic between amateurs of the two nations

**ANSWER A:** Licensed amateur operators from other countries traveling within the U.S. and its possessions may apply for a reciprocal permit. If the U.S. has a reciprocal operating agreement with their country, the foreign operators may use their equipment in the U.S. on a temporary basis. [97.5c,d,e, 97.107]

**Countries Holding U.S. Reciprocal Agreements**

| | | | | |
|---|---|---|---|---|
| Antigua, Barbuda | Chile | Greece | Liberia | Seychelles |
| Argentina | Colombia | Greenland | Luxembourg | Sierra Leone |
| Australia | Costa Rica | Grenada | Macedonia | Solomon Islands |
| Austria | Croatia | Guatemala | Marshall Is. | South Africa |
| Bahamas | Cyprus | Guyana | Mexico | Spain |
| Barbados | Denmark | Haiti | Micronesia | St. Lucia |
| Belgium | Dominica | Honduras | Monaco | St. Vincent and Grenadines |
| Belize | Dominican Rep. | Iceland | Netherlands | Surinam |
| Bolivia | Ecuador | India | Netherlands Ant. | Sweden |
| Bosnia-Herzegovina | El Salvador | Indonesia | New Zealand | Switzerland |
| Botswana | Fiji | Ireland | Nicaragua | Thailand |
| Brazil | Finland | Israel | Norway | Trinidad, Tobago |
| Canada[1] | France[2] | Italy | Panama | Turkey |
| | Germany | Jamaica | Paraguay | Tuvalu |
| | | Japan | Papua New Guinea | United Kingdom[3] |
| | | Jordan | Peru | Uruguay |
| | | Kiribati | Philippines | Venezuela |
| | | Kuwait | Portugal | |

1. Do not need reciprocal permit
2. Includes all French Territories
3. Includes all British Territories

**E1C02 Who is authorized for alien reciprocal operation in places where the FCC regulates the amateur service?**

A. Anyone holding a valid amateur service license issued by a foreign government
B. Any non-US citizen holding an amateur service license issued by their government with which the US has a reciprocal operating arrangement
C. Anyone holding a valid amateur service license issued by a foreign government with which the US has a reciprocal operating arrangement
D. Any non-US citizen holding a valid amateur license issued by a foreign government, as long as the person is a citizen of that country

**ANSWER B:** Reciprocal operating permits are only available to foreigners, and a U.S. FCC alien amateur reciprocal permit may not be held by a U.S. citizen. [97.107]

**E1C03 What are the frequency privileges authorized for alien reciprocal operation?**

A. Those authorized to a holder of the equivalent US amateur operator license
B. Those that the alien has in his or her own country
C. Those authorized to the alien by his country of citizenship, but not to exceed those authorized to Amateur Extra Class operators
D. Those approved by the International Amateur Radio Union

**ANSWER C:** If you are licensed in another country and are operating as an alien reciprocal, you must stick with the privileges of your own country of citizenship, not to exceed those authorized by the FCC to U.S. Extra Class operators. [97.107]

**E1C04 What indicator must a Canadian amateur station include with the assigned call sign in the station identification announcement when operating in the US?**

A. No indicator is required
B. The grid-square locator number for the location of the station must be included after the call sign
C. The permit number and the call-letter district number of the station location must be included before the Canadian-assigned call sign
D. The letter-numeral indicating the station location after the Canadian call sign and the closest city and state once during the communication

**ANSWER D:** Canadian operators operating mobile or base here in the USA will say their Canadian call sign and then a letter-numeral indicating their approximate location. If a Canadian is operating in Chicago, they would first give their Canadian call sign, followed by "W9, near Chicago, Illinois." [97.119(g)]

**E1C05 When may a US citizen holding a foreign amateur service license be authorized for alien reciprocal operation in places where the FCC regulates the amateur service?**

A. Never; US citizens are not eligible for alien reciprocal operation
B. When the US citizen also holds citizenship in the foreign country
C. When the US citizen was born in the foreign country
D. When the US citizen has no current FCC amateur service license

**ANSWER A:** U.S. citizens are not eligible for a reciprocal permit, even though they may hold a call sign from another foreign country. [97.107]

**E1C06 Which of the following would disqualify a foreign amateur operator from being authorized for alien reciprocal operation in places where the FCC regulates the amateur service?**

A. Not being a citizen of the country that issued the amateur service license
B. Having citizenship in their own country but not US citizenship
C. Holding only an amateur license issued by their own country but holding no FCC amateur service license grant
D. Holding an amateur service license issued by their own country authorizing privileges beyond Amateur Extra Class operator privileges

**ANSWER A:** You must be a citizen of the country that issued the amateur service license in order to qualify as an alien reciprocal operator here in the U.S. [97.107]

**E1C07 What special document is required before a Canadian citizen holding a Canadian amateur service license may reciprocal operate in the US?**

A. A written FCC authorization for alien reciprocal operation
B. No special document is required
C. The citizen must have an FCC-issued validation of their Canadian license
D. The citizen must have an FCC-issued Certificate of US License Grant without Examination to operate for a period longer than 10 days

**ANSWER B:** There are no additional special documents necessary or required, other than the original Canadian license. U.S. hams may travel into Canada without special Canadian licensing. [97.107a]

---

**E1C08 What operating privileges does a properly licensed alien amateur have in the US, if the US and the alien amateur's home country have a multilateral or bilateral reciprocal operating agreement?**

A. All privileges of their home license
B. All privileges of an Amateur Extra Class operator license
C. Those authorized by their home license, not to exceed the operating privileges of an Amateur Extra Class operator license
D. Those granted by the home license that match US privileges authorized to amateur operators in ITU Region 1

**ANSWER C:** Recent, worldwide agreements are streamlining the process of being able to operate amateur radio in a foreign country, both for us as well as aliens visiting us. An alien amateur visiting our country may enjoy many of the same band privileges that he or she enjoyed abroad, but those privileges may NOT exceed the operating privileges of an amateur Extra license. If they have extended frequency privileges that we don't have, they may not use them here in the U.S. Some countries allow voice within our CW sub-bands, but when operating here, they would need to stay strictly within the voice sub-bands for USA operators. [97.107b]

---

**E1C09 From which locations may a licensed alien amateur operator be the control operator of an amateur station?**

A. Only locations within the boundaries of the 50 United States
B. Only locations listed as the primary station location on an FCC amateur service license
C. Only locations on ground within the US and its territories; no shipboard or aeronautical mobile operation is permitted
D. Any location where the amateur service is regulated by the FCC

**ANSWER D:** Our reciprocal operating agreements may allow the alien to be a control operator of a U.S. ham station, just as long as the location is within jurisdiction of our Federal Communications Commission. [97.5c]

---

**E1C10 Which of the following operating arrangements allow an FCC licensed US citizen to operate in many European countries and alien amateurs from many European countries to operate in the US?**

A. CEPT agreement
B. IARP agreement
C. ITU agreement
D. All of these choices are correct

**ANSWER A:** The CEPT agreement (European Conference of Postal & Telecommunications Administrations – CEPT) is great news for all U.S. amateur operators when visiting many countries in Europe – NO MORE RECIPROCAL PAPERWORK! Before leaving for Europe, obtain a copy of FCC Public Notice DA99-1098, released June 7, 1999, along with the original and copies of your U.S.

ham license. This paperwork is necessary in case you are asked about the equipment you are bringing into that country at customs. Have fun in Europe with ham radio! Don't forget your paperwork and the required FCC Public Notice. [97.5d]

**Participating CEPT countries, as of March, 2002, are:**

| | | | |
|---|---|---|---|
| Austria | Estonia | Latvia | Romania |
| Belgium | Finland | Liechtenstein | Slovak Republic |
| Bosnia & Herzegovina | France & its possessions | Lithuania | Slovenia |
| Bulgaria | Germany | Luxembourg | Spain |
| Croatia | Hungary | Monaco | Sweden |
| Cyprus | Iceland | Netherlands | Switzerland |
| Czech Republic | Ireland | Netherland Antilles | Turkey |
| Denmark | Italy | Norway | United Kingdom & its possessions |
| | | Portugal | |

**E1C11 Which of the following multilateral or bilateral operating arrangements allow an FCC licensed US citizen and many Central and South American amateur operators to operate in each other's countries?**

A. CEPT agreement
B. IARP agreement
C. ITU agreement
D. All of these choices are correct

**ANSWER B:** Heading south for a vacation and want to take along your ham equipment to Central and South America? In addition to Canada, the following countries and the U.S. have an IARP (International Amateur Radio Permit) agreement: Argentina, Brazil, Peru, Uruguay, and Venezuela. [97.5e]

South American IARP Countries

**E1C12 What additional station identification, in addition to his or her own call sign, does an alien operator supply when operating in the US under an FCC authorization for alien reciprocal operation?**

A. No additional operation is required
B. The grid-square locator closest to his or her present location is included before the call

C. The serial number of the permit and the call-letter district number of the station location is included before the call
D. The letter-numeral indicating the station location in the US included before their call and the closest city and state given once during the communication

**ANSWER D:** "Letter-numeral" and "city and state" are the key words to identify the correct answer to this question. FCC Rule 97.119(g) reads, ". . .an indicator consisting of the appropriate letter-numeral designating the station location must be included before the call sign issued to the station by the licensing country...." "At least once during each intercommunication, the identification announcement must include the geographical location as nearly as possible by city and state...." [97.119g]

***E1D Radio Amateur Civil Emergency Service (RACES): definition; purpose; station registration; station license required; control operator requirements; control operator privileges; frequencies available; limitations on use of RACES frequencies; points of communication for RACES operation; permissible communications***

**E1D01 What is the Radio Amateur Civil Emergency Service (RACES)?**

A. A radio service using amateur service frequencies on a regular basis for communications that can reasonably be furnished through other radio services
B. A radio service using amateur stations for civil defense communications during periods of local, regional, or national civil emergencies
C. A radio service using amateur service frequencies for broadcasting to the public
D. A radio service using local government frequencies by Amateur Radio operators for emergency communications

**ANSWER B:** Remember Civil Defense? The C in RACES is for civil, and the correct answer is Radio Amateur Civil Emergency Service. [97.3a37]

The RACES Logo

**E1D02 What is the purpose of RACES?**

A. To provide civil-defense communications during emergencies
B. To provide emergency communications for boat or aircraft races
C. To provide routine and emergency communications for athletic races
D. To provide routine and emergency military communications

**ANSWER A:** Most RACES organizations are looking for volunteers to provide Civil Defense communications in emergencies. [97.3a37]

**E1D03 With what organization must an amateur station be registered before participating in RACES?**

A. The Amateur Radio Emergency Service
B. The US Department of Defense
C. A civil defense organization
D. The FCC Enforcement Bureau

**ANSWER C:** Some members of RACES are issued a distinctive uniform, and may be required to put in a certain amount of civil defense drill time each month. [97.407a]

**E1D04 Which amateur stations may be operated in RACES?**

A. Only those licensed to Amateur Extra class operators
B. Any FCC-licensed amateur station except a station licensed to a Technician class operator
C. Any FCC-licensed amateur station certified by the responsible civil defense organization for the area served
D. Any FCC licensed amateur station participating in the Military Affiliate Radio System (MARS)

**ANSWER C:** All amateurs are eligible for RACES membership. [97.407a]

---

**E1D05 What frequencies are authorized normally to an amateur station participating in RACES?**

A. All amateur service frequencies otherwise authorized to the control operator
B. Specific segments in the amateur service MF, HF, VHF and UHF bands
C. Specific local government channels
D. Military Affiliate Radio System (MARS) channels

**ANSWER A:** Anyone with a valid amateur license, including a Novice license, is eligible to be certified by a civil defense organization as a RACES operator. However, you do not gain any out-of-band privileges as a RACES operator. [97.407(b)]

---

**E1D06 What are the frequencies authorized to an amateur station participating in RACES during a period when the President's War Emergency Powers are in force?**

A. All frequencies in the amateur service authorized to the control operator
B. Specific segments in the amateur service MF, HF, VHF and UHF bands
C. Specific local government channels
D. Military Affiliate Radio System (MARS) channels

**ANSWER B:** If a war should break out, RACES operators may be authorized specific segments in the amateur service MF, HF, VHF, and UHF bands. Look at the incorrect answers, and see how good they look, and then memorize the correct answer! [97.407(b)]

---

**E1D07 What frequencies are normally available for RACES operation?**

A. Only those authorized to the civil defense organization
B. Only those authorized to federal government communications
C. Only the top 25 kHz of each amateur service band
D. All frequencies authorized to the amateur service

**ANSWER D:** RACES members are granted all frequencies available to the amateur service. It's only during war time that RACES stations may be restricted to specific frequencies found in Part 97.407(b)(1). FCC Part 97.407(b) states, "...in the event of an emergency which necessitates the invoking of the President's War Emergency Powers under the provisions of Section 706 of the Communications Act...." [97.407b]

---

**E1D08 What type of emergency can cause limits to be placed on the frequencies available for RACES operation?**

A. An emergency during which the President's War Emergency Powers are invoked
B. An emergency in only one of the United States would limit RACES operations to a single HF band
C. An emergency confined to a 25-mile area would limit RACES operations to a single VHF band
D. An emergency involving no immediate danger of loss of life

**ANSWER A:** Only during war times, when the President invokes War Emergency Powers, are RACES members limited to special frequencies. [97.407b]

**E1D09 Who may be the control operator of a RACES station?**

A. Anyone holding an FCC-issued amateur operator license other than Novice
B. Only an Amateur Extra Class operator licensee
C. Anyone who holds an FCC-issued amateur operator license and is certified by a civil defense organization
D. Any person certified as a RACES radio operator by a civil defense organization and who holds an FCC issued GMRS license

**ANSWER C:** Be sure to read the complete answer all the way through – Answer D goes wrong when it talks about another type of radio license. Anyone who holds an FCC-issued amateur operator license and is certified by a Civil Defense organization may be the control operator of a RACES station. Don't speed read the exam! [97.407(a)]

**E1D10 With which stations may amateur stations participating in RACES communicate?**

A. Any amateur station
B. Amateur stations participating in RACES and specific other stations authorized by the responsible civil defense official
C. Any amateur station or a station in the Disaster Communications Service
D. Any Citizens Band station that is also registered in RACES

**ANSWER B:** As a RACES station, you may communicate with other RACES stations and certain other stations authorized by the responsible Civil Defense official. [97.407c,d]

**In an emergency, authorized hams participating in a RACES organization may communicate from a police helicopter.**

**E1D11 What communications are permissible in RACES?**

A. Any type of communications when there is no emergency
B. Any Amateur Radio Emergency Service communications
C. National defense or immediate safety of people and property and communications authorized by the area civil defense organization
D. National defense and security or immediate safety of people and property communications authorized by the President

**ANSWER C:** For communications concerning national defense and security, or immediate safety of life and property, your area civil defense organization may decide on its own when RACES operation begins. They may also allow specific periods for testing and drills. [97.407e]

***E1E Amateur Satellite Service: definition; purpose; station license required for space station; frequencies available; telecommand operation: definition; eligibility; telecommand station (definition); space telecommand station; special provisions; telemetry: definition; special provisions; space station: definition; eligibility; special provisions; authorized frequencies (space station); notification requirements; earth operation: definition; eligibility; authorized frequencies (Earth station)***

**E1E01 What is the amateur-satellite service?**

A. A radio navigation service using satellites for the purpose of self-training, intercommunication and technical studies carried out by amateurs
B. A spacecraft launching service for amateur-built satellites
C. A service using amateur stations on satellites for the purpose of self-training, intercommunication and technical investigations
D. A radio communications service using stations on Earth satellites for weather information gathering

**ANSWER C:** The amateur satellite service allows hams to communicate through satellite orbiting repeaters and translators. The purpose is exactly the same as the amateur service. Amateurs are not limited to navigation, not limited to only weather information, and not limited to just radar experimentation. [97.3a3]

**E1E02 What is a space station in the amateur-satellite service?**

A. An amateur station located more than 50 km above the Earth's surface
B. An amateur station designed for communications with other amateur stations by means of Earth satellites
C. An amateur station that transmits communications to initiate, modify or terminate functions of an Earth station
D. An amateur station designed for communications with other amateur stations by reflecting signals off objects in space

**ANSWER A:** If your station is 50 kilometers or higher above the Earth's surface, you become an amateur satellite or space station. [97.3a40]

**E1E03 What is a telecommand station in the amateur-satellite service?**

A. An amateur station that transmits communications to initiate, modify or terminate functions of a space station
B. An amateur station located on the Earth's surface for communications with other Earth stations by means of Earth satellites
C. An amateur station located more than 50 km above the Earth's surface
D. An amateur station that transmits telemetry consisting of measurements of upper atmosphere data from space

**ANSWER A:** Remember this question from before? Telecommand means initiating, modifying, or terminating a function of another station, such as a space station. [97.3a44]

**Tracking and communicating through amateur satellites can be done with a cross-polarized satellite antenna**

---

**E1E04 What is an Earth station in the amateur-satellite service?**

A. An amateur station within 50 km of the Earth's surface for communications with Amateur stations by means of objects in space
B. An amateur station that is not able to communicate using amateur satellites
C. An amateur station that transmits telemetry consisting of measurement of upper atmosphere data from space
D. Any amateur station on the surface of the Earth

**ANSWER A:** Below 50 kilometers, you're considered an earth station when your are communicating via a space station. [97.3a16]

---

**E1E05 Which of the following types of communications may space stations transmit?**

A. Automatic retransmission of signals from Earth stations and other space stations
B. One-way communications
C. Telemetry consisting of specially coded messages
D. All of these choices are correct

**ANSWER D:** Our new International Space Station, plus the space shuttles, may transmit many types of amateur radio communications, such as repeater-type automatic transmissions, one-way signaling, and telemetry consisting of on-board coded messages. [97.207]

Want to learn more about ham radio in space?
We suggest you visit the AMSAT website at: www.AMSAT.org.

---

**E1E06 Which amateur stations are eligible to operate as a space station?**

A. Any except those of Technician Class operators
B. Only those of General, Advanced or Amateur Extra Class operators
C. Only those of Amateur Extra Class operators
D. Any FCC-licensed amateur station

**ANSWER D:** There is no longer a specific higher grade license requirement for space operation. Any grade of license may now qualify. Originally, only an Extra Class operator could transmit from space. [97.207a]

---

**E1E07 What special provision must a space station incorporate in order to comply with space station requirements?**

A. The space station must be capable of effecting a cessation of transmissions by telecommand whenever so ordered by the FCC
B. The space station must cease all transmissions after 5 years

C. The space station must be capable of changing its orbit whenever such a change is ordered by NASA
D. The station call sign must appear on all sides of the spacecraft

**ANSWER A:** All space stations must be capable of receiving telecommand to shut them down if ordered to do so by the FCC. [97.207b]

---

**E1E08 When must the licensee of a space station give the FCC International Bureau the first written pre-space notification?**

A. Any time before initiating the launch countdown for the spacecraft
B. No less than 3 months after initiating construction of the space station
C. No less that 12 months before launch of the space station platform
D. No less than 27 months prior to initiating space station transmissions

**ANSWER D:** It takes a whopping 27 months of written notification to the FCC before initiating space station transmissions. A licensee also must send in a second notification 5 months prior to initiating space station operation. [97.207g1]

---

**E1E09 Which amateur service HF bands have frequencies authorized to space stations?**

A. Only 40m, 20m, 17m, 15m, 12m and 10m
B. Only 40 m, 20 m, 17m, 15 m and 10 m bands
C. 40 m, 30 m, 20 m, 15 m, 12 m and 10 m bands
D. All HF bands

**ANSWER A:** By international agreement, some high-frequency voice bands are available for space operation. This includes 40 meters, 20 meters, 17 meters ITU, 15 meters, 12 meters ITU, and 10 meters. [97.207]

---

**E1E10 Which VHF amateur service bands have frequencies available for space stations?**

A. 2 meters
B. 2 meters and 1.25 meters
C. 6 meters, 2 meters, and 1.25 meters
D. 6 meters and 2 meters

**ANSWER A:** On VHF (30 MHz-300 MHz), only the 2-meter band is allowed for space station operation. Operating on 1.25 meters (the 222 MHz band) and 6 meters are prohibited. You can find the right answer by looking at the wrong answers. [97.207]

---

**E1E11 Which amateur service UHF bands have frequencies available for a space station?**

A. 70 cm, 23 cm, 13 cm
B. 70 cm
C. 70 cm and 33 cm
D. 33 cm and 13 cm

**ANSWER A:** On UHF, 300 MHz-3000 MHz, the 33 cm (900 MHz) band is NOT allowed for space station, but all other bands are allowed. This includes 70 cm (435 MHz), 23 cm (1260 MHz), and 13 cm (2400 MHz). [97.207]

---

**E1E12 Which amateur stations are eligible to be telecommand stations?**

A. Any amateur station designated by NASA
B. Any amateur station so designated by the space station licensee
C. Any amateur station so designated by the ITU
D. All of these choices are correct

**ANSWER B:** Any amateur station so designated by the space station licensee may be a telecommand station. [97.211a]

**E1E13 What unique privilege is afforded a telecommand station?**

A. A telecommand station may transmit command messages to the space station using codes intended to obscure their meaning
B. A telecommand station may transmit music to the space station
C. A telecommand station may transmit with a PEP output of 5000 watts
D. A telecommand station is not required to transmit its call sign at the end of the communication

**ANSWER A:** Telecommand stations that may turn on and turn off satellites ARE PERMITTED to use special codes that are secret for this operation. [97.211b]

---

**E1E14 What is the term for space-to-Earth transmissions used to communicate the results of measurements made by a space station?**

A. Data transmission
B. Frame check sequence
C. Telemetry
D. Space-to-Earth telemetry indicator (SETI) transmissions

**ANSWER C:** Remember, "telemetry" is communications of the results of measurements made by the space station. [97.207 (f)]

---

**E1E15 Which amateur stations are eligible to operate as Earth stations?**

A. Any amateur station whose licensee has filed a pre-space notification with the FCC International Bureau
B. Only those of General, Advanced or Amateur Extra Class operators
C. Only those of Amateur Extra Class operators
D. Any amateur station, subject to the privileges of the class of operator license held by the control operator

**ANSWER D:** An amateur Earth station is located down here on planet Earth. Operating as an Earth station for space communications is available to all amateur stations. Be sure to use only those frequency privileges allowed by your license class. [97.209 (a)]

---

***E1F Volunteer Examiner Coordinators (VECs): definition; VEC qualifications; VEC agreement; scheduling examinations; coordinating VEs; reimbursement for expenses; accrediting VEs; question pools; Volunteer Examiners (VEs): definition; requirements; accreditation; reimbursement for expenses; VE conduct; preparing an examination; examination elements; definition of code and written elements; preparation responsibility; examination requirements; examination credit; examination procedure; examination administration; temporary operating authority***

---

**E1F01 Who may prepare an Element 4 amateur operator license examination?**

A. The VEC Question Pool Committee, which selects questions from the appropriate VEC question pool
B. A VEC that selects questions from the appropriate FCC bulletin
C. An Extra class VE that selects questions from the appropriate FCC bulletin
D. An Extra class VE or a qualified supplier who selects questions from the appropriate VEC question pool

**ANSWER D:** These 28 questions reflect your responsibility as an Extra Class operator to know how to administer and prepare all levels of examinations. For the Element 4 exam, an Extra Class operator or a "qualified supplier" will select questions from the VEC question pool. [97.507a, b, c; 97.523]

**E1F02 Where are the questions listed that must be used in all written US amateur license examinations?**
A. In the instructions that each VEC give to their VEs
B. In an FCC-maintained question pool
C. In the VEC-maintained question pool
D. In the appropriate FCC Report and Order
**ANSWER C:** The Extra Class exam, like all ham tests, may only take questions verbatim from the VEC-maintained question pool. [97.507b]

---

**E1F03 Who is responsible for maintaining the question pools from which all amateur license examination questions must be taken?**
A. All of the VECs
B. The VE team
C. The VE question pool team
D. The FCC Wireless Telecommunications Bureau
**ANSWER A:** All of the Volunteer Examiner Coordinators contribute to the question pool and insure it is up to date. If a rule change occurs that disqualifies a question, all VEC's will eliminate that question from the question pool. [97.523]

---

**E1F04 Who must select from the VEC question pool the set of questions that are administered in an Element 3 examination?**
A. Only a VE holding an Amateur Extra Class operator license grant
B. The VEC coordinating the examination session
C. A VE holding an FCC-issued Amateur Extra or Advanced Class operator license grant
D. The FCC Enforcement Bureau
**ANSWER C:** Only a VE holding an Extra Class or an old Advanced Class license may prepare a General Class Element 3 exam. [97.507a1]

---

**E1F05 Who must select from the VEC question pool the set of questions that are administered in an Element 2 examination?**
A. The VEC coordinating the examination session
B. A VE holding an FCC-issued Technician, General, Advanced or Amateur Extra Class operator license grant
C. Only a VE holding an Amateur Extra or Advanced Class operator license grant
D. The FCC Office of Engineering and Technology
**ANSWER B:** For the entry-level Element 2 Tech exam, this particular question asks, "Who must select?" – NOT who can administer the test. The selection of questions could be a VE holding any license other than the old Novice. This includes the Technician Class license, but keep in mind, we are talking about selecting test questions, NOT administering them – that requires a higher-grade license. [97.507a2]

---

**E1F06 What is the purpose of an amateur operator telegraphy examination?**
A. It determines the examinee's level of commitment to the amateur service
B. All of these choices are correct
C. It proves that the examinee has the ability to send correctly by hand and to receive correctly by ear texts in the International Morse Code
D. It helps preserve the proud tradition of radiotelegraphy skill in the amateur service

**ANSWER C:** This is a good one – "Proves the ability to send and receive" Morse Code. At most exams, the applicant is only asked to receive the code. But here at Gordon West Radio School, we also like to see our exam applicants for Element 1 send CW as well. And it's right here in the rules: "...to send correctly by hand." [97.503a]

**E1F07 What is the purpose of an Element 4 examination?**

A. It proves the examinee has the qualifications necessary to perform properly the duties of an Amateur Extra Class operator
B. It proves the examinee is qualified as an electronics technician
C. It proves the examinee is an electronics expert
D. It proves that the examinee is an expert radio operator

**ANSWER A:** Passing Element 4 proves the examinee has the qualifications necessary to perform the duties of an amateur Extra Class operator. Well, memorize this for the test, but know that the real proving is when that new Extra Class operator gets on the air and begins transmitting and receiving. This is where the real learning of Extra Class begins! [97.503b]

**E1F08 What is a Volunteer-Examiner Coordinator?**

A. A person who has volunteered to administer amateur operator license examinations
B. A person who has volunteered to prepare amateur operator license examinations
C. An organization that has entered into an agreement with the FCC to coordinate amateur operator license examinations
D. The person that has entered into an agreement with the FCC to be the VE session manager

**ANSWER C:** A Volunteer Examiner Coordinator is an organization that has an agreement with the FCC to coordinate amateur radio license exams. Two of the biggest VEC's are the American Radio Relay League, and the W5YI VEC. [97.521]

**Working with kids as an Elmer or VE is both fun and a great way to help ham radio grow.**

**E1F09 What is an accredited Volunteer Examiner?**

A. An amateur operator who is approved by three or more fellow volunteer examiners to administer amateur license examinations
B. An amateur operator who is approved by a VEC to administer amateur operator license examinations
C. An amateur operator who administers amateur license examinations for a fee
D. An amateur operator who is approved by an FCC staff member to administer amateur operator license examinations

**ANSWER B:** A volunteer examiner coordinator (VEC) will accredit amateur operators as volunteer examiners (VEs). [97.3a48]

---

**E1F10 What is a VE Team?**

A. A group of at least three VEs who administer examinations for an amateur operator license
B. The VEC staff
C. One or two VEs who administer examinations for an amateur operator license
D. A group of FCC Volunteer Enforcers who investigate Amateur Rules violations

**ANSWER A:** It takes a minimum of 3 accredited volunteer examiners to conduct an examination session. [97.509(a)]

---

**E1F11 Which persons seeking to be VEs cannot be accredited?**

A. Persons holding less than an Advanced Class operator license
B. Persons less than 21 years of age
C. Persons who have ever had an amateur operator or amateur station license suspended or revoked
D. Persons who are employees of the federal government

**ANSWER C:** Any amateur who has ever had his or her license suspended or revoked may not serve as an accredited volunteer examiner. [97.509b4]

---

**E1F12 What is the VE accreditation process?**

A. Each General, Advanced and Amateur Extra Class operator is automatically accredited as a VE when the license is granted
B. The amateur operator must pass a VE examination administered by the FCC Enforcement Bureau
C. The prospective VE obtains accreditation from a VE team
D. Each VEC ensures that its Volunteer Examiner applicants meet FCC requirements to serve as VEs

**ANSWER D:** VE accreditation is received from a volunteer examiner coordinator (VEC) who makes sure that each VE meets specific FCC requirements, and has no conflict of interest between their VE duties and their normal occupation. [97.509b1, 97.525]

---

**E1F13 Where must the VE team be stationed while administering an examination?**

A. All administering VEs must be present and observing the examinees throughout the entire examination
B. The VEs must leave the room after handing out the exam(s) to allow the examinees to concentrate on the exam material
C. The VEs may be elsewhere provided at least one VE is present and is observing the examinees throughout the entire examination
D. The VEs may be anywhere as long as they each certify in writing that examination was administered properly

**ANSWER A:** As a VE, you must be present and in a location to physically observe the candidate taking the test throughout the entire examination. [97.509c]

---

**E1F14 Who is responsible for the proper conduct and necessary supervision during an amateur operator license examination session?**

A. The VEC coordinating the session
B. The FCC
C. The administering VEs
D. The VE session manager

---

**ANSWER C:** The administering Volunteer Examiners must continuously supervise everyone in the room taking the tests. Administering VEs should not talk amongst themselves as the examinees are pouring over the test questions. [97.509c]

**E1F15 What should a VE do if a candidate fails to comply with the examiner's instructions during an amateur operator license examination?**

A. Warn the candidate that continued failure to comply will result in termination of the examination
B. Immediately terminate the candidate's examination
C. Allow the candidate to complete the examination, but invalidate the results
D. Immediately terminate everyone's examination and close the session

**ANSWER B:** If an examinee fails to comply with your instructions, you should immediately terminate that candidate's examination. [97.509c]

**E1F16 What special procedures must a VE team follow for an examinee with a physical disability?**

A. A special procedure that accommodates the disability
B. A special procedure specified by the coordinating VEC
C. A special procedure specified by a physician
D. None; the VE team does not have to provide special procedures

**ANSWER A:** Volunteer examiner teams take great pride in developing special procedures to accommodate a disability. These procedures vary with the degree of disability, and there is no set, specific procedure by the coordinating VEC. The procedures are adopted by the VE team themselves. There also is no list of specific disabilities that require special accommodation – again, it will be the call of the individual VE team. [97.509k]

**E1F17 To which of the following examinees may a VE not administer an examination?**

A. The VE's close relatives as listed in the FCC rules
B. Acquaintances of the VE
C. Friends of the VE
D. There are no restrictions as to whom a VE may administer an examination

**ANSWER A:** As a VE, you may NOT test your own close relatives. [97.509(d)]

**E1F18 What may be the penalty for a VE who fraudulently administers or certifies an examination?**

A. Revocation of the VE's amateur station license grant and the suspension of the VE's amateur operator license grant
B. A fine of up to $1000 per occurrence
C. A sentence of up to one year in prison
D. All of these choices are correct

**ANSWER A:** Any VE caught administering a fraudulent exam may have his or her station license revoked and the VE's amateur radio license grant suspended. [97.509e]

**E1F19 What must the VE team do with your test papers when you have finished this examination?**

A. The VE team must collect them for grading at a later date
B. The VE team must collect and send them to the coordinating VEC for grading
C. The VE team must collect and grade them immediately
D. The VE team must collect and send them to the FCC for grading

**ANSWER C:** After the exam, the VE team must collect and grade all papers immediately. [97.509h]

---

**E1F20 What action must the coordinating VEC complete within 10 days of collecting the information from an examination session?**

A. Screen collected information
B. Resolve all discrepancies and verify that the VEs' certifications are properly completed
C. For qualified examinees, forward electronically all required data to the FCC
D. All of these choices are correct

**ANSWER D:** The job of the Volunteer Examiner Coordinator is time-sensitive; they have only 10 days after receiving an examination session to screen it and check it, resolve any discrepancies, and electronically forward the results to the FCC. [97.519b]

---

**E1F21 What must the VE team do if an examinee scores a passing grade on all examination elements needed for an upgrade or new license?**

A. Photocopy all examination documents and forward them to the FCC for processing
B. Notify the FCC that the examinee is eligible for a license grant
C. Issue the examinee the new or upgrade license
D. Three VEs must certify that the examinee is qualified for the license grant and that they have complied with the VE requirements

**ANSWER D:** When an examinee upgrades to a new, higher license, the 3-member VE team must certify the exam paperwork and sign the CSCE (Certificate of Successful Completion of Examination) showing that they have complied with the VE requirements. The applicant holds onto the CSCE until such time as the license is issued by the FCC. [97.509i]

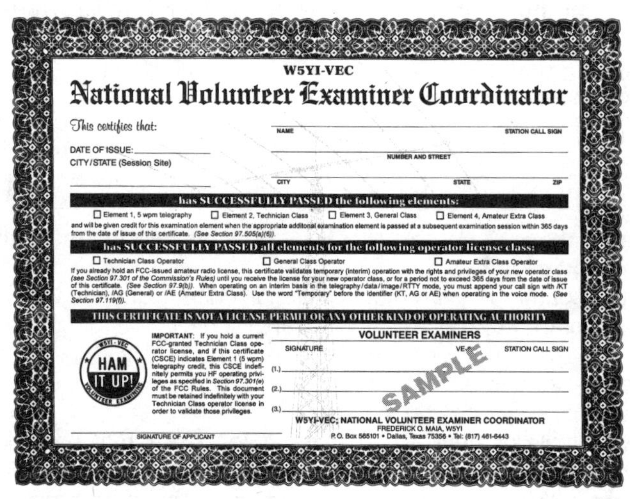

W5YI-VEC

National Volunteer Examiner Coordinator

This certifies that:

NAME — STATION CALL SIGN

DATE OF ISSUE: ____

CITY/STATE (Session Site)

NUMBER AND STREET

CITY — STATE — ZIP

**has SUCCESSFULLY PASSED the following elements:**

☐ Element 1, 5 wpm telegraphy ☐ Element 2, Technician Class ☐ Element 3, General Class ☐ Element 4, Amateur Extra Class

and will be given credit for this examination element when the appropriate additonal examination element is passed at a subsequent examination session within 365 days from the date of issue of this certificate. *(See Section 97.505(a)(6)).*

**has SUCCESSFULLY PASSED all elements for the following operator license class:**

☐ Technician Class Operator ☐ General Class Operator ☐ Amateur Extra Class Operator

If you already hold an FCC-issued amateur radio license, this certificate validates temporary (interim) operation with the rights and privileges of your new operator class *(see Section 97.301 of the Commission's Rules)* until you receive the license for your new operator class, or for a period not to exceed 365 days from the date of issue of this certificate. *(See Section 97.9(b)).* When operating on an interim basis in the telegraphy/data/image/RTTY mode, you must append your call sign with /KT (Technician), /AG (General) or /AE (Amateur Extra Class). Use the word "Temporary" before the identifier (KT, AG or AE) when operating in the voice mode. *(See Section 97.119(f)).*

**THIS CERTIFICATE IS NOT A LICENSE PERMIT OR ANY OTHER KIND OF OPERATING AUTHORITY**



**IMPORTANT:** If you hold a current FCC-granted Technician Class operator license, and if this certificate (CSCE) indicates Element 1 (5 wpm) telegraphy credit, this CSCE indefinitely permits you HF operating privileges as specified in *Section 97.301(e)* of the FCC Rules. This document must be retained indefinitely with your Technician Class operator license in order to validate those privileges.

VOLUNTEER EXAMINERS

SIGNATURE — VE-# — STATION CALL SIGN

(1.) ____

(2.) ____

(3.) ____

SAMPLE

SIGNATURE OF APPLICANT

W5YI-VEC; NATIONAL VOLUNTEER EXAMINER COORDINATOR
FREDERICK O. MAIA, W5YI
P.O. Box 565101 • Dallas, Texas 75356 • Tel: (817) 461-6443

**CSCE (Certificate of Successful Completion of Examination)**

---

**E1F22 What must the VE team do if the examinee does not score a passing grade on the examination?**

A. Return the application document to the examinee and inform the examinee of the grade
B. Return the application document to the examinee

C. Inform the examinee that he or she did not pass
D. Explain how the incorrect questions should have been answered

**ANSWER A:** You will return the application to the examinee and inform the candidate of the failing grade. Try to do this compassionately, urging the candidate to continue to study and try again soon. [97.509j]

---

**E1F23 What are the consequences of failing to appear for readministration of an examination when so directed by the FCC?**

A. The licensee's license will be cancelled and a new license will be issued that is consistent with examination elements not invalidated
B. The licensee must pay a monetary fine
C. The licensee is disqualified from any future examination for an amateur operator license grant
D. The person may be sentenced to incarceration

**ANSWER A:** Occasionally, the FCC asks an applicant to appear for readministration of an exam. This occurs if the exam session appears to have been flawed. If the licensee fails to appear for the re-exam, the licensee's license will be canceled and a new license will be issued that is consistent with examination elements not invalidated. [97.519d3]

---

**E1F24 What are the types of out-of-pocket expenses for which the FCC rules authorize a VE and VEC to accept reimbursement?**

A. Preparing, processing, administering and coordinating an examination for an amateur radio license
B. Teaching an amateur operator license examination preparation course
C. None; a VE must never accept any type of reimbursement
D. Providing amateur operator license examination preparation training materials

**ANSWER A:** Volunteer examiners and VEC's are allowed to recoup out-of-pocket expenses for the preparing, processing, administering, and coordinating an examination for an amateur radio license. There is a set fee that each applicant pays for the entire exam session. [97.527]

---

**E1F25 How much reimbursement may the VE team and VEC accept for preparing, processing, administering and coordinating an examination?**

A. Actual out-of-pocket expenses
B. Up to the national minimum hourly wage times the number of hours spent providing the services
C. Up to the maximum fee per examinee announced by the FCC annually
D. As much as the examinee is willing to donate

**ANSWER A:** The fee collected is for actual out-of-pocket expenses. [97.509e, 97.527]

---

**E1F26 What amateur operator license examination credit must be given for a valid Certificate of Successful Completion of Examination (CSCE)?**

A. Only the written elements the CSCE indicates the examinee passed with in the previous 365 days
B. Only the telegraphy elements the CSCE indicates the examinee passed within the previous 365 days
C. Each element the CSCE indicates the examinee passed within the previous 365 days
D. None

**ANSWER C:** If an applicant passes an element, but doesn't have all of the elements necessary for an upgrade, they will be issued a Certificate of Successful Completion of Examination (CSCE) showing each element passed. The CSCE is valid for 365 days. You may test for General theory, but the applicant still has not completed the code. The passed theory element receives 365-day CSCE paperwork. [97.505a6]

**E1F27 For what period of time does a Technician class licensee, who has just been issued a CSCE for having passed a 5 WPM Morse code examination, have authority to operate on the Novice/Technician HF subbands?**

A. 365 days from the examination date as indicated on the CSCE
B. 1 year from the examination date as indicated on the CSCE
C. Indefinitely, so long as the Technician license remains valid
D. 5 years plus a 5-year grace period from the examination date as indicated on the CSCE

**ANSWER C:** This is a great Q & A! You will be asked this question a lot at the exam site. If a Technician comes in and passes the 5-wpm code test, they will be issued a CSCE for the test-passing process and are encouraged to get their General theory passed within the next 365 days. However, this code-passing Technician may operate on Novice/Technician HF CW and voice sub-bands indefinitely, so long as the Technician license remains valid, even though the 365-day CSCE "appears" to expire. Now read my explanation for the next question. [97.301(e)]

**E1F28 What period of time does a Technician class licensee, who has just been issued a CSCE for having passed a 5 WPM Morse code examination, have in order to use this credit toward a license upgrade?**

A. 365 days from the examination date as indicated on the CSCE
B. 15 months from the examination date as indicated on the CSCE
C. There is no time limit, so long as the Technician license remains valid
D. 5 years plus a 5-year grace period from the examination date as indicated on the CSCE

**ANSWER A:** The Technician class licensee, having passed the 5-wpm code test, has only 365 days to study my General book and General theory audio course and pass the written exam. If they wait any longer than 365 days, they will need to retake the code test. But again, remember that their operating privileges having passed the code test remain valid for the duration of their Technician class license on the HF sub-bands. [97.505a6]

***E1G Certification of external RF power amplifiers and external RF power amplifier kits; Line A; National Radio Quiet Zone; business communications; definition and operation of spread spectrum; auxiliary station operation***

**E1G01 What does it mean if an external RF amplifier is listed on the FCC database as certificated for use in the amateur service?**

A. An RF amplifier of that model may be used in any radio service
B. That particular RF amplifier model may be marketed for use in the amateur service
C. All similar models of RF amplifiers produced by other manufacturers may be marketed
D. All models of RF amplifiers produced by that manufacturer may be marketed

**ANSWER B:** Do you own a linear amplifier? If so, look at the certification label and see that it is marked as approved for use in the amateur service. All commercially-made RF amplifiers must be listed on the FCC database. [97.315c]

**E1G02 Which of the following is one of the standards that must be met by an external RF power amplifier if it is to qualify for a grant of Certification?**

A. It must have a time-delay to prevent it from operating continuously for more than ten minutes
B. It must satisfy the spurious emission standards when driven with at least 50W mean RF power (unless a higher drive level is specified)
C. It must not be capable of modification without voiding the warranty
D. It must exhibit no more than 6dB of gain over its entire operating range

**ANSWER B:** The FCC-certified, high-frequency power amplifier must require a drive level greater than 50 watts in order to make it virtually unusable with a 5-watt CB radio. It must also satisfy the spurious emission standards at the 50-watt level. This again keeps the casual CB radio operator from using a modified ham amplifier since it won't cycle on below 50 watts input power. [97.317a3]

This little 2-meter linear amplifier meets stringent FCC spurious emission standards.

**2-Meter Power Amplifier**

---

**E1G03 Under what condition may an equipment dealer sell an external RF power amplifier capable of operation below 144 MHz if it has not been granted FCC certification?**

A. It was purchased in used condition from an amateur operator and is sold to another amateur operator for use at that operator's station
B. The equipment dealer assembled it from a kit
C. It was imported from a manufacturer in a country that does not require certification of RF power amplifiers
D. It was imported from a manufacturer in another country, and it was certificated by that country's government

**ANSWER A:** Only "home brew" amplifiers for frequencies below 144 MHz could be purchased in used condition by an equipment dealer and then sold to another amateur operator. This rule prevents the sale of amplifiers really intended for CB use that are constructed without the FCC blessing. [97.315b5]

---

**E1G04 Which of the following geographic descriptions approximately describes Line A?**

A. A line roughly parallel to, and south of, the US-Canadian border
B. A line roughly parallel to, and west of, the US Atlantic coastline
C. A line roughly parallel to, and north of, the US-Mexican border and Gulf coastline
D. A line roughly parallel to, and east of, the US Pacific coastline

**ANSWER A:** The United States and Canada have adopted the "Line A" as a buffer zone for certain types of radio stations. This keeps normal "line-of-sight" radio energy from interfering with the other country's communications. Amateurs are restricted on the 70-cm band between 420 MHz-430 MHz from transmitting if they are located north of "Line A". This is because Canada has a different type of radio allocation in the bottom part of the 70-cm band. [97.3a32]

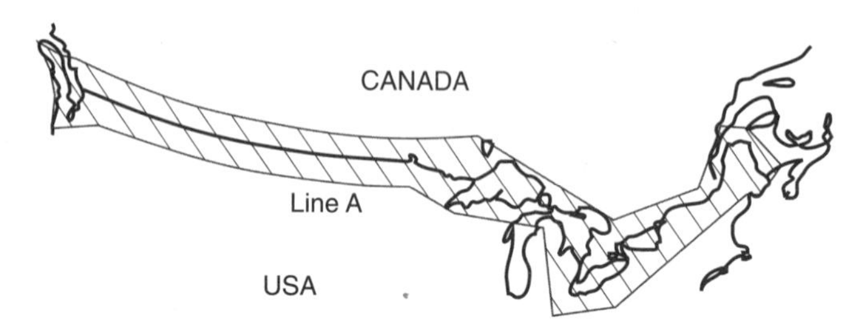


**Line A Amateur Radio Restrictions**

**E1G05 Amateur stations may not transmit in which frequency segment if they are located north of Line A?**

A. 21.225-21.300 MHz
B. 53-54 MHz
C. 222-223 MHz
D. 420-430 MHz

**ANSWER D:** Frequencies between 420 MHz-430 MHz are normally used for digital modes, fast-scan television, and auxiliary links. Within the "Line A" area next to the Canadian border, U.S. hams are prohibited from operating below 430 MHz. [97.303f1]

**E1G06 What is the National Radio Quiet Zone?**

A. An area in Puerto Rico surrounding the Aricebo Radio Telescope
B. An area in New Mexico surrounding the White Sands Test Area
C. An Area in Maryland, West Virginia and Virginia surrounding the National Radio Astronomy Observatory
D. An area in Florida surrounding Cape Canaveral

**ANSWER C:** A national radio quiet zone has been established near the National Radio Astronomy Observatory in Maryland, West Virginia, and Virginia. The quiet zone restriction applies to "constant-on" stations like beacons and repeaters. Operation from your home or vehicle would not be affected unless you actually enter the property of the National Radio Astronomy Observatory in Greenbank, West Virginia. [97.3a32]

**E1G07 What type of automatically controlled amateur station must not be established in the National Radio Quiet Zone before the licensee gives written notification to the National Radio Astronomy Observatory?**

A. Beacon station
B. Auxiliary station
C. Repeater station
D. Earth station

**ANSWER A:** An amateur beacon station continuously transmits, and could create problems in a radio quiet zone. You would not be able to establish a beacon station in a national radio quiet zone before you have received written notification from the National Radio Astronomy Observatory. [97.203e]

**E1G08 When may the control operator of a repeater accept payment for providing communication services to another party?**

A. When the repeater is operating under portable power
B. When the repeater is operating under local control
C. During Red Cross or other emergency service drills
D. Under no circumstances

**ANSWER D:** Under no circumstances can someone pay you for handling a ham radio call. Even if they want to give you a tip, tell the stranded motorist to learn more about the great hobby of Amateur Radio. [97.113a2]

---

**E1G09 When may an amateur station send a message to a business?**

A. When the total money involved does not exceed $25
B. When the control operator is employed by the FCC or another government agency
C. When transmitting international third-party communications
D. When neither the amateur nor his or her employer has a pecuniary interest in the communications

**ANSWER D:** In 1993, the prohibited business communications rules were redefined by the FCC, allowing an amateur to send a message to a business as long as neither the amateur nor his or her employer has a pecuniary interest in the communications. If you run a tree-trimming business it would not be legal to use an autopatch to call your answering machine to find out where your next customer is waiting. On the other hand, if you are a school teacher and discover that a big pine tree has fallen down across the school's driveway, it would be perfectly acceptable to place an autopatch call to a tree-trimmer and have them come out to do their job. But could the school teacher call into the school district and tell them this teacher is going to be late for work? Probably NOT legal. [97.113a3]

---

**E1G10 Which of the following types of amateur operator-to-amateur operator communication are prohibited?**

A. Communications transmitted for hire or material compensation, except as otherwise provided in the rules
B. Communication that has a political content
C. Communication that has a religious content
D. Communication in a language other English

**ANSWER A:** This is a new question, dealing with permissible communications from one ham to another. Political, religious, and foreign language communications ARE permitted, but communications transmitted for hire or material compensation are normally prohibited. [97.113]

---

**E1G11 What is the term for emissions using bandwidth-expansion modulation?**

A. RTTY
B. Image
C. Spread spectrum
D. Pulse

**ANSWER C:** Spread spectrum may operate over an entire amateur radio band by split-second transmission over hundreds and thousands of "slices" within that band. When two spread spectrum stations are in QSO, their signals are all but invisible to the casual operator. The two stations can stay locked-on without interruption even though certain parts of the amateur band may be in use by conventional emission modes. New digital 900 MHz cordless phones operate spread spectrum, so you may already have a spread spectrum system around your household! [97.3c8]

**E1G12 FCC-licensed amateur stations may use spread spectrum (SS) emissions to communicate under which of the following conditions?**

A. When the other station is in an area regulated by the FCC
B. When the other station is in a country permitting SS communications
C. When the transmission is not used to obscure the meaning of any communication
D. All of these choices are correct

**ANSWER D:** Spread-spectrum communications are allowed to any other station under FCC regulation, and to stations in foreign countries that have specifically permitted spread-spectrum communications to their country from ours. Spread-spectrum may not be used to obscure the meaning of any communication. If these 3 requirements are met, and as long as you stay below 100 watts, you are entitled to use spread spectrum. [97.311(a)]

---

**E1G13 Under any circumstance, what is the maximum transmitter power for an amateur station transmitting emission type SS communications?**

A. 1 W
B. 1.5 W
C. 100 W
D. 1.5 kW

**ANSWER C:** Since hams are still in the early stages of experimenting with spread spectrum, the FCC limits maximum power output to 100 watts. [97.311d]

---

**E1G14 What of the following is a use for an auxiliary station?**

A. To provide a point-to-point communications uplink between a control point and its associated remotely controlled station
B. To provide a point-to-point communications downlink between a remotely controlled station and its control point
C. To provide a point-to-point control link between a control point and its associated remotely controlled station
D. All of these choices are correct

**ANSWER D:** Auxiliary stations are a great way to provide an uplink between a control point and the remotely controlled station for a downlink between that station and your control point. You also can provide a point-to-point control link between a control point and its associated remotely controlled station, and this is usually accomplished on UHF or microwave frequencies. [97.109c]

## Subelement E2 — Operating Procedures [5 Exam Questions — 5 Groups]

***E2A Amateur Satellites: orbital mechanics; frequencies available for satellite operation; satellite hardware; satellite operations***

**E2A01 What is the direction of an ascending pass for an amateur satellite?**

A. From west to east
B. From east to west
C. From south to north
D. From north to south

**ANSWER C:** Ascending... going up! From south to north.

---

**E2A02 What is the direction of a descending pass for an amateur satellite?**

A. From north to south
B. From west to east
C. From east to west
D. From south to north

**ANSWER A:** Descending... going down! From north to south.

**E2A03 What is the period of an amateur satellite?**

A. The point of maximum height of a satellite's orbit
B. The point of minimum height of a satellite's orbit
C. The amount of time it takes for a satellite to complete one orbit
D. The time it takes a satellite to travel from perigee to apogee

**ANSWER C:** One complete orbit is the period of a satellite.

---

**E2A04 What are the receiving and retransmitting frequency bands used for Mode V/H in amateur satellite operations?**

A. Satellite receiving on Amateur bands in the range of 21 to 30 MHz and retransmitting on 144 to 148 MHz
B. Satellite receiving on 435 to 438 MHz and retransmitting on 144 to 148 MHz
C. Satellite receiving on 435 to 438 MHz and retransmitting on Amateur bands in the range of 21 to 30 MHz
D. Satellite receiving on 144 to 148 MHz and retransmitting on Amateur bands in the range of 21 to 30 MHz

**ANSWER D:** This question reflects a change in satellite mode "lingo." Instead of listing the different modes as "Mode A, Mode J, etc.," we now list them as VHF, UHF, or high-frequency modes. Here is what you need to remember:

**Amateur Satellite Frequency Bands**

| Mode | Frequency Range |
|---|---|
| Mode H – high-frequency | 21-30 MHz |
| Mode V – very-high frequency | 144-148 MHz |
| Mode U – ultra-high frequency | 435-438 MHz |
| Mode L – L-band | 1.260-1.270 GHz |

Further, you need to carefully read the question, because they always ask satellite receiving frequency first, and transmitting frequency second. Thus, for receiving and transmitting using Mode V/H, the satellite is receiving V on 144-148 MHz, and transmitting on H from 21-30 MHz.

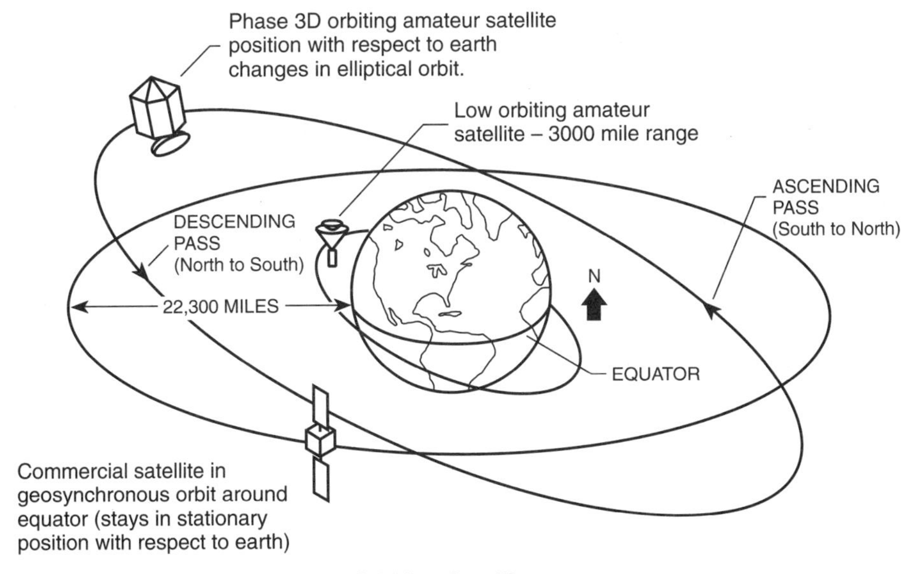


**Orbiting Satellites**

**E2A05 What are the receiving and retransmitting frequency bands used for Mode U/V in amateur satellite operations?**

A. Satellite receiving on Amateur bands in the range of 21 to 30 MHz and retransmitting on 144 to 148 MHz
B. Satellite receiving on 435 to 438 MHz and retransmitting on 144 to 148 MHz
C. Satellite receiving on 435 to 438 MHz and retransmitting on Amateur bands in the range of 21 to 30 MHz
D. Satellite receiving on 144 to 148 MHz and retransmitting on Amateur bands in the range of 21 to 30 MHz

**ANSWER B:** Mode U/V, the satellite is receiving on U, 435-438 MHz, and transmitting on V, 144-148 MHz.

---

**E2A06 What are the receiving and retransmitting frequency bands used for Mode V/U in amateur satellite operations?**

A. Satellite receiving on 435 to 438 MHz and retransmitting on 144 to 148 MHz
B. Satellite receiving on 144 to 148 MHz and retransmitting on Amateur bands in the range of 21 to 30 MHz
C. Satellite receiving on 144 to 148 MHz and retransmitting on 435 to 438 MHz
D. Satellite receiving on 435 to 438 MHz and transmitting on 21 to 30 MHz

**ANSWER C:** For Mode V/U, the satellite is receiving on V, 144-148 MHz, and transmitting on U, 435-438 MHz.

---

**E2A07 What are the receiving and retransmitting frequency bands used for Mode L/U in amateur satellite operations?**

A. Satellite receiving on 435 to 438 MHz and retransmitting on 21 to 30 MHz
B. Satellite receiving on Amateur bands in the range of 21 to 30 MHz and retransmitting on 435 to 438 MHz
C. Satellite receiving on 435 to 438 MHz and retransmitting on 1.26 to 1.27 GHz
D. Satellite receiving on 1.26 to 1.27 GHz and retransmitting on 435 to 438 MHz

**ANSWER D:** On Mode L/U, the satellite is receiving L on 1.26-1.27 GHz, and transmitting U on 435-438 MHz. Got it? Remember: H for HF, V for VHF, U for UHF, and L for the GHz L band. Review these 4 questions. One of them will probably be on your upcoming Extra Class exam.

---

**E2A08 What is a linear transponder?**

A. A repeater that passes only linear or CW signals
B. A device that receives and retransmits signals of any mode in a certain passband
C. An amplifier that varies its output linearly in response to input signals
D. A device that responds to satellite telecommands and is used to activate a linear sequence of events

**ANSWER B:** A linear transponder in a satellite will retransmit signals of many different modes through the satellite, including CW, SSB and, while not recommended, even FM.

---

**E2A09 What is the name of the effect that causes the downlink frequency of a satellite to vary by several kHz during a low-earth orbit?**

A. The Kepler effect
B. The Bernoulli effect
C. The Einstein effect
D. The Doppler effect

**ANSWER D:** As a satellite is speeding toward you, the Doppler effect will make its frequencies several hundred hertz higher. As the satellite passes overhead and speeds away from you, the signals will begin to decrease in frequency.

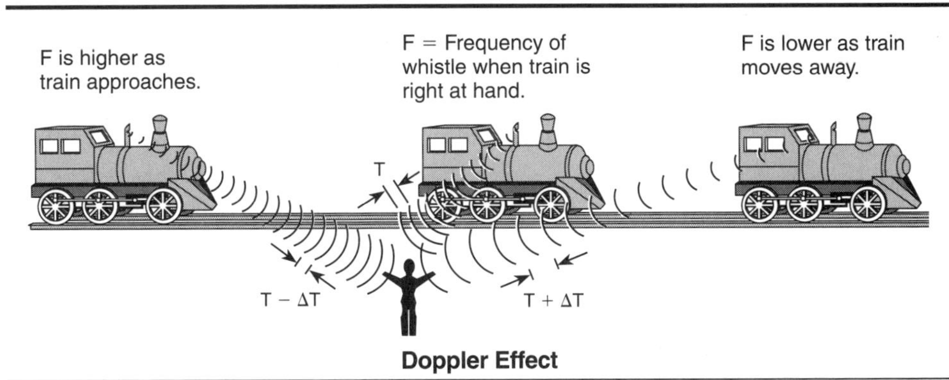


**Doppler Effect**

**E2A10 Why may the received signal from an amateur satellite exhibit a fairly rapid pulsed fading effect?**

A. Because the satellite is rotating
B. Because of ionospheric absorption
C. Because of the satellite's low orbital altitude
D. Because of the Doppler effect

**ANSWER A:** Many amateur satellites continuously rotate to keep all sides of the "bird" equally bathed in sunlight. This distributes the heat build-up, and allows all of the solar panels to receive energy from the sun. Because of the rotation, we hear a characteristic rapid pulse in and out on the downlink.

**E2A11 What type of antenna can be used to minimize the effects of spin modulation and Faraday rotation?**

A. A nonpolarized antenna
B. A circularly polarized antenna
C. An isotropic antenna
D. A log-periodic dipole array

**ANSWER B:** The circular-polarized antenna will give you best results when working amateur satellites. Although you lose approximately 3 dB gain over a horizontal antenna, circular polarization minimizes fading.

**E2A12 How may the location of a satellite at a given time be predicted?**

A. By means of the Doppler data for the specified satellite
B. By subtracting the mean anomaly from the orbital inclination
C. By adding the mean anomaly to the orbital inclination
D. By means of the Keplerian elements for the specified satellite

**ANSWER D:** You can log onto many satellite web pages and learn the Keplerian elements for a specific satellite or a specific day and time for the location of the satellite. You can find this information on the AMSAT website at: www.AMSAT.org.

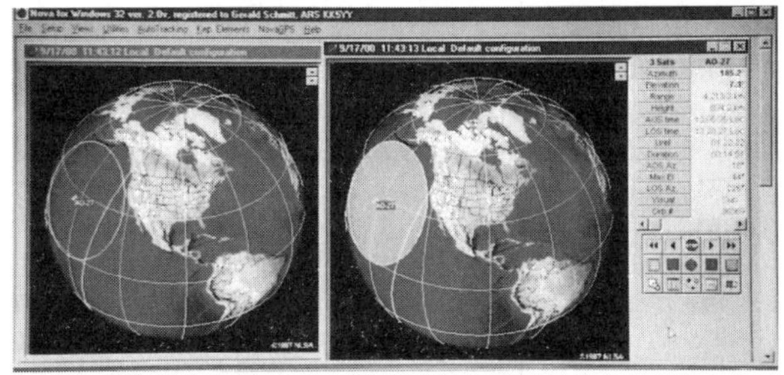

**Computer programs and websites can show you where and when an amateur satellite or the Space Station will be in range of your ham station**

***E2B Television: fast scan television (FSTV) standards; slow scan television (SSTV) standards; facsimile (fax) communications***

**E2B01 How many times per second is a new frame transmitted in a fast-scan television system?**

A. 30 C. 90
B. 60 D. 120

**ANSWER A:** Operating fast-scan television is fun on the 430-MHz, 900-MHz, and 1200-MHz bands. A new frame is created 30 times per second by two complete fields for each frame. This gives you a picture just like home television.

**Amateur Television**

**E2B02 How many horizontal lines make up a fast-scan television frame?**

A. 30 C. 525
B. 60 D. 1050

**ANSWER C:** A total of 525 lines from two fields make up a complete frame.

**E2B03 How is the interlace scanning pattern generated in a fast-scan television system?**

A. By scanning the field from top to bottom
B. By scanning the field from bottom to top
C. By scanning from left to right in one field and right to left in the next
D. By scanning odd numbered lines in one field and even numbered ones in the next

**ANSWER D:** Each field is 262.5 lines, and the frame is created by scanning even numbered lines in one field and odd numbered lines in the second field.

**E2B04 What is blanking in a video signal?**

A. Synchronization of the horizontal and vertical sync pulses
B. Turning off the scanning beam while it is traveling from right to left and from bottom to top
C. Turning off the scanning beam at the conclusion of a transmission
D. Transmitting a black and white test pattern

**ANSWER B:** The scanning beam is turned off when it completes its travel from left to right, and is returning from right to left to start another line. The scanning beam is also blanked when the beam reaches the bottom and is returning to the top to begin the next field.

**E2B05 What is the bandwidth of a vestigial sideband AM fast-scan television transmission?**

A. 3 kHz
B. 10 kHz
C. 25 kHz
D. 6 MHz

**ANSWER D:** Amateur television requires a wide bandwidth and thus occupies a tremendous amount of spectrum – 6 MHz. This is why you won't find fast-scan amateur television below the 420-MHz ham band.

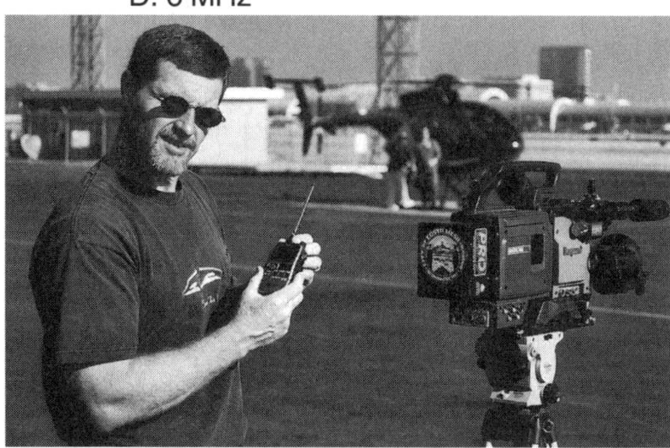

**Amateur TV signals can be received on a variety of equipment – even a small hand-held monitor**

---

**E2B06 What is the standard video level, in percent PEV, for black in amateur fast scan television?**

A. 0%
B. 12.5%
C. 70%
D. 100%

**ANSWER C:** The standard video level voltage for black is 70% of the peak envelope voltage (PEV).

---

**E2B07 What is the standard video level, in percent PEV, for blanking in amateur fast scan television?**

A. 0%
B. 12.5%
C. 75%
D. 100%

**ANSWER C:** The standard video level voltage for blanking is in the "blacker-than-black" range at 75% of the peak envelope voltage (PEV).

---

**E2B08 Which of the following is NOT a common method of transmitting accompanying audio with amateur fast-scan television?**

A. Amplitude modulation of the video carrier
B. Frequency-modulated sub-carrier
C. A separate VHF or UHF audio link
D. Frequency modulation of the video carrier

**ANSWER A:** We would not modulate the video carrier, using amplitude modulation, for amateur fast-scan television. We frequency modulate the sub-carrier, or provide a separate audio channel link, usually on the 2-meter band. We also could frequency modulate the video carrier, but NEVER amplitude modulate it.

---

**E2B09 What is facsimile?**

A. The transmission of characters by radioteletype that form a picture when printed
B. The transmission of still pictures by slow-scan television
C. The transmission of video by amateur television
D. The transmission of printed pictures for permanent display on paper

**ANSWER D:** When you pass your examination, you will probably all send your author a facsimile of your Certificate of Successful Completion of Examination. This is a *printed picture* (key words), and he will keep it on file at his Radio School and send you a graduation certificate and some nice gift certificates. (See page 245 for free gift certificate and graduation certificate offer.)

**Facsimile System**
**Courtesy of Alden Electronics**

---

**E2B10 What is the modern standard scan rate for a fax image transmitted by an amateur station?**

A. 240 lines per minute
B. 50 lines per minute
C. 150 lines per second
D. 60 lines per second

**ANSWER A:** Amateur radio facsimile stations transmit a standard 240 lines per minute. Remember this key – out of all the answers, 240 is the highest resolution.

---

**E2B11 What is the approximate transmission time per frame for a fax picture transmitted by an amateur station at 240 lpm?**

A. 6 minutes
B. 3.3 minutes
C. 6 seconds
D. 1/60 second

**ANSWER B:** It takes about as much time to cook a 3-minute egg as it does to receive an amateur station facsimile picture. Exactly 3.3 minutes will allow for a single frame at 240 lines per minute (lpm).

---

**E2B12 What information is sent by slow-scan television transmissions?**

A. Baudot or ASCII characters that form a picture when printed
B. Pictures for permanent display on paper
C. Moving pictures
D. Still pictures

**ANSWER D:** Another interesting area of amateur radio operation is slow-scan TV. The pictures are displayed on a tube or an LCD readout (not on paper as answer B has it). Slow-scan television can be in color, too, *sending still pictures by radio* (key words). It takes approximately 45 seconds to receive a color slow-scan picture.

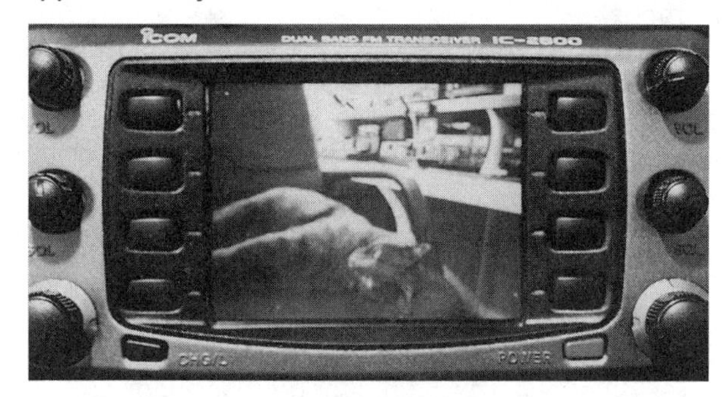


**Amateur slow-scan TV pictures can be displayed on a tube, or an LCD display like this one**

**E2B13 How many lines are commonly used in each frame on an amateur slow-scan color television picture?**

A. 30 to 60 C. 128 or 256
B. 60 or 100 D. 180 or 360

**ANSWER C:** It takes 128 lines to make up a complete frame. 256 lines increases the resolution of the color.

---

**E2B14 What is the audio frequency for black in an amateur slow-scan television picture?**

A. 2300 Hz C. 1500 Hz
B. 2000 Hz D. 120 Hz

**ANSWER C:** You can hear the audio sounds of slow-scan television over the airwaves with low tones around 1500 Hz, and high tones around 2300 Hz. The lowest audio frequency for black is 1500. The audio frequency in Answer D would be too low to be accurately picked up by a receiver. Listen on the 20 Meter Band around 14.235 MHz for examples.

---

**E2B15 What is the audio frequency for white in an amateur slow-scan television picture?**

A. 120 Hz C. 2000 Hz
B. 1500 Hz D. 2300 Hz

**ANSWER D:** Slow-scan white is the highest audio tone, 2300 Hz. Slow-scan television is making a major comeback from the early days of the high-persistence CRT. Now pictures are sent in color using your home computer and a "sound-blaster" card. You can tune into slow-scan activity on the 20-meter band at 14.230 MHz and 14.234 MHz. You will select "receive picture" on your PC slow-scan computer program (available by mail order in most ham magazines). Digitization captures the incoming signal into the buffer. It always starts at position zero, and digitizes until either the buffer is full or a key is pressed. Once the incoming signal is digitized, the program will attempt to display the image according to the present format and calibration of values. Try slow-scan on your computer. It's fun!

---

**E2B16 What is the standard video level, in percent PEV, for white in an amateur fast-scan television transmission?**

A. 0% C. 70%
B. 12.5% D. 100%

**ANSWER B:** For fast-scan video, the percent PEV (peak envelope voltage) for WHITE is 12.5%.

---

**E2B17 Which of the following is NOT a characteristic of FMTV (Frequency-Modulated Amateur Television) as compared to vestigial sideband AM television?**

A. Immunity from fading due to limiting
B. Poor weak signal performance
C. Greater signal bandwidth
D. Greater complexity of receiving equipment

**ANSWER A:** Here is one of those mind-boggling questions where they want you to know a "not" characteristic of FM amateur television compared to AM amateur TV. "Immunity from fading due to limiting" is not a characteristic of FM TV – it WILL fade due to limiting.

**E2B18 What is the approximate bandwidth of a slow-scan TV signal?**

A. 600 Hz
B. 2 kHz
C. 2 MHz
D. 6 MHz

**ANSWER B:** Have you tried one of those computer programs or slow-scan camera systems on VHF or high frequency? Yup, slow-scan IS allowed on high frequency as well as the higher bands because it only occupies 2 kHz of bandwidth – no more than a typical SSB signal. It takes about 30 seconds to send a color image.

---

**E2B19 Which of the following systems is used to transmit high-quality still images by radio?**

A. AMTOR
B. Baudot RTTY
C. AMTEX
D. Facsimile

**ANSWER D:** Another method of transmitting high-quality still images by radio is facsimile. All of the other modes listed as possible answers are NOT for sending still images by radio.

---

**E2B20 What special restrictions are imposed on fax transmissions?**

A. None; they are allowed on all amateur frequencies
B. They are restricted to 7.245 MHz, 14.245 MHz, 21.345, MHz, and 28.945 MHz
C. They are allowed in phone band segments if their bandwidth is no greater than that of a voice signal of the same modulation type
D. They are not permitted above 54 MHz

**ANSWER C:** Facsimile transmissions, as well as slow-scan TV, are allowed in phone band segments as long as their bandwidth is no greater than that of a voice signal of the same modulation type. You can easily hear 20-meter slow-scan signals by tuning in 14.230 and 14.233, a hot-bed of SSTV activity that is easily seen on a computer program or on your little handheld SSTV transmit/receive equipment.

---

***E2C Contest and DX operating; spread-spectrum transmissions; automatic HF forwarding; selecting your operating frequency***

---

**E2C01 When operating during a contest, which of these standards should you generally follow?**

A. Always listen before transmitting, be courteous and do not cause harmful interference to other communications
B. Always reply to other stations calling CQ at least as many times as you call CQ
C. When initiating a contact, always reply with the call sign of the station you are calling followed by your own call sign
D. Always include your signal report, name and transmitter power output in any exchange with another station

**ANSWER A:** "Always listen before transmitting" is good amateur radio practice at anytime – contesting, or not.

---

**E2C02 What is one of the main purposes for holding on-the-air operating contests?**

A. To test the dollar-to-feature value of station equipment during difficult operating circumstances
B. To enhance the communicating and operating skills of amateur operators in readiness for an emergency

C. To measure the ionosphere's capacity for refracting RF signals under varying conditions
D. To demonstrate to the FCC that amateur station operation is possible during difficult operating circumstances

**ANSWER B:** Almost every weekend there is some type of local amateur radio contest over the airwaves. While some amateur operators may be critical of contesting activities over "prime time" weekends, contests are a good way to increase the number of amateur operators coming onto the airwaves and enhancing their fast-paced communication skills. Fast-paced skills are necessary to efficiently handle emergency communications.

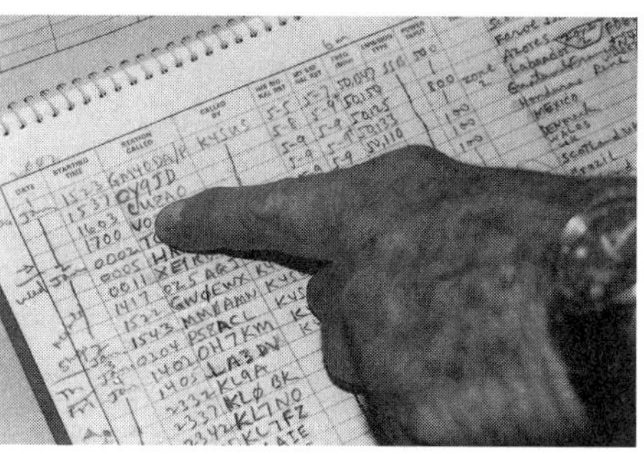

**A good contester keeps an accurate log – either on paper or with software**

**E2C03 Which of the following is typical of operations during an international amateur DX contest?**

A. Calling CQ is always done on an odd minute and listening is always done on an even minute
B. Contacting a DX station is best accomplished when the WWV K index is above a reading of 8
C. Some DX operators use split frequency operations (transmitting on a frequency different from the receiving frequency)
D. DX contacts during the day are never possible because of known band attenuation from the sun

**ANSWER C:** During international DX contests, a rare DX station may transmit just outside of our normal phone band, such as 14.145 phone, and listen "up 10," meaning we would transmit back to them on phone at 14.155, an area reserved EXCLUSIVELY for Extra class operators. See how important passing the Extra class examination will be for working some rare DX? Split operation is accomplished by running both VFO's on your rig – and on the real large transceivers, tracking separate receive from your transmit.

**E2C04 If a DX station asks for your grid square locator, what should be your reply?**

A. The square of the power fed to the grid of your final amplifier and your current city, state and country
B. The DX station's call sign followed by your call sign and your RST signal report
C. The subsection of the IARU region in which you are located based upon dividing the entire region into a grid of squares 10 km wide
D. Your geographic Maidenhead grid location (e.g., FN31AA) based on your current latitude and longitude

**ANSWER D:** VHF and UHF radio enthusiasts usually identify their location by grid squares. A longitude x latitude grid square is formed for every 20° of longitude (E-W) and 10° of latitude (N-S). The 20° x 10° square is assigned a 2-letter identification. Each 20° x 10° square is further divided into 100 2° x 1° squares, and identified with a 2-digit number from 00 to 99. A grid-square map for the U.S.A. is shown on page 257. Look at it and identify your grid square. Your author's is DM13. Each 2° x 1° grid square is also subdivided into even smaller squares for those operators making contact on microwave frequencies, such as the 10-GHz band.

**E2C05 What does a Maidenhead gridsquare refer to?**

A. A two-degree longitude by one-degree latitude square, as part of a world wide numbering system
B. A one-degree longitude by one-degree latitude square, beginning at the South Pole
C. An antenna made of wire grid used to amplify low-angle incoming signals while reducing high-angle incoming signals
D. An antenna consisting of a screen or grid positioned directly beneath the radiating element

**ANSWER A:** Amateur operators on frequencies from 50 MHz and higher usually refer to their location by grid squares – one-degree latitude by two-degrees longitude. Be careful you don't get Answer A and B mixed up – Answer A simply lists longitude first as 2 degrees, and latitude second as 1 degree.

---

**E2C06 During a VHF/UHF contest, in which band section would you expect to find the highest level of contest activity?**

A. At the top of each band, usually in a segment reserved for contests
B. In the middle of each band, usually on the national calling frequency
C. In the weak signal segment of the band, with most of the activity near the calling frequency
D. In the middle of the band, usually 25 kHz above the national calling frequency

**ANSWER C:** On the VHF and UHF bands, contest activity is usually single sideband and CW. This takes place at the bottom of each band in the weak signal segment. No FM!

---

**E2C07 If you are in the US calling a station in Texas on a frequency of 1832 kHz and a station replies that you are in the window, what does this mean?**

A. You are operating out of the band privileges of your license
B. You are calling at the wrong time of day to be within the window of frequencies that can be received in Texas at that time
C. You are transmitting in a frequency segment that is reserved for international DX contacts by gentlemen's agreement
D. Your modulation has reached an undesirable level and you are interfering with another contact

**ANSWER C:** It's tough to keep track of all of the "gentlemen's agreement" band plans on worldwide. For that matter, VHF and UHF, too! If someone asks you to move off of "the window," they mean please QSY because you are transmitting in an area where local stations are listening for international DX.

---

**E2C08 Why are received spread-spectrum signals so resistant to interference?**

A. Signals not using the spectrum-spreading algorithm are suppressed in the receiver
B. The high power used by a spread-spectrum transmitter keeps its signal from being easily overpowered
C. The receiver is always equipped with a special digital signal processor (DSP) interference filter
D. If interference is detected by the receiver it will signal the transmitter to change frequencies

**ANSWER A:** Amateur operators transmitting and receiving spread-spectrum signals are just pioneering one more way that hams can serve our valuable frequency resources. Although spread-spectrum may utilize an entire band, the hopping technique won't disturb other signals on the band, and all other signals not using the same spread-spectrum algorithm are suppressed in the receiver. So just when you thought there was no more room for any more signals on a particular band, spread-spectrum gets through!

---

**E2C09 How does the spread-spectrum technique of frequency hopping (FH) work?**

A. If interference is detected by the receiver it will signal the transmitter to change frequencies
B. If interference is detected by the receiver it will signal the transmitter to wait until the frequency is clear
C. A pseudo-random binary bit stream is used to shift the phase of an RF carrier very rapidly in a particular sequence
D. The frequency of the transmitted signal is changed very rapidly according to a particular sequence also used by the receiving station

**ANSWER D:** Frequency hopping depends on a particular signaling sequence, and the frequency of the RF carrier is changed so quickly that it is almost indistinguishable except to other spread-spectrum receivers locked on to the same signaling sequence.

---

**E2C10 While participating in an HF contest, how should you attempt to contact a station calling CQ and stating that he is listening on another specific frequency?**

A. By sending your full call sign on the listening frequency specified
B. By sending only the suffix of your call sign on the listening Frequency
C. By sending your full call sign on the frequency on which you heard the station calling CQ
D. By sending only the suffix of your call sign on the frequency on which you heard the station calling CQ

**ANSWER A:** If the DX station is operating "split," as in this question, only transmit on the frequency where they indicate they are listening. Always make sure to send your full call sign when operating DX, not just the last letter or 2 in your call sign, which is not legal. Full call sign. Rules also require that you momentarily listen to your offset transmit frequency to ensure the frequency is not in use.

---

**E2C11 When operating SSB in a VHF contest, how should you attempt to contact a station calling CQ while a pileup of other stations are also trying to contact the same station?**

A. By sending your full call sign after the distant station transmits QRZ
B. By sending only the last letters of your call sign after the distant station transmits QRZ
C. By sending your full call sign and grid square as soon as you hear the distant station transmit QRZ
D. By sending the call sign of the distant station three times, the words "this is", then your call sign three times

**ANSWER A:** When operating in a VHF SSB contest, getting a signal through a pile-up is a matter of TIMING. Just like on HF, always send your full call sign, and do so quickly – right after the distant station transmits "QRZ," which means he is waiting for someone to say their call sign. A little trick I use is to wait a few seconds until everyone else piles on top of each other, and then give my full call sign when I hear the pile-up quit in the moment of silence before the distant station goes "QRZ" again! Just send your own call sign; the DX station knows its own call sign. Don't send your grid square until after they've acknowledged your call.

**E2C12 In North America during low sunspot activity, signals from Europe become weak and fluttery across an entire HF band two to three hours after sunset, what might help to contact other European DX stations?**

A. Switch to a higher frequency HF band, because the MUF has increased
B. Switch to a lower frequency HF band because the MUF has decreased
C. Wait 90 minutes or so for the signal degradation to pass
D. Wait 24 hours before attempting another communication on the band

**ANSWER B:** We are on the downward slide of solar cycle 23, and the 20-meter band begins to die down two to three hours after sunset. When signals begin to become too weak to work, switch to a lower frequency HF band to compensate for the lower maximum useable frequency (MUF).

***E2D Operating VHF / UHF digital modes: packet clusters; digital bulletin boards; Automatic Position Reporting System (APRS)***

**E2D01 What does "CMD:" mean when it is displayed on the video monitor of a packet station?**

A. The TNC is ready to exit the packet terminal program
B. The TNC is in command mode, ready to receive instructions from the keyboard
C. The TNC will exit to the command mode on the next keystroke
D. The TNC is in KISS mode running TCP/IP, ready for the next command

**ANSWER B:** When the prompt "CMD" shows up on the monitor when you are in the packet mode, it indicates that the terminal node controller is ready to receive keyboard instructions.

**Packet Terminal Node Controller (TNC)**
Courtesy of MFJ Enterprises, Inc.

**E2D02 What is a Packet Cluster Bulletin Board?**

A. A packet bulletin board devoted primarily to serving a special interest group
B. A group of general-purpose packet bulletin boards linked together in a cluster
C. A special interest cluster of packet bulletin boards devoted entirely to packet radio computer communications
D. A special interest telephone/modem bulletin board devoted to amateur DX operations

**ANSWER A:** The Packet Cluster Bulletin Board serves specific interest groups in ham radio. The most popular are the DX packet cluster bulletin boards. As soon as a rare DX station is monitored on the air, the Packet Cluster Bulletin Board system alerts everyone tuned in that there is a rare DX – and who it is, and where to tune in. Almost like shooting fish in a barrel, huh?

---

**E2D03 In comparing HF and 2-meter packet operations, which of the following is NOT true?**

A. HF packet typically uses an FSK signal with a data rate of 300 bauds; 2-meter packet uses an AFSK signal with a data rate of 1200 bauds
B. HF packet and 2-meter packet operations use the same code for information exchange
C. HF packet is limited to Amateur Extra class amateur licensees; 2-meter packet is open to all but Novice Class amateur licensees
D. HF packet operations are limited to CW/Data-only band segments; 2-meter packet is allowed wherever FM operations are allowed

**ANSWER C:** High-frequency and VHF packet operation is open for everyone, regardless of license. The question asks which is NOT true – HF packet is NOT limited to just one class of license; and since we know that Novice class licensees do not have access to the 2-meter band, Answer C is your best choice for the answer that contains statements that are NOT true.

---

**E2D04 What is the purpose of a digital store and forward on an Amateur satellite?**

A. To stockpile packet TNCs and other digital hardware to be distributed to RACES operators in the event of an emergency
B. To relay messages across the country via a network of HF digital stations
C. To store messages in an amateur satellite for later download by other stations
D. To store messages in a packet digipeater for relay via the Internet

**ANSWER C:** Amateur orbiting satellites make great store-and-forward messengers to hold messages that may be downloaded later by other stations.

---

**E2D05 Which of the following techniques is normally used by low-earth orbiting digital satellites to relay messages around the world?**

A. Digipeating
B. Store and forward
C. Multi-satellite relaying
D. Node hopping

**ANSWER B:** Store-and-forward is a popular way of sending a wireless e-mail that may be received halfway around the world when the satellite passes over the distant station.

---

**E2D06 What is the common 2-meter APRS frequency?**

A. 144.20 MHz
B. 144.39 MHz
C. 145.02 MHz
D. 146.52 MHz

**ANSWER B:** APRS stands for Automatic Position Reporting System. APRS links a portable or fixed-mount Global Positioning System (GPS) receiver to a ham transmitter or transceiver. APRS allows the ham radio to transmit pre-set timed packet bursts of the GPS position, either on worldwide high frequencies, or on the VHF and UHF bands. On 2 meters, the most common APRS frequency is 144.390 MHz.

---

**E2D07 Which of the following digital protocols does APRS use?**

A. AX.25
B. 802.11
C. PACTOR
D. AMTOR

**ANSWER A:** The digital packet protocol in use for APRS is described as AX.25, the ham version of commercial packet networks operating X.25. The AX.25 protocol is recognized and authorized by the Federal Communications Commission with specific speed limitations depending on what band you are operating on.

**APRS links GPS to a ham transmitter to report the GPS position.**

**E2D08 Which of the following types of packet frames is used to transmit APRS beacon data?**

A. Connect frames
B. Disconnect frames
C. Acknowledgement frames
D. Un-numbered Information frames

**ANSWER D:** The packet frames used to transmit APRS beacon data are called UIF – Un-numbered Information Frames. Networking systems could allow an APRS position and short message to be transmitted and relayed throughout the country. One manufacturer produces a dual-band handheld with the APRS terminal node controller already built-in, and handheld GPS equipment that output NMEA data are available for under $99.

**E2D09 Under clear communications conditions, which of these digital communications modes has the fastest data throughput?**

A. AMTOR
B. 170-Hz shift, 45 baud RTTY
C. PSK31
D. 300-baud packet

**ANSWER D:** Although 300-baud packet sounds relatively slow, on the worldwide bands it's faster than most other modes.

**E2D10 How can an APRS station be used to help support a public service communications activity, such as a walk-a-thon?**

A. An APRS station with an emergency medical technician can automatically transmit medical data to the nearest hospital
B. APRS stations with General Personnel Scanners can automatically relay the participant numbers and time as they pass the check points
C. An APRS station with a GPS unit can automatically transmit information to show the station's position along the course route
D. All of these choices are correct

**ANSWER C:** Many marathons are supported by ham operators on foot, bicycles, and motorcycles with an APRS station tied into a GPS to automatically transmit position information along the course route. All this can be done without the operator needing to do a thing – simply set the time interval you want your APRS packets transmitted, and presto, everyone with the right equipment can easily see the APRS position information.

---

**E2D11 Which of the following data sources are needed to accurately transmit your geographical location over the APRS network?**

A. The NMEA-0183 formatted data from a Global Positioning System (GPS) satellite receiver
B. The latitude and longitude of your location, preferably in degrees, minutes and seconds, entered into the APRS computer software
C. The NMEA-0183 formatted data from a LORAN navigation system
D. All of these choices are correct

**ANSWER D:** Hooking up your APRS transmitter to a Global Positioning System receiver will give you much better performance over the older Loran navigation system, but Loran could also work. In real-world APRS, Answer B should be refined to show that the latitude and longitude is sent in degrees, minutes, and FRACTIONS OF A MINUTE. Not many stations transit degrees, minutes, and seconds.

---

***E2E Operating HF digital modes***

---

**E2E01 What is the most common method of transmitting data emissions below 30 MHz?**

A. DTMF tones modulating an FM signal
B. FSK (frequency-shift keying) of an RF carrier
C. AFSK (audio frequency-shift keying) of an FM signal
D. Key-operated on/off switching of an RF carrier

**ANSWER B:** On high frequencies, data emissions are sent using frequency-shift keying without the frequency being modulated. We normally operate lower sideband, and 1600 Hz and 1800 Hz are the most common shifts for mark and space.

---

**E2E02 What do the letters FEC mean as they relate to AMTOR operation?**

A. Forward Error Correction
B. First Error Correction
C. Fatal Error Correction
D. Final Error Correction

**ANSWER A:** The word "AMTOR" stands for amateur teleprinting over radio. This is a digital mode with error correction leading to perfect copy when band conditions are favorable on worldwide frequencies. "FEC" stands for forward-error correction, AMTOR Mode B, where each character is sent twice.

---

**E2E03 How is Forward Error Correction implemented?**

A. By transmitting blocks of 3 data characters from the sending station to the receiving station, which the receiving station acknowledges
B. By transmitting a special FEC algorithm which the receiving station uses for data validation
C. By transmitting extra data that may be used to detect and correct transmission errors
D. By varying the frequency shift of the transmitted signal according to a predefined algorithm

**ANSWER C:** During AMTOR FEC, each data character is sent twice with no acknowledgment by the receiving station.

---

**E2E04 If an oscilloscope is connected to a TNC or terminal unit and is displaying two crossed ellipses, one of which suddenly disappears, what would this indicate about the observed signal?**

A. The phenomenon known as selective fading has occurred
B. One of the signal filters has saturated
C. The receiver should be retuned, as it has probably moved at least 5 kHz from the desired receive frequency
D. The mark and space signal have been inverted and the receiving equipment has not yet responded to the change

**ANSWER A:** Most high-frequency stations feature a companion station monitor oscilloscope that will display two crossed ellipses, indicating a properly tuned-in digital signal. If one of the ellipses disappears, the phenomenon is known as selective fading and will cause your received copy to stall until the signal comes out of the fade and is then printed out on the screen or on paper. Don't touch the receiver during selective fading – you are on frequency; it's just that propagation is causing the signal to fade in and out.

**Don't touch that dial! Selective fading is caused by propagation conditions – not your transceiver.**

---

**E2E05 What is the name for a bulletin transmission system that includes a special header to allow receiving stations to determine if the bulletin has been previously received?**

A. ARQ mode A
B. FEC mode B
C. AMTOR
D. AMTEX

**ANSWER D:** A relatively unknown word for your new Extra Class vocabulary is "AMTEX," the name of a bulletin that contains a special header or address to allow receiving stations to check their memory to see whether or not this bulletin has been previously received. If the bulletin has been received before, it will not be decoded until such time as a new header shows a new bulletin.

---

**E2E06 What is the most common data rate used for HF packet communications?**

A. 48 bauds
B. 110 bauds
C. 300 bauds
D. 1200 bauds

**ANSWER C:** High frequency packet must labor along at a relatively slow 300 baud so as not to exceed signal bandwidth limits below 30 MHz. 300 baud seems relatively slow compared to higher speeds on VHF and UHF, and high-frequency packet is not necessarily a "right now" transmission method. However, it is highly accurate, and even binary files may be sent around the world via HF packet!

**E2E07 What is the typical bandwidth of a properly modulated MFSK16 signal?**

A. 31 Hz
B. 316 Hz
C. 550 Hz
D. 2 kHz

**ANSWER B:** MFSK refers to a super-charged RTTY signal, Millennium Frequency Shift Keying. Instead of simple RTTY with two tones, MFSK 16 uses 16 tones, one at a time at 15.625 baud, spaced at 15.625 Hertz apart. This works out to a relatively narrow 316 Hertz bandwidth, just like RTTY, and offers higher non-error rates with improved reception under noisy conditions. It is an ideal mode for 40 meters, 80 meters, and 160 meters.

---

**E2E08 Which of the following HF digital modes can be used to transfer binary files?**

A. Hellschreiber
B. PACTOR
C. RTTY
D. AMTOR

**ANSWER B:** PACTOR is more than 11 years old, and the transmission sounds like 2 birds chirping where data is sent in blocks and the receiving station sends ACK for acknowledging everything sent, or NAK indicating the request to send the block again. PACTOR may be used to transfer binary files, allowing you to download programs from a BBS and then work the program on your laptop or home computer.

---

**E2E09 – Question deleted from Element 4 Pool by QPC.**

---

**E2E10 – Question deleted from Element 4 Pool by QPC.**

---

**E2E11 What is the Baudot code?**

A. A code used to transmit data only in modern computer-based data systems using seven data bits
B. A binary code consisting of eight data bits
C. An alternate name for Morse code
D. The International Telegraph Alphabet Number 2 (ITA2) which uses five data bits

**ANSWER D:** The original name for transmitting and receiving letters and numbers over the air was a code called International Teletype Alphabet Number 2. We now call it Baudot, giving us the on-air capability to produce both letters and numbers, plus punctuation symbols, with the 5-bit code. Everything is sent upper case.

---

**E2E12 Which of these digital communications modes has the narrowest bandwidth?**

A. AMTOR
B. 170-Hz shift, 45 baud RTTY
C. PSK31
D. 300-baud packet

**ANSWER C:** Phase Shift Keying is the type of modulation that generates PSK, and the "31" means the bit rate at 31.25. There is no mistaking the whistle sound of PSK 31, and the modern transceiver with adaptive digital signal processing can pull a signal out of the noise when you swear you might not get any reception at all. If you have a modern ham set and a computer, working PSK 31 at around 50 wpm requires only a relatively inexpensive matching device and the readily-available software and a sound card. You will be fascinated when you eavesdrop on two operators running PSK 31, and I expect that it will ultimately become one of the most popular computer-to-computer digital modes, going well beyond RTTY.

## Subelement E3 – Radio Wave Propagation [3 Exam Questions - 3 Groups]

### *E3A Earth-Moon-Earth (EME or moonbounce) communications; meteor scatter*

**E3A01 What is the maximum separation between two stations communicating by moonbounce?**

A. 500 miles maximum, if the moon is at perigee
B. 2000 miles maximum, if the moon is at apogee
C. 5000 miles maximum, if the moon is at perigee
D. Any distance as long as the stations have a mutual lunar window

**ANSWER D:** It takes a minimum of a pair of long-boom Yagis, plus the maximum legal power output, to establish a moonbounce QSO on CW or SSB phone. If you are working a station with a huge dish antenna, you might get by with a single long boom Yagi and a good VHF or UHF kilowatt amplifier. Moonbounce is most common on the bottom edge of the 2-meter band, and to communicate over any distance, both stations must be able to see the moon in the sky. Monthly ham magazines publish the best EME (earth-moon-earth) days.

---

**E3A02 What characterizes libration fading of an earth-moon-earth signal?**

A. A slow change in the pitch of the CW signal
B. A fluttery irregular fading
C. A gradual loss of signal as the sun rises
D. The returning echo is several hertz lower in frequency than the transmitted signal

**ANSWER B:** Remember EME means earth-moon-earth and that the moon is not a perfect reflector of radio signals. All those craters on the moon will cause your return signal to sound fluttery, with rapid, irregular fading.

---

**E3A03 When scheduling EME contacts, which of these conditions will generally result in the least path loss?**

A. When the moon is at perigee
B. When the moon is full
C. When the moon is at apogee
D. When the MUF is above 30 MHz

**ANSWER A:** PERigee is the time when the moon is closest to the earth. This is the PERfect condition for EME to occur, and a PERfect day to schedule an EME contact. A moon at perigee, just appearing above the horizon, will give you the best opportunity for moon-bounce.

---

**E3A04 What type of receiving system is desirable for EME communications?**

A. Equipment with very low power output
B. Equipment with very low dynamic range
C. Equipment with very low gain
D. Equipment with very low noise figures

**ANSWER D:** For EME communications, it's best to use a receiver with front-end transistors that have extremely high gain and extremely low noise (a "hot front end"). For his station, your author uses GaAsFET transistors for maximum gain and minimum noise.

**E3A05 What transmit and receive time sequencing is normally used on 144 MHz when attempting an earth-moon-earth contact?**

A. Two-minute sequences, where one station transmits for a full two minutes and then receives for the following two minutes
B. One-minute sequences, where one station transmits for one minute and then receives for the following one minute
C. Two-and-one-half minute sequences, where one station transmits for a full 2.5 minutes and then receives for the following 2.5 minutes
D. Five-minute sequences, where one station transmits for five minutes and then receives for the following five minutes

**ANSWER A:** Moon-bounce communication takes plenty of power, a hot receiver with digital signal processing, big antennas, and plenty of patience. Most successful contacts are made by a prearranged schedule. One station transmits two minutes, and then listens to the other station that will transmit for two minutes.

---

**E3A06 What transmit and receive time sequencing is normally used on 432 MHz when attempting an EME contact?**

A. Two-minute sequences, where one station transmits for a full two minutes and then receives for the following two minutes
B. One-minute sequences, where one station transmits for one minute and then receives for the following one minute
C. Two and one half minute sequences, where one station transmits for a full 2.5 minutes and then receives for the following 2.5 minutes
D. Five minute sequences, where one station transmits for five minutes and then receives for the following five minutes

**ANSWER C:** At 432-MHz band, on the 70-cm band, conditions are even more rough than what you would find on 2 meters, so you need to send two-and-one-half-minute sequences, rather than just two minutes.

---

**E3A07 What frequency range would you normally tune to find EME stations in the 2-meter band?**

A. 144.000 - 144.001 MHz
B. 144.000 - 144.100 MHz
C. 144.100 - 144.300 MHz
D. 145.000 - 145.100 MHz

**ANSWER B:** Moon-bounce is found in the CW weak-signal portion of the band. Your author prefers 144.074 MHz, but anywhere between 144.000 to 144.100 is the spot for moon-bounce.

---

**E3A08 What frequency range would you normally tune to find EME stations in the 70-cm band?**

A. 430.000 - 430.150 MHz
B. 430.100 - 431.100 MHz
C. 431.100 - 431.200 MHz
D. 432.000 - 432.100 MHz

**ANSWER D:** The 70-cm band is again the 432-MHz band – everything below 432.100 is within the weak-signal (only) portion of the band where CW and SSB are found.

---

**E3A09 When a meteor strikes the earth's atmosphere, a cylindrical region of free electrons is formed at what layer of the ionosphere?**

A. The E layer
B. The F1 layer
C. The F2 layer
D. The D layer

**ANSWER A:** It is within the E-layer of the ionosphere that 6- and 2-meter excitement like Sporadic-E and meteor scatter contacts take place.

**E3A10 Which range of frequencies is well suited for meteor-scatter communications?**

A. 1.8 - 1.9 MHz
B. 10 - 14 MHz
C. 28 - 148 MHz
D. 220 - 450 MHz

**ANSWER C:** During a recent meteor shower, your author was able to maintain 30-second meteor-scatter QSO's on the 10-meter band, 10-second QSO's on the 6-meter band, and 3-second QSO's on the 2-meter band. The higher you go in frequency, the shorter the contact time. 28-148 MHz are the best areas to hear the effects of meteor-scatter. If you live within 100 miles of an airport, listen to 75 MHz on a scanner for airport radio navigation signals to bounce off of meteor trails.

---

**E3A11 What transmit and receive time sequencing is normally used on 144 MHz when attempting a meteor-scatter contact?**

A. Two-minute sequences, where one station transmits for a full two minutes and then receives for the following two minutes
B. One-minute sequences, where one station transmits for one minute and then receives for the following one minute
C. 15-second sequences, where one station transmits for 15 seconds and then receives for the following 15 seconds
D. 30-second sequences, where one station transmits for 30 seconds and then receives for the following 30 seconds

**ANSWER C:** On meteor-scatter, stations transmitting from the west to the east go for the first 15 seconds and the third 15 seconds of a minute. East calls west on the second and fourth 15 seconds of the minute. This is the pattern used for scheduled contacts. This pattern is not used for random meteor-scatter calls.

---

***E3B Transequatorial; long path; gray line***

---

**E3B01 What is transequatorial propagation?**

A. Propagation between two points at approximately the same distance north and south of the magnetic equator
B. Propagation between two points at approximately the same latitude on the magnetic equator
C. Propagation between two continents by way of ducts along the magnetic equator
D. Propagation between two stations at the same latitude

**ANSWER A:** A TE(transequatorial) signal is propagated between stations north and south of the equator. Some of the best contacts have been between Florida and countries in South America. Any stations in contact with each other are about the same distance away from the equator – ideal conditions for a TE contact. Both CW as well as SSB are the normal modes for TE on VHF and UHF bands.

---

**E3B02 What is the approximate maximum range for signals using transequatorial propagation?**

A. 1000 miles
B. 2500 miles
C. 5000 miles
D. 7500 miles

**ANSWER C:** The total distance for a TE(transequatorial) contact is about 5,000 miles.

---

**E3B03 What is the best time of day for transequatorial propagation?**

A. Morning
B. Noon
C. Afternoon or early evening
D. Late at night

**ANSWER C:** Similar to sporadic E, the best time of day for TE propagation is mid-afternoon or early evening. The time of day has allowed ultraviolet radiation from the sun too "warm up" the possible path.

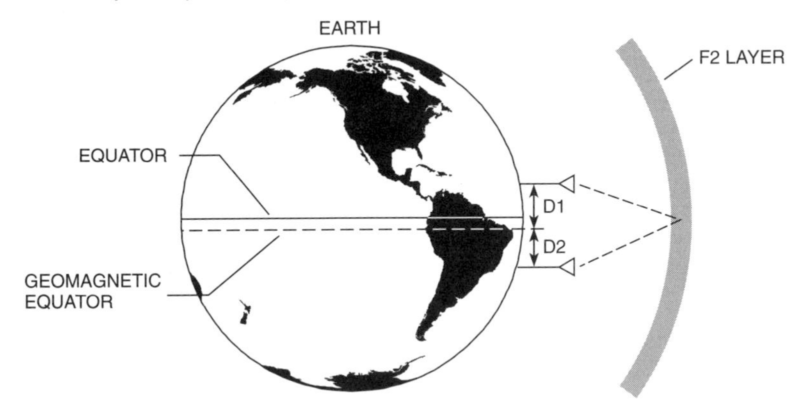


For transequatorial propagation to occur, stations at D1 and D2 must be equidistance from geomagnetic equator. Popular theory is that reflections occur in and off the F2 layer during peak sunspot activity.

**Transequatorial Propagation**

---

**E3B04 What type of propagation is probably occurring if an HF beam antenna must be pointed in a direction 180 degrees away from a station to receive the strongest signals?**

A. Long-path
B. Sporadic-E
C. Transequatorial
D. Auroral

**ANSWER A:** If two stations wish to communicate on high frequency when the sun is ionizing the F1 and F2 layers, both stations would normally aim their antennas at each other for an excellent "short path" contact. But if the sun is on the other side of the earth to both stations, the stations may wish to try long-path propagation by aiming their antennas in the opposite direction – 180 degrees away from the station with which they wish to communicate. They are actually sending their signals all the way around the world, the long way, to get a signal to the other station. Many times you can hear a tell-tale echo when a station is being received via a long path.

---

**E3B05 On what amateur bands can long-path propagation provide signal enhancement?**

A. 160 to 40 meters
B. 30 to 10 meters
C. 160 to 10 meters
D. 6 meters to 2 meters

**ANSWER C:** Long-range, long-path propagation can be found on the worldwide bands from 10 meters down to 160 meters. Twenty meters is always the best band to expect almost daily long-path propagation, which is best in the late evening or early morning hours.

---

**E3B06 What amateur band consistently yields long-path enhancement using a modest antenna of relatively high gain?**

A. 80 meters
B. 20 meters
C. 10 meters
D. 6 meters

**ANSWER B:** The amateur radio 20-meter band is the best one for long-path DX with a relatively modest directional antenna system. When you pass your upcoming Extra Class exam, you will gain an exclusive 20-meter CW window from 14-14.025 MHz, and an exclusive Extra Class window from 14.150 MHz to 14.175 MHz, just for working rare 20-meter DX on a band shared only by other Extra Class operators in the United States.

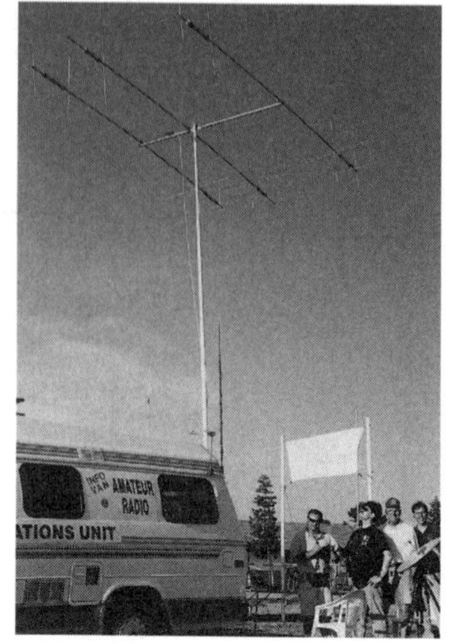


**A 20-meter antenna system for DX communications during Field Day**

---

**E3B07 What is the typical reason for hearing an echo on the received signal of a station in Europe while directing your HF antenna toward the station?**

A. The station's transmitter has poor frequency stability
B. The station's transmitter is producing spurious emissions
C. Auroral conditions are causing a direct and a long-path reflected signal to be received
D. There are two signals being received, one from the most direct path and one from long-path propagation

**ANSWER D:** When long-path conditions exist, European ham stations and even European shortwave stations will be received with a noticeable echo on the signal. The echo is the slight delay of the signal coming in long-path from what you are also hearing sooner as a short-path signal.

---

**E3B08 What type of propagation is probably occurring if radio signals travel along the terminator between daylight and darkness?**

A. Transequatorial
B. Sporadic-E
C. Long-path
D. Gray-line

**ANSWER D:** Once you earn your Extra Class license, you will stay up for long hours working rare DX. Just as the sun is rising here, it may be setting over Europe and Asia, and working stations in the twilight at both ends of the circuit may lead to some extraordinary DX contacts that can last up to 30 minutes.

---

**E3B09 At what time of day is gray-line propagation most prevalent?**

A. Twilight, at sunrise and sunset
B. When the sun is directly above the location of the transmitting station
C. When the sun is directly overhead at the middle of the communications path between the two stations
D. When the sun is directly above the location of the receiving station

**ANSWER A:** There are excellent computer programs that will illustrate the area of sunshine, darkness, and twilight. Go for the twilight contact!

**E3B10 What is the cause of gray-line propagation?**

A. At midday the sun, being directly overhead, superheats the ionosphere causing increased refraction of radio waves
B. At twilight solar absorption drops greatly while atmospheric ionization is not weakened enough to reduce the MUF
C. At darkness solar absorption drops greatly while atmospheric ionization remains steady
D. At mid afternoon the sun heats the ionosphere, increasing radio wave refraction and the MUF

**ANSWER B:** During twilight, the D layer quickly disappears yielding to less absorption, while the E and F layers continue relatively strong. This same condition holds true during solar eclipses.

---

**E3B11 What communications are possible during gray-line propagation?**

A. Contacts up to 2,000 miles only on the 10-meter band
B. Contacts up to 750 miles on the 6- and 2-meter bands
C. Contacts up to 8,000 to 10,000 miles on three or four HF bands
D. Contacts up to 12,000 to 15,000 miles on the 2 meter and 70 centimeter bands

**ANSWER C:** It is not uncommon to reach beyond 8,000 miles to 10,000 miles on three or four high-frequency bands using gray-line propagation.

---

***E3C Auroral propagation; selective fading; radio-path horizon; take-off angle over flat or sloping terrain; earth effects on propagation***

---

**E3C01 What effect does auroral activity have upon radio communications?**

A. The readability of SSB signals increases
B. FM communications are clearer
C. CW signals have a clearer tone
D. CW signals have a fluttery tone

**ANSWER D:** If you live in the Northwest, Northeast, or along the Great Lakes, chances are you have seen some auroral activity. The aurora is a visual glowing of the lower ionospheric layers, and will cause VHF and UHF signals to reflect off of this shimmering curtain. Since the aurora is always changing, CW signals will sound like a fluttery tone.

---

**E3C02 What is the cause of auroral activity?**

A. A high sunspot level
B. A low sunspot level
C. The emission of charged particles from the sun
D. Meteor showers concentrated in the northern latitudes

**ANSWER C:** During periods of major sunspots, auroral activity will be high. When a giant solar flare occurs on the face of the sun, we see it here on earth about eight minutes later and feel the effects of its ultraviolet radiation. About two days later, slower-moving, charged particles begin to pump up the ionosphere, many times creating spectacular auroral light shows. It's the charged particles that take a couple of days to finally reach us that give us an aurora borealis.

---

**E3C03 Where in the ionosphere does auroral activity occur?**

A. At F-region height
B. In the equatorial band
C. At D-region height
D. At E-region height

**ANSWER D:** It's up in the E-layer that we see most of the effects of auroral activity. Now be careful of your answer – even though we are talking about the E-layer, the correct answer is answer D. If you went for the D-layer answer, that would really be answer C. I think you get what I mean – when there is a letter in the correct answer, make sure you don't transpose that letter over to the A-B-C-D multiple-choice answer!

---

**E3C04 Which emission mode is best for auroral propagation?**

A. CW
B. SSB
C. FM
D. RTTY

**ANSWER A:** Bouncing a signal off of an aurora is difficult. CW always provides the best opportunity for someone hearing raspy sounds coming into their receiver. On many auroral contacts on CW, we don't even hear a tone – all that can be heard are just short and long rushes of super-imposed noise. Play my audiotape Extra Class course and listen for yourself.

**CW is the best emission mode for auroral contacts.**

Courtesy of Chip Margelli, K7JA

---

**E3C05 What causes selective fading?**

A. Small changes in beam heading at the receiving station
B. Phase differences between radio-wave components of the same transmission, as experienced at the receiving station
C. Large changes in the height of the ionosphere at the receiving station ordinarily occurring shortly after either sunrise or sunset
D. Time differences between the receiving and transmitting stations

**ANSWER B:** The key words *phase differences* depict an incoming signal that will sound distorted due to selective fading of the sidebands.

---

**E3C06 How does the bandwidth of a transmitted signal affect selective fading?**

A. It is more pronounced at wide bandwidths
B. It is more pronounced at narrow bandwidths
C. It is the same for both narrow and wide bandwidths
D. The receiver bandwidth determines the selective fading effect

**ANSWER A:** The wider the bandwidth, the more severely the signal will suffer from selective fading.

**E3C07 How much farther does the VHF/UHF radio-path horizon distance exceed the geometric horizon?**

A. By approximately 15% of the distance
B. By approximately twice the distance
C. By approximately one-half the distance
D. By approximately four times the distance

**ANSWER A:** Radio waves bend slightly over the horizon because of the difference in the air's refractive index at higher altitudes. You can calculate your approximate VHF range to the horizon with the formula: D (miles) = 1.415 × $\sqrt{2H}$ (in feet). Determine the height (in feet) of the station you are wishing to communicate with, calculate its distance to the horizon, and add the two answers together for your rock-solid communications range. Depending on local weather conditions, a 15% to 30% range enhancement over the optical horizon will usually take place at VHF and UHF radio frequency bands.

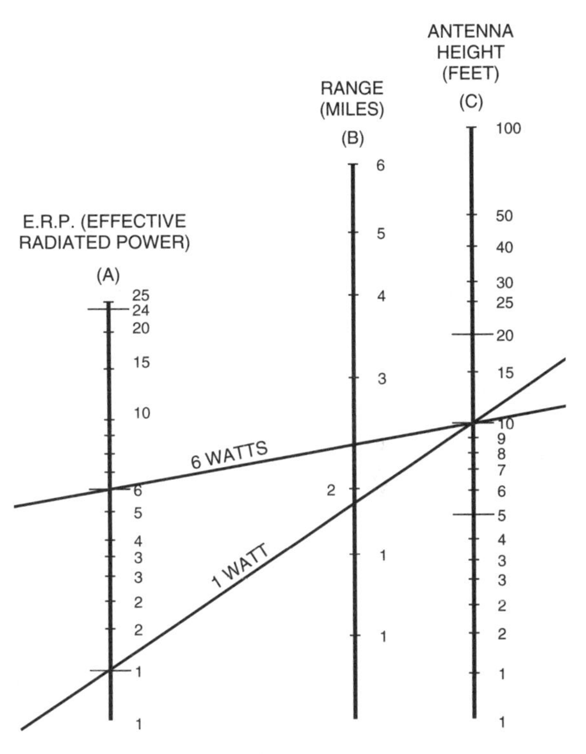


**VHF Range Nomograph**

**E3C08 For a 3-element beam antenna with horizontally mounted elements, how does the main lobe takeoff angle vary with height above flat ground?**

A. It increases with increasing height
B. It decreases with increasing height
C. It does not vary with height
D. It depends on E-region height, not antenna height

**ANSWER B:** When you put up a 3-element Yagi antenna, the high frequency Yagi is usually mounted horizontally and the antenna should be mounted at least one-half wavelength above the ground on the lowest band of operation. Most 3-element Yagis feature tri-band operation for 10 meters, 15 meters and 20 meters. 20-meter band operation would like to see the antenna at a half wavelength off the earth, or about 32 feet in the air. Anything lower than a half wavelength from the operating frequency dramatically raises the main lobe takeoff angle, and the higher the takeoff angle, the shorter the skip distance. You want to get your beam up at least one-half wavelength above the earth for a nice low takeoff angle for long-range DX.

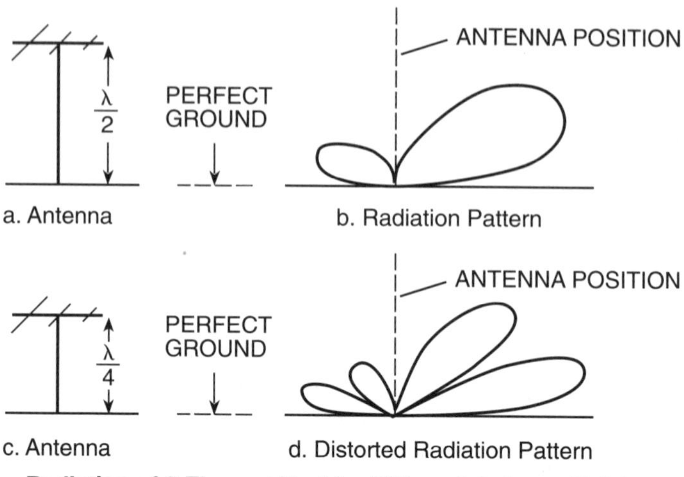


**Radiation of 3-Element Yagi for Different Antenna Heights**

**E3C09 What is the name of the high-angle wave in HF propagation that travels for some distance within the F2 region?**

A. Oblique-angle ray
B. Pedersen ray
C. Ordinary ray
D. Heaviside ray

**ANSWER B:** The wave that propagates for some distance within the F2 region is called the Pedersen ray.

**E3C10 What effect is usually responsible for propagating a VHF signal over 500 miles?**

A. D-region absorption
B. Faraday rotation
C. Tropospheric ducting
D. Moonbounce

**ANSWER C:** Some incredible extra-long-range VHF and UHF propagation may take place in the presence of a high-pressure system. Sinking warm air, called a subsidence, overlays cool air just above the surface of the earth and over water. A well-defined inversion layer occurs when there is a sharp boundary between the sinking warm air and the cool air beneath. Over big cities, we call it smog. To the VHF and UHF DX enthusiast, we call it a for-sure key for VHF *tropospheric ducting* (key words) super range. Your author has communicated many times from his station in California to Hawaii on the 2-meter band during summer months when tropospheric ducting is at its best!

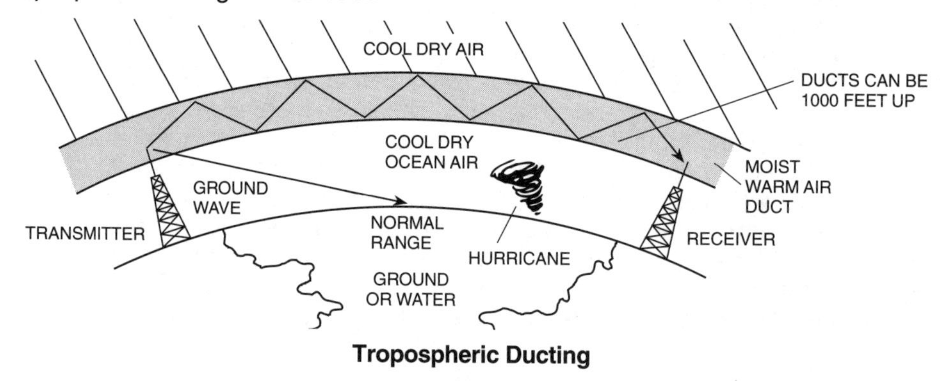


**Tropospheric Ducting**

**E3C11 For a 3-element beam antenna with horizontally mounted elements, how does the main lobe takeoff angle vary with the downward slope of the ground (moving away from the antenna)?**

A. It increases as the slope gets steeper
B. It decreases as the slope gets steeper
C. It does not depend on the ground slope
D. It depends of the F-region height

**ANSWER B:** I hope you live on the side of a hill. Hopefully, the side that drops off abruptly is in the direction of long-range DX. The steeper the drop-off, the lower the take-off angle from your 3-element beam. The lower the angle, the further your skip distance.

---

**E3C12 In the northern hemisphere, in which direction should a directional antenna be pointed to take maximum advantage of auroral propagation?**

A. South
B. North
C. East
D. West

**ANSWER B:** You would aim your directional antenna directly at the aurora in order to pick up VHF and UHF fluttery auroral CW sounds. Sometimes with big auroras, you can even transmit SSB. Forget about FM, it usually will not work off of an aurora.

---

**E3C13 As the frequency of a signal is increased, how does its ground wave propagation change?**

A. It increases
B. It decreases
C. It stays the same
D. Radio waves don't propagate along the earth's surface

**ANSWER B:** Ground waves travel further on lower frequencies, so as the frequency of a signal is increased, ground wave propagation decreases. This same propagation phenomena occurs with audio frequencies. Lower audio frequencies travel further than higher audio frequencies. That's why fog horns always blast out a very low audio tone.

---

**E3C14 What type of polarization does ground-wave propagation have?**

A. Vertical
B. Horizontal
C. Circular
D. Elliptical

**ANSWER A:** We use vertical polarization for best ground-wave propagation. Knife-edge bending is a good example of ground waves, vertically polarized, that drag the bottom half of the wave over a mountain top, resulting in the ground-wave signal getting into valleys where normally there would be no ground-wave reception.

---

**E3C15 Why does the radio-path horizon distance exceed the geometric horizon?**

A. E-region skip
B. D-region skip
C. Auroral skip
D. Radio waves may be bent

**ANSWER D:** Because of the difference in the air's refractive index at higher altitudes, radio waves will bend slightly giving them a longer horizon than the optical horizon. See the explanation at E3C07.

## Subelement E4 – Amateur Radio Practices [5 Exam Questions — 5 Groups]

***E4A Test equipment: spectrum analyzers (interpreting spectrum analyzer displays; transmitter output spectrum), logic probes (indications of high and low states in digital circuits; indications of pulse conditions in digital circuits)***

**E4A01 How does a spectrum analyzer differ from a conventional time-domain oscilloscope?**

A. A spectrum analyzer measures ionospheric reflection; an oscilloscope displays electrical signals
B. A spectrum analyzer displays signals in the time domain; an oscilloscope displays signals in the frequency domain
C. A spectrum analyzer displays signals in the frequency domain; an oscilloscope displays signals in the time domain
D. A spectrum analyzer displays radio frequencies; an oscilloscope displays audio frequencies

**ANSWER C:** The spectrum analyzer displays the strength of signals at particular frequencies according to frequency (in the frequency domain) along a horizontal axis. A signal can be locked in and centered in the middle of the screen with 25 kHz to the left and 25 kHz to the right for a close examination of any off-frequency spurs it might have. A simple oscilloscope cannot display the signal in the frequency domain – it displays the signals on a time axis (in the time domain) going from left to right. With your output signal displayed on a spectrum analyzer, you may adjust your transmitter output network for minimum spurious signals.

**E4A02 What parameter does the horizontal axis of a spectrum analyzer display?**

A. Amplitude
B. Voltage
C. Resonance
D. Frequency

**ANSWER D:** The spectrum analyzer allows you to analyze the frequency of signals by displaying the frequencies of the signals on the horizontal axis.

**Spectrum Analyzer**
**Courtesy of Hewlett-Packard Company**

**E4A03 What parameter does the vertical axis of a spectrum analyzer display?**

A. Amplitude
B. Duration
C. Frequency
D. Time

**ANSWER A:** The vertical axis of the spectrum analyzer displays the amplitude of the signal.

**E4A04 Which test instrument is used to display spurious signals from a radio transmitter?**

A. A spectrum analyzer
B. A wattmeter
C. A logic analyzer
D. A time-domain reflectometer

**ANSWER A:** You can check for spurious signals using a spectrum analyzer.

---

**E4A05 Which test instrument is used to display intermodulation distortion products in an SSB transmission?**

A. A wattmeter
B. A spectrum analyzer
C. A logic analyzer
D. A time-domain reflectometer

**ANSWER B:** Only the spectrum analyzer allows you to see your output versus frequency by displaying it in the frequency domain. This allows you to check for harmonics that cause distortion in your SSB transmitter.

---

**E4A06 Which of the following is NOT something that could be determined with a spectrum analyzer?**

A. The degree of isolation between the input and output ports of a 2 meter duplexer
B. Whether a crystal is operating on its fundamental or overtone frequency
C. The speed at which a transceiver switches from transmit to receive when being used for packet radio
D. The spectral output of a transmitter

**ANSWER C:** The spectrum analyzer measures the frequency domain of signals. It does not measure the time nor speed at which a transceiver switches. This is something you would do with a conventional oscilloscope. The spectrum analyzer is now an affordable piece of test equipment available for under $1,500.

---

**E4A07 What is an advantage of using a spectrum analyzer to observe the output from a VHF transmitter?**

A. There are no advantages; an inexpensive oscilloscope can display the same information
B. It displays all frequency components of the transmitted signal
C. It displays a time-varying representation of the modulation envelope
D. It costs much less than any other instrumentation useful for such measurements

**ANSWER B:** The big advantage of a spectrum analyzer on VHF and UHF frequencies is to check for spurious output while adjusting the transmitter section. In the answer, they call the spurious outputs "frequency components," and this is what you hope you don't have when transmitting on VHF.

---

**E4A08 What advantage does a logic probe have over a voltmeter for monitoring the status of a logic circuit?**

A. It has many more leads to connect to the circuit than a voltmeter
B. It can be used to test analog and digital circuits
C. It can read logic circuit voltage more accurately than a voltmeter
D. It is smaller and shows a simplified readout

**ANSWER D:** Now that most transceivers incorporate numerous digital circuits, the simple logic probe indicating that circuits are either ON (low state) or OFF (high state) is a quick way to compare circuit operation to that listed in the technical manuals. Incidentally, technical manuals are normally not included with each new piece of equipment – you must buy them separately if you plan to service your own equipment.

**E4A09 Which test instrument is used to directly indicate high and low digital voltage states?**

A. An ohmmeter
B. An electroscope
C. A logic probe
D. A Wheatstone bridge

**ANSWER C:** The logic probe, about the same size as a fountain pen, indicates high and low logic states. High is a 1, and low is a 0.

**Logic Probe**
Courtesy of RadioShack

---

**E4A10 What can a logic probe indicate about a digital logic circuit?**

A. A short-circuit fault
B. An open-circuit fault
C. The resistance between logic modules
D. The high and low logic states

**ANSWER D:** You can use a logic probe to help determine whether or not an integrated circuit (IC) chip is good or bad.

---

**E4A11 Which of the following test instruments can be used to indicate pulse conditions in a digital logic circuit?**

A. A logic probe
B. An ohmmeter
C. An electroscope
D. A Wheatstone bridge

**ANSWER A:** An oscilloscope is always your best piece of test equipment – but the logic probe is a close second because of all the electronic functions now being performed digitally.

---

**E4A12 Which of the following procedures should you follow when connecting a spectrum analyzer to a transmitter output?**

A. Use high quality coaxial lines
B. Attenuate the transmitter output going to the spectrum analyzer
C. Use a signal divider
D. Match the antenna to the load

**ANSWER B:** Never transmit directly into a spectrum analyzer. You will blow it up, for sure. Your transmitter signal must be attenuated to just a fraction of a watt in order to protect the spectrum analyzer input. I make it a practice to NEVER direct couple to a spectrum analyzer. Rather, I use a little pigtail and “sniff” my transmit RF.

***E4B Frequency measurement devices (i.e., frequency counter, oscilloscope Lissajous figures, dip meter); meter performance limitations; oscilloscope performance limitations; frequency counter performance limitations***

**E4B01 What is a frequency standard?**

A. A frequency chosen by a net control operator for net operations
B. A device used to produce a highly accurate reference frequency
C. A device for accurately measuring frequency to within 1 Hz
D. A device used to generate wide-band random frequencies

**ANSWER B:** You probably already have one or two frequency standards in your ham shack, but guess what is the world's most expensive frequency standard? It's the time signals from WWV and WWVH. You can pick them up on 5 MHz, 10 MHz, 15 MHz, and 20 MHz. These signals are absolutely on frequency, 24 hours a day, for your equipment and test apparatus calibration.

**E4B02 What factors limit the accuracy, frequency response and stability of a frequency counter?**

A. Phase comparator slew rate, speed of the logic and time base stability
B. Time base accuracy, speed of the logic and time base stability
C. Time base accuracy, temperature coefficient of the logic and time base reactance
D. Number of digits in the readout, external frequency reference and temperature coefficient of the logic

**ANSWER B:** Everyone should have frequency counter in their ham shack. Be sure to get one that has *accuracy, speed and stability* (key words).

**E4B03 How can the accuracy of a frequency counter be improved?**

A. By using slower digital logic
B. By improving the accuracy of the frequency response
C. By increasing the accuracy of the time base
D. By using faster digital logic

**ANSWER C:** Once you master all of the brochures for frequency counters, you will find that the *time base* (key words) is the key element for increased accuracy of the frequency readout.

**Frequency counter and SWR indicator**
Courtesy of MFJ

**E4B04 If a frequency counter with a specified accuracy of +/−1.0 ppm reads 146,520,000 Hz, what is the most the actual frequency being measured could differ from the reading?**

A. 165.2 Hz
B. 14.652 kHz
C. 146.52 Hz
D. 1.4652 MHz

**ANSWER C:** You may have a question on your test regarding the readout error on a frequency counter with a specific part per million (ppm) time base. It's easy to solve these problems with a simple calculator. Since they're asking parts per million, first convert the frequency given in Hz to MHz. You do this simply by moving the decimal point 6 places to the left. Now multiply the frequency in MHz times the parts per million time base accuracy (in this case 1 ppm). This will give you the error they are looking for in the answer. Try this on your calculator: Clear 146.52 × 1 = 146.52 Hz. Here is the formula:

**Counter Readout Error**
Readout Error = f × a

f = Frequency in **MHz** being measured
a = Counter accuracy in **parts per million**

**E4B05 If a frequency counter with a specified accuracy of +/−0.1 ppm reads 146,520,000 Hz, what is the most the actual frequency being measured could differ from the reading?**

A. 14.652 Hz
B. 0.1 MHz
C. 1.4652 Hz
D. 1.4652 kHz

**ANSWER A:** Hertz to MHz will give you a frequency of 146.52. Multiple that by 0.1 parts per million, and see if your calculator reads 14.652 Hz. The keystrokes are: Clear 146.52 × 0.1 = 14.652.

**E4B06 If a frequency counter with a specified accuracy of +/−10 ppm reads 146,520,000 Hz, what is the most the actual frequency being measured could differ from the reading?**

A. 146.52 Hz
B. 10 Hz
C. 146.52 kHz
D. 1465.20 Hz

**ANSWER D:** Keystroke the following on your calculator: Clear 146.52 × 10 = 1465.2 Hz. You are multiplying the frequency in MHz by the time base accuracy of 10 ppm.

**E4B07 If a frequency counter with a specified accuracy of +/−1.0 ppm reads 432,100,000 Hz, what is the most the actual frequency being measured could differ from the reading?**

A. 43.21 MHz
B. 10 Hz
C. 1.0 MHz
D. 432.1 Hz

**ANSWER D**: How refreshing, a new frequency. From Hz to MHz, we end up with 432.1 MHz, which multiplied by 1.0 ppm gives you the answer. The following are the keystrokes: Clear 432.1 × 1 = 432.1 Hz.

**E4B08 If a frequency counter with a specified accuracy of +/−0.1 ppm reads 432,100,000 Hz, what is the most the actual frequency being measured could differ from the reading?**

A. 43.21 Hz
B. 0.1 MHz
C. 432.1 Hz
D. 0.2 MHz

**ANSWER A:** Remember, multiply the frequency in MHz times the time base accuracy. Try this on your calculator: Clear 432.1 × 0.1 = 43.21.

**E4B09 If a frequency counter with a specified accuracy of +/−10 ppm reads 432,100,000 Hz, what is the most the actual frequency being measured could differ from the reading?**

A. 10 MHz
B. 10 Hz
C. 4321 Hz
D. 432.1 Hz

**ANSWER C:** Here is the last problem. Again, frequency in MHz times the time base accuracy is performed on your calculator like this: Clear 432.1 × 10 = 4321 Hz. When you are shopping for a frequency counter, remember that a counter with 0.1 ppm time base will give you a much more accurate reading than a counter with a 10 ppm time base.

**E4B10 If a 100 Hz signal is fed to the horizontal input of an oscilloscope and a 150 Hz signal is fed to the vertical input, what type of Lissajous figure will be displayed on the screen?**

A. A looping pattern with 100 loops horizontally and 150 loops vertically
B. A rectangular pattern 100 mm wide and 150 mm high
C. A looping pattern with 3 loops horizontally and 2 loops vertically
D. An oval pattern 100 mm wide and 150 mm high

**ANSWER C:** This is a 3/2 relationship, so you will see a looping pattern with 3 loops across for 150 Hz, and 2 loops vertically for 100 Hz.

---

**E4B11 What is a dip-meter?**

A. A field-strength meter
B. An SWR meter
C. A device consisting of a variable frequency LC oscillator and an indicator showing the metered feedback current
D. A marker generator

**ANSWER C:** When you tune your mobile antenna system, you watch for a dip on your SWR bridge. But you don't necessarily need a transmitter to generate a signal – you could do it with a portable, variable LC oscillator that features a little SWR meter that will measure the feedback current (reflective power). Your *variable* (key word) oscillator may be hooked up directly to a mobile antenna, and you can sweep through the frequencies to see where the antenna is resonant.

**SWR Analyzer**
Courtesy of MFJ Enterprises, Inc.

---

**E4B12 What does a dip-meter do?**

A. It accurately indicates signal strength
B. It measures frequency accurately
C. It measures transmitter output power accurately
D. It gives an indication of the resonant frequency of a nearby circuit

**ANSWER D:** Many senior hams will remember winding their own coils around lead pencils or oatmeal boxes. They used a dip-meter to give an indication of the *resonant frequency* (key words) of the circuit.

---

**E4B13 How does a dip-meter function?**

A. Reflected waves at a specific frequency desensitize a detector coil
B. Power coupled from an oscillator causes a decrease in metered current
C. Power from a transmitter cancels feedback current
D. Harmonics from an oscillator cause an increase in resonant circuit Q

**ANSWER B:** Your little portable dip-meter, updated for the '90s as an SWR analyzer, couples power from its built-in oscillator to your antenna. At resonance, the antenna grabs all of the current, and this causes a decrease in metered current, seen as a dip on the indicating meter. The meter dips because all of the current is getting soaked up at resonance within the antenna system.

**E4B14 What two ways could a dip-meter be used in an amateur station?**

A. To measure resonant frequency of antenna traps and to measure percentage of modulation
B. To measure antenna resonance and to measure percentage of modulation
C. To measure antenna resonance and to measure antenna impedance
D. To measure resonant frequency of antenna traps and to measure a tuned circuit resonant frequency

**ANSWER D:** The dip-meter is a handy way to determine the resonant frequency of a box full of unmarked antenna traps. You can also measure a tuned circuit resonant frequency, such as an antenna coupler. The dip-meter is a great one to measure a *tuned circuit* (key words).

---

**E4B15 For best accuracy, how tightly should a dip-meter be coupled with the LC circuit being checked?**

A. As loosely as possible
B. As tightly as possible
C. First loosely, then tightly
D. With a jumper wire between the meter and the circuit to be checked

**ANSWER A:** If you get too close to the circuit you are testing, you could de-tune it with your dip-meter coil. This is why, for best accuracy, you want to stay as loosely coupled as possible.

---

**E4B16 What factors limit the accuracy, frequency response and stability of an oscilloscope?**

A. Accuracy and linearity of the time base and the linearity and bandwidth of the deflection amplifiers
B. Tube face voltage increments and deflection amplifier voltage
C. Accuracy and linearity of the time base and tube face voltage increments
D. Deflection amplifier output impedance and tube face frequency increments

**ANSWER A:** The accuracy and faithful linearity of both the time base and linearity of the bandwidth of the deflection amplifiers determine the accuracy and stability of your oscilloscope.

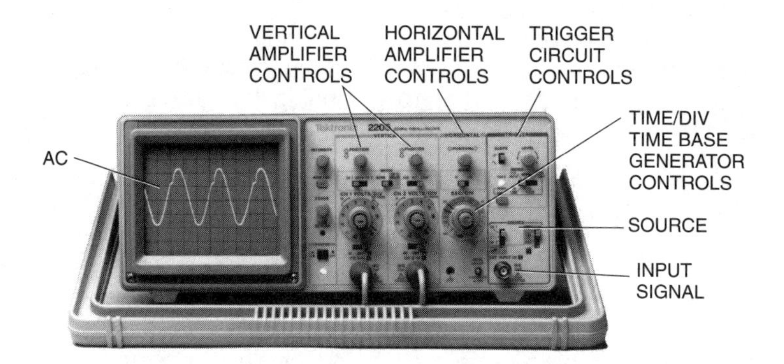


**A Typical Oscilloscope showing an AC Waveform**

Source: *Basic Electronics* © 1994, Master Publishing, Inc., Lincolnwood, Illinois

**E4B17 What happens in a dip-meter when it is too tightly coupled with a tuned circuit being checked?**

A. Harmonics are generated
B. A less accurate reading results
C. Cross modulation occurs
D. Intermodulation distortion occurs

**ANSWER B:** If you put your dip-meter right on top of a coil, you will have a less *accurate reading* (key words) – only get close enough to see a smooth dip on your readout. On the SWR analyzer type of dip-meters, everything is done for you automatically inside the box. All you need to do is to screw on the antenna cable, and sweep the frequency dial to see where the antenna is resonant.

---

**E4B18 What factors limit the accuracy, frequency response and stability of a D'Arsonval-type meter?**

A. Calibration, coil impedance and meter size
B. Calibration, mechanical tolerance and coil impedance
C. Coil impedance, electromagnetic voltage and movement mass
D. Calibration, series resistance and electromagnet current

**ANSWER B:** Ham sets are now being produced without the traditional needle movement meter. Your author is sad to see them go, but the new colored LED readouts replace the traditional mechanical meter movement. Because the traditional mechanical (key word) meter was prone to vibration failure, the jeweled needle movement could go out of calibration. In addition, the mechanical movement is not as fail-safe as the newer LED displays.

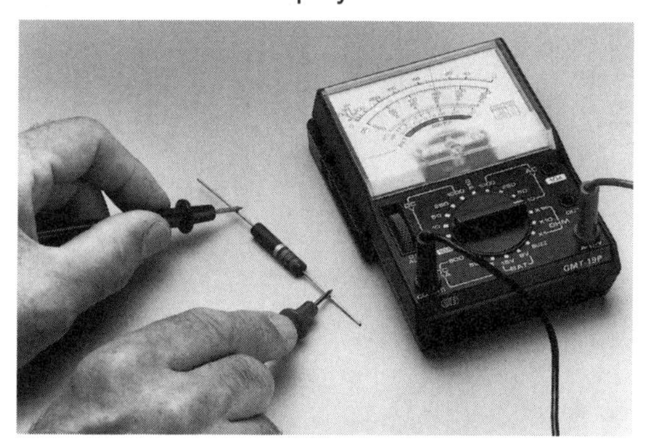

D'Arsonval-type Meter

---

**E4B19 How can the frequency response of an oscilloscope be improved?**

A. By using a triggered sweep and a crystal oscillator as the time base
B. By using a crystal oscillator as the time base and increasing the vertical sweep rate
C. By increasing the vertical sweep rate and the horizontal amplifier frequency response
D. By increasing the horizontal sweep rate and the vertical amplifier frequency response

**ANSWER D:** You can improve the frequency response of your oscilloscope by INCREASING the horizontal sweep rate and INCREASING the vertical amplifier frequency response.

***E4C Receiver performance characteristics (i.e., phase noise, desensitization, capture effect, intercept point, noise floor, dynamic range {blocking and IMD}, image rejection, MDS, signal-to-noise-ratio); intermodulation and cross-modulation interference***

**E4C01 What is the effect of excessive phase noise in the local oscillator section of a receiver?**

A. It limits the receiver ability to receive strong signals
B. It reduces the receiver sensitivity
C. It decreases the receiver third-order intermodulation distortion dynamic range
D. It allows strong signals on nearby frequencies to interfere with reception of weak signals

**ANSWER D:** The latest amateur transceivers all boast minimum phase noise in the receiver local oscillator. Any worldwide set with excessive phase noise in the LO might allow adjacent frequency signals to interfere with on-frequency weak signals.

---

**E4C02 What is the term for the reduction in receiver sensitivity caused by a strong signal near the received frequency?**

A. Desensitization
B. Quieting
C. Cross-modulation interference
D. Squelch gain rollback

**ANSWER A:** If signals mysteriously come in strong, then abruptly get weak, and then come back strong again, chances are there is someone nearby transmitting on an adjacent frequency, desensitizing your receiver. The best way to check for adjacent signals is through the use of a frequency counter.

---

**E4C03 Which of the following can cause receiver desensitization?**

A. Audio gain adjusted too low
B. Strong adjacent-channel signals
C. Audio bias adjusted too high
D. Squelch gain adjusted too low

**ANSWER B:** Adjacent frequency signals may overload a receiver's "front-end" causing a receiver to momentarily lose sensitivity.

---

**E4C04 Which of the following is one way receiver desensitization can be reduced?**

A. Improve the shielding between the receiver and the transmitter causing the problem
B. Increase the transmitter audio gain
C. Decrease the receiver squelch level
D. Increase the receiver bandwidth

**ANSWER A:** If you have a dual-band handheld transceiver, you will notice that the insides are all enclosed in tiny metal cans. This shielding is necessary to keep the receiver from desensitizing on one band when you are transmitting on another band. *Shielding* (key word) is the best way to minimize internal desensitization.

---

**E4C05 What is the FM capture effect?**

A. All signals on a frequency are demodulated by an FM receiver
B. All signals on a frequency are demodulated by an AM receiver
C. The strongest signal received is the only demodulated signal
D. The weakest signal received is the only demodulated signal

**ANSWER C:** Capture effect occurs in an FM receiver when a stronger signal may override a weaker signal, causing only the strongest signal to be received. This phenomena occurs when powerful signals overtake the limiter and detector, making weaker signals virtually disappear. However, capture effect is less noticeable when two FM signals are relatively close in signal strength, resulting in a garbled tone with neither signal getting through. This is referred to as "doubling" – the unintentional transmission of two FM operators at the same time on the same frequency.

---

**E4C06 What is the term for the blocking of one FM phone signal by another, stronger FM phone signal?**

A. Desensitization
B. Cross-modulation interference
C. Capture effect
D. Frequency discrimination

**ANSWER C:** A unique property of your FM receiver is the ability to hear only the strongest incoming FM signal. If a distant station is transmitting, and a strong local station comes on the air, you will only hear the local station. The stronger signal is "capturing" your receiver.

---

**E4C07 What is meant by the noise floor of a receiver?**

A. The weakest signal that can be detected under noisy atmospheric conditions
B. The amount of phase noise generated by the receiver local oscillator
C. The minimum level of noise that will overload the receiver RF amplifier stage
D. The weakest signal that can be detected above the receiver internal noise

**ANSWER D:** For VHF and UHF weak signal work on SSB and CW, you want to operate a receiver with an extremely low noise floor. The GaAsFET transistor is found in quality VHF/UHF transceivers for weak signal work because of its extremely low noise floor that will allow weak signals to be detected above the receiver's internal noise.

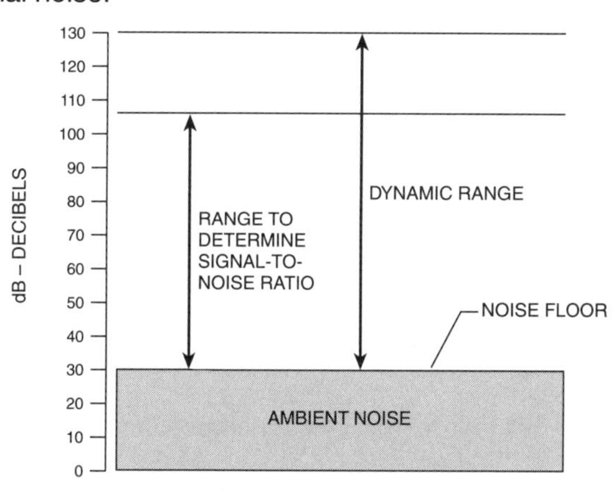


**Noise Floor in a Receiver**

Source: *Installing and Maintaining Sound Systems,* G. McComb, ©1996, Master Publishing, Inc.

---

**E4C08 What is the blocking dynamic range for a receiver that has an 8-dB noise figure and an IF bandwidth of 500 Hz when the blocking level (1-dB compression point) is -20 dBm?**

A. -119 dBm
B. 119 dB
C. 146 dB
D. -146 dBm

**ANSWER B:** Blocking dynamic range is the difference, in dB, between the noise floor and an incoming signal that will cause one dB of gain compression in the receiver. If you look over the high frequency amateur radio specification sheets, you will find that most receivers have a blocking dynamic range between 117 dB to 119 dB, 119 dB being the correct answer here. We always express the dynamic range value as an absolute value with no minus sign.

E ▶ $P_N = KT_0B$ where: $P_N$ = Noise Power

k = Bolzmann's Constant, ($1.38 \times 10^{-23}$ volts/degrees)

$T_0$ = 290 Kelvins

B = Bandwidth in Hz

Shortcut:

Noise floor = -174dBm (in 1Hz bandwidth)

\+ 27dB (10 log 500Hz) our bandwidth

\+ 8dB (our noise figure)

– 139dBm

BDR = –139dBm –20dBm (1dB compression)

=> –119dBm but always expressed as positive number in dB

=> 119dB

---

**E4C09 What is meant by the dynamic range of a communications receiver?**

A. The number of kHz between the lowest and the highest frequency to which the receiver can be tuned

B. The maximum possible undistorted audio output of the receiver, referenced to one milliwatt

C. The ratio between the minimum discernible signal and the largest tolerable signal without causing audible distortion products

D. The difference between the lowest-frequency signal and the highest-frequency signal detectable without moving the frequency control

**ANSWER C:** The dynamic range of a communications receiver is a ratio of the minimum signal to the maximum signal without distortion. Look for the key word *ratio* in the answer to this question.

---

**E4C10 What type of problems are caused by poor dynamic range in a communications receiver?**

A. Cross modulation of the desired signal and desensitization from strong adjacent signals

B. Oscillator instability requiring frequent retuning, and loss of ability to recover the opposite sideband, should it be transmitted

C. Cross modulation of the desired signal and insufficient audio power to operate the speaker

D. Oscillator instability and severe audio distortion of all but the strongest received signals

**ANSWER A:** In a receiver that does not have good dynamic range, cross-modulation of the desired signal can be encountered, and the receiver can drop in sensitivity (be desensitized) by strong adjacent frequency signals.

**E4C11 If you measured the MDS of a receiver, what would you be measuring?**

A. The meter display sensitivity (MDS), or the responsiveness of the receiver S-meter to all signals
B. The minimum discernible signal (MDS), or the weakest signal that the receiver can detect
C. The minimum distorting signal (MDS), or the strongest signal the receiver can detect without overloading
D. The maximum detectable spectrum (MDS), or the lowest to highest frequency range of the receiver

**ANSWER B:** The most common sensitivity measurement for a CW or SSB receiver is the minimum discernible signal (MDS), more commonly called the noise floor. Most worldwide amateur radio sets proudly proclaim their specifications for this important consideration on your part when selecting the equipment. But to gain full advantage of a minimum discernible signal, you must go with an antenna system in an area where there is extremely low atmospheric and man-made noise, too. Move to the desert, or live aboard a boat out on the high seas with the engine turned off!

---

**E4C12 How does intermodulation interference between two repeater transmitters usually occur?**

A. When the signals from the transmitters are reflected out of phase from airplanes passing overhead
B. When they are in close proximity and the signals mix in one or both of their final amplifiers
C. When they are in close proximity and the signals cause feedback in one or both of their final amplifiers
D. When the signals from the transmitters are reflected in phase from airplanes passing overhead

**ANSWER B:** "Intermod," in its true sense, occurs within the repeater final amplifier. When signals mix in one or both repeater amplifiers, sum and difference signals on many different frequencies can be heard.

---

**E4C13 How can intermodulation interference between two repeater transmitters in close proximity often be reduced or eliminated?**

A. By using a Class C final amplifier with high driving power
B. By installing a terminated circulator or ferrite isolator in the feed line to the transmitter and duplexer
C. By installing a band-pass filter in the antenna feed line
D. By installing a low-pass filter in the antenna feed line

**ANSWER B:** Hams who maintain repeaters will install *circulators and isolators* (key words) in the feed line to the transmitter and duplexer to minimize other nearby repeaters getting into their system. This is an ongoing problem because most mountain tops will have different radio systems that will be coming and going, causing the repeater operator a continuous headache when any new tenants begin transmitting from a nearby antenna.

---

**E4C14 If a receiver tuned to 146.70 MHz receives an intermodulation-product signal whenever a nearby transmitter transmits on 146.52 MHz, what are the two most likely frequencies for the other interfering signal?**

A. 146.34 MHz and 146.61 MHz
B. 146.88 MHz and 146.34 MHz
C. 146.10 MHz and 147.30 MHz
D. 73.35 MHz and 239.40 MHz

**ANSWER A:** To calculate the possible sources of intermodulation interference, we check for the signals that lead to this type of a common problem at base and repeater stations.

**Finding the Intermodulation Interference Frequencies**

There are four possible calculations to find the intermodulation interfering frequencies — two where the frequencies are added and two where the frequencies are subtracted. The calculations where the frequencies are subtracted are the only ones used because only these calculations produce frequencies close enough to cause significant interference. As the question states, the intermodulation product $f_{IMD}$ is received on 146.70 MHz when a signal at 146.52 MHz is transmitted. The calculations when the frequencies are subtracted are:

E

$$f_{IMD(2)} = 2f_1 - f_2$$

Rearranging,

$$f_2 = 2f_1 - f_{IMD(2)}$$

$$f_2 = 2(146.52) - 146.70$$

$$f_2 = 146.34 \text{ MHz}$$

$$f_{IMD(4)} = 2f_2 - f_1$$

Rearranging,

$$2f_2 = f_{IMD(4)} + f_1$$

$$f_2 = \frac{f_{IMD(4)} + f_1}{2}$$

$$f_2 = \frac{146.70 + 146.52}{2}$$

$$f_2 = 146.61 \text{ MHz}$$

---

**E4C15 If the signals of two transmitters mix together in one or both of their final amplifiers and unwanted signals at the sum and difference frequencies of the original signals are generated, what is this called?**

A. Amplifier desensitization
B. Neutralization
C. Adjacent channel interference
D. Intermodulation interference

**ANSWER D:** Repeaters are fun to operate through, but a real headache to keep on the air with a clean signal. The problem is intermodulation caused by other close proximity transmitters. What you may hear are two signals coming in at once, and this is the effect commonly called "intermod."

---

**E4C16 What is cross-modulation interference?**

A. Interference between two transmitters of different modulation type
B. Interference caused by audio rectification in the receiver preamp
C. Harmonic distortion of the transmitted signal
D. Modulation from an unwanted signal is heard in addition to the desired signal

**ANSWER D:** If you have driven your little handheld transceiver down to a big city, chances are you have heard the effects of cross-modulation interference. "Cross-mod" brings in modulation from an *unwanted signal* (key words) that rides on top of the signal you are presently listening to. This mix occurs inside your receiver with handheld transceivers. Do not confuse cross-modulation with intermodulation. "Intermod" occurs in the transmitters, and cross-modulation occurs in the receivers.

---

**E4C17 What causes intermodulation in an electronic circuit?**

A. Too little gain
B. Lack of neutralizaton
C. Nonlinear circuits or devices
D. Positive feedback

**ANSWER C:** Intermodulation interference is minimized by a careful selection of linear circuits and components. If that newly-designed transceiver for 2 meters or 440 MHz, or that repaired repeater is experiencing high levels of intermodulation, the problem could be non-linear circuits, devices, and components inside the receiver.

**E4C18 What two factors determine the sensitivity of a receiver?**

A. Dynamic range and third-order intercept
B. Cost and availability
C. Intermodulation distortion and dynamic range
D. Bandwidth and noise figure

**ANSWER D:** VHF and UHF transceivers with top sensitivity use low-noise front-end transistors. When choosing a receiver, look closely at the specifications for the bandwidth and the noise figure.

**E4C19 What is the limiting condition for sensitivity in a communications receiver?**

A. The noise floor of the receiver
B. The power-supply output ripple
C. The two-tone intermodulation distortion
D. The input impedance to the detector

**ANSWER A:** Very weak signals get masked by the background noise of a receiver – the so-called noise floor of the receiver. The sensitivity of a receiver is limited by the noise generated by the devices inside which contribute to the receiver's noise floor.

**E4C20 Selectivity can be achieved in the front-end circuitry of a communications receiver by using what means?**

A. An audio filter
B. An additional RF amplifier stage
C. A preselector
D. An additional IF amplifier stage

**ANSWER C:** A preselector achieves selectivity in the front-end circuitry of a communications receiver. When you change bands, different relay-switched filters are automatically brought on line.

**E4C21 What degree of selectivity is desirable in the IF circuitry of an amateur RTTY receiver?**

A. 100 Hz
B. 300 Hz
C. 6000 Hz
D. 2400 Hz

**ANSWER B:** For RTTY, 300 Hz is good selectivity.

**E4C22 What degree of selectivity is desirable in the IF circuitry of a single-sideband phone receiver?**

A. 1 kHz
B. 2.4 kHz
C. 4.2 kHz
D. 4.8 kHz

**ANSWER B:** For SSB, a 2.4-kHz filter bandwidth is the most popular for good fidelity audio.

**E4C23 What is an undesirable effect of using too wide a filter bandwidth in the IF section of a receiver?**

A. Output-offset overshoot
B. Filter ringing
C. Thermal-noise distortion
D. Undesired signals will reach the audio stage

**ANSWER D:** Using an AM 10-kHz IF filter for pulling in 2.3 kHz SSB signals would result in undesired signals beyond the normal bandwidth of the desired signal.

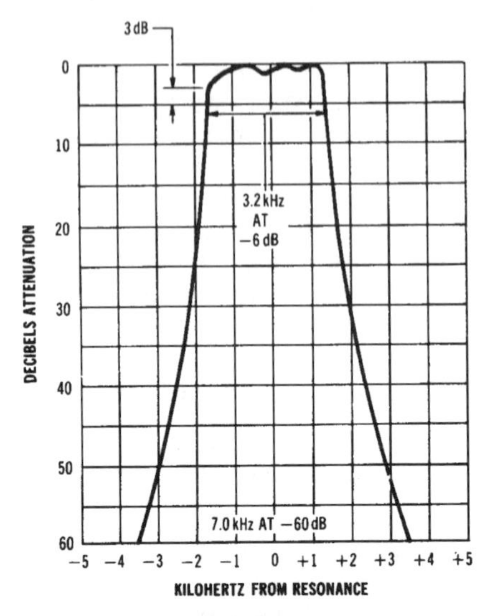


**SBB Filter**

---

**E4C24 How should the filter bandwidth of a receiver IF section compare with the bandwidth of a received signal?**

A. It should be slightly greater than the received-signal bandwidth
B. It should be approximately half the received-signal bandwidth
C. It should be approximately twice the received-signal bandwidth
D. It should be approximately four times the received-signal bandwidth

**ANSWER A:** For best fidelity, filter bandwidth should be slightly greater than the received signal bandwidth. If you are using one of those new SSB base stations, you have plenty of choices!

---

**E4C25 What degree of selectivity is desirable in the IF section of an FM phone receiver?**

A. 1 kHz
B. 2.4 kHz
C. 4.2 kHz
D. 15 kHz

**ANSWER D:** In your FM equipment, the filter must be a minimum of 15 kHz wide to accommodate 10-kHz bandwidth typical deviation.

---

**E4C26 In a receiver, if the third-order intermodulation products have a power of -70 dBm when using two test tones at -30 dBm, what is the third-order intercept point?**

A. -20 dBm
B. -10 dBm
C. 0 dBm
D +10 dBm

**ANSWER B:** For the third-order intercept point, add the difference between -30 dBm and -70 dBm, which is +40, then take half the difference which will be +20 and add it to the 2 test tones at -30 dBm, and +20 added to -30 gives us -10 dBm. Remember, for third-order intermodulation products, always half the difference before you start your addition.

---

**E4C27 In a receiver, if the second-order intermodulation products have a power of -70 dBm when using two test tones at -30 dBm, what is the second-order intercept point?**

A. -20 dBm
B. -10 dBm
C. 0 dBm
D. +10 dBm

**ANSWER D:** For second-order intermodulation, it is simply the difference between -30 dBm and -70 dBm, which is +40 dBm. Add +40 dBm to -30 dBm, and you end up with +10 dBm. For second order, it is simply the difference, NOT half the difference that we would use with third-order intermodulation calculations. Got it?

***E4D Noise suppression: vehicular system noise; electronic motor noise; static; line noise***

**E4D01 What is one of the most significant problems associated with reception in HF transceivers?**

A. Ignition noise
B. Doppler shift
C. Radar interference
D. Mechanical vibrations

**ANSWER A:** Ignition noise is a significant problem with mobile high-frequency installations. After you start up your engine, if the noise is worse than 1 S-unit, you may need to build some simple noise filters to reduce the problem. Special noise-suppression high-voltage wiring, plus special spark plugs, may help eliminate the problem.

**Mobil HF installations usually require noise suppression filters**

**E4D02 What is the proper procedure for suppressing electrical noise in a mobile transceiver?**

A. Follow the vehicle manufacturer's recommended procedures
B. Insulate all plane sheet metal surfaces from each other
C. Apply antistatic spray liberally to all non-metallic surfaces
D. Install filter capacitors in series with all DC wiring

**ANSWER A:** Today's modern motor vehicle has sophisticated computer components that are not easily serviced by a ham trying to lick electrical noise. Electrical noise might be phantom carriers, called birdies, found on your favorite high-frequency band, or on VHF and UHF. Contact the vehicle manufacturer, and see whether or not they have a service bulletin on how to remedy, or at least move, the offending phantom signals. There is no easy solution to this common problem in newer vehicles.

**CAUTION:** Today's automobiles, SUVs, and trucks contains very sophisticated computerized control systems. We recommend that you always consult with the manufacturer before making any modifications to your vehicle. If not done correctly, such modifications could void your warranty and cause serious damage to your car's electrical system.

**E4D03 Where should ferrite beads be installed to suppress ignition noise in a mobile transceiver?**

A. In the resistive high-voltage cable
B. Between the starter solenoid and the starter motor
C. In the primary and secondary ignition leads
D. In the antenna lead to the transceiver

**ANSWER C:** It is the primary and secondary ignition leads which radiate strong spark pulses that create your ignition noise. Ferrite beads strung on the leads may help reduce the higher-frequency noises.

---

**E4D04 How can alternator whine be minimized?**

A. By connecting the radio's power leads to the battery by the longest possible path
B. By connecting the radio's power leads to the battery by the shortest possible path
C. By installing a high-pass filter in series with the radio's DC power lead to the vehicle's electrical system
D. By installing filter capacitors in series with the DC power lead

**ANSWER B:** When you keep the radio power leads as short as possible, you use your battery as the main power filter point to minimize alternator whine. The battery acts like a giant capacitor to filter out the singing-type sound that may be heard on both receive as well as transmit on mobile units.

---

**E4D05 How can conducted and radiated noise caused by an automobile alternator be suppressed?**

A. By installing filter capacitors in series with the DC power lead and by installing a blocking capacitor in the field lead
B. By connecting the radio to the battery by the longest possible path and installing a blocking capacitor in both leads
C. By installing a high-pass filter in series with the radio's power lead and a low-pass filter in parallel with the field lead
D. By connecting the radio's power leads directly to the battery and by installing coaxial capacitors in the alternator leads

**ANSWER D:** Wire your radio's red and black leads directly to your battery positive and negative terminals using the shortest possible run. Be sure you put a fuse right next to the battery positive pick-off point for safety. Also, it is recommended that you install coaxial capacitors in the alternator leads.

---

**E4D06 How can noise from an electric motor be suppressed?**

A. Install a ferrite bead on the AC line used to power the motor
B. Install a brute-force, AC-line filter in series with the motor leads
C. Install a bypass capacitor in series with the motor leads
D. Use a ground-fault current interrupter in the circuit used to power the motor

**ANSWER B:** One way of reducing noise coming up the line from an electric motor is to install an AC-line filter in series with the motor leads. The incorrect answers have got the right components, but the wrong usage. We use RF beads to keep RF off of specific circuits, and we use bypass capacitors in parallel to bypass noise. As for the ground-fault interrupter, this won't help with motor noise. Go with the brute-force AC-line filter in series.

**E4D07 What is a major cause of atmospheric static?**

A. Sunspots
B. Thunderstorms
C. Airplanes
D. Meteor showers

**ANSWER B:** Thunderstorms carry lightning activity, and lightning noise can be received as far away as 800 miles on the 80- and 40-meter bands.

---

**E4D08 How can it be determined if line-noise interference is being generated within your home?**

A. By checking the power-line voltage with a time-domain reflectometer
B. By observing the AC power line waveform with an oscilloscope
C. By turning off the AC power line main circuit breaker and listening on a battery-operated radio
D. By observing the AC power line voltage with a spectrum analyzer

**ANSWER C:** Home-office equipment like FAX machines or telephones with intercom, fish-tank heaters, and ultrasonic bug repellers all can create whistles and rhythmical pulsing sounds on both high-frequency and VHF/UHF station outputs. To find out whether the noise is being generated by your own home electronics, shut off your AC power mains and continue listening as you run your rig off of a 12-volt battery source. If the noise instantly disappears, use your 2-meter handi-talkie and start "sniffing" for the noise source when the power is turned on again. Of course, many home appliances will blink 12:00 after you turn the power back on!

---

**E4D09 What type of signal is picked up by electrical wiring near a radio transmitter?**

A. A common-mode signal at the frequency of the radio transmitter
B. An electrical-sparking signal
C. A differential-mode signal at the AC power line frequency
D. Harmonics of the AC power line frequency

**ANSWER A:** Electrical wiring will many times pick up common-mode radio signals at the frequency of the radio transmitter. The question is slightly in error – the wiring near the radio antenna is where most of the signal gets picked up, not necessarily at the radio transmitter. Answer A is correct.

---

**E4D10 Which of the following types of equipment would be least useful in locating power line noise?**

A. An AM receiver with a directional antenna
B. An FM receiver with a directional antenna
C. A hand-held RF sniffer
D. An ultrasonic transducer, amplifier and parabolic reflector

**ANSWER B:** An FM receiver is specifically designed not to pick up AM signals and AM noise. To trace power line noise, a simple AM receiver, or a handheld sniffer, or even an ultrasonic transducer tied into an amplifier and parabolic reflector will help you home in on arcing wires around a power-line insulator, even though you can't see where the snap, crackle, and pop are coming from.

***E4E Component mounting techniques (i.e., surface, dead bug (raised), circuit board; direction finding: techniques and equipment; fox hunting***

**E4E01 What circuit construction technique uses leadless components mounted between circuit board pads?**

A. Raised mounting
B. Integrated circuit mounting
C. Hybrid device mounting
D. Surface mounting

**ANSWER D:** Amateur radio equipment has never been more compact. Many sections of your new worldwide transceiver or your VHF/UHF equipment have boards standing on end where the components are "SMT" – surface mount technology. Surface-mounted components are mounted on one side of the board, not through holes in the board. Working on these tiny components requires a steady hand, a magnifying glass, and a fan to blow away the heat.

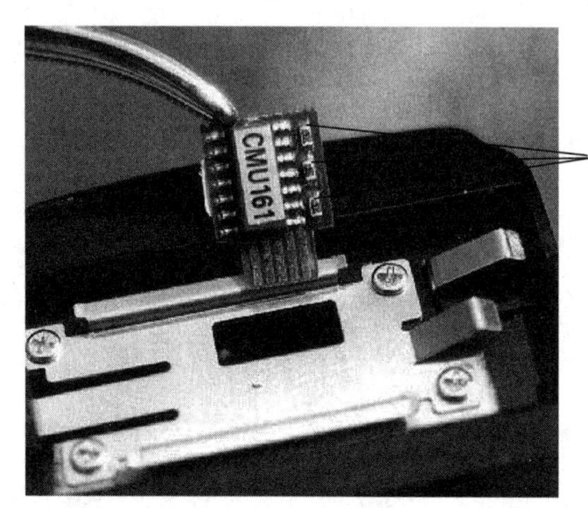


**VHF/UHF Equipment Board with Surface-Mount Devices**
(The board with chip set is from a Standard Radio transceiver.)

**E4E02 What is the main drawback of a wire-loop antenna for direction finding?**

A. It has a bi-directional pattern broadside to the loop
B. It is non-rotatable
C. It receives equally well in all directions
D. It is practical for use only on VHF bands

**ANSWER A:** Wire loop direction finding antennas will pick up maximum signal strength when the loop is broadside to the signal source. It is bi-directional – it will pick up noise sources on either side of the loop.

**E4E03 What pattern is desirable for a direction-finding antenna?**

A. One which is non-cardioid
B. One with good front-to-back and front-to-side ratio
C. One with good top-to-bottom and side-to-side ratio
D. One with shallow nulls

**ANSWER B:** A lightweight quad antenna with three or more elements offers excellent direction-finding capabilities.

**E4E04 What is the triangulation method of direction finding?**

A. The geometric angle of ground waves and sky waves from the signal source are used to locate the source
B. A fixed receiving station plots three beam headings from the signal source on a map
C. Beam antenna headings from several receiving stations are used to plot the signal source on a map
D. A fixed receiving station uses three different antennas to plot the location of the signal source

**ANSWER C:** Amateur radio operators participating in a joint "T-hunt" will swing their beams and record a magnetic bearing to where signal strength is the greatest. Through triangulation, they can then determine the approximate location of the transmitting station. All participating stations should agree to use either magnetic north or true north for their heading directions.

---

**E4E05 Why is an RF attenuator desirable in a receiver used for direction finding?**

A. It narrows the bandwidth of the received signal
B. It eliminates the effects of isotropic radiation
C. It reduces loss of received signals caused by antenna pattern nulls
D. It prevents receiver overload from extremely strong signals

**ANSWER D:** When using a receiver for direction finding, a step adjustable or variable adjustable attenuator is necessary on the input when you get extremely close to the signal source. You add additional attenuation in order to keep your receiver from overloading as you maintain maximum sensitivity for signal variations as the antenna direction is changed.

---

**E4E06 What is a sense antenna?**

A. A vertical antenna added to a loop antenna to produce a cardioid reception pattern
B. A horizontal antenna added to a loop antenna to produce a cardioid reception pattern
C. A vertical antenna added to an Adcock antenna to produce a omnidirectional reception pattern
D. A horizontal antenna added to an Adcock antenna to produce a omnidirectional reception pattern

**ANSWER A:** Older aviation and marine loop antennas were coupled with a "sense antenna" in order for the operator to resolve bearing ambiguities. There was usually a button that the operator would push to "phase-in" the signal from the vertical sense antenna.

---

**E4E07 What is a loop antenna?**

A. A large circularly-polarized antenna
B. A small coil of wire tightly wound around a toroidal ferrite core
C. Several turns of wire wound in the shape of a large open coil
D. Any antenna coupled to a feed line through an inductive loop of wire

**ANSWER C**: A wire loop antenna is normally used for direction finding.

**E4E08 How can the output voltage of a loop antenna be increased?**

A. By reducing the permeability of the loop shield
B. By increasing the number of wire turns in the loop and reducing the area of the loop structure
C. By reducing either the number of wire turns in the loop or the area of the loop structure
D. By increasing either the number of wire turns in the loop or the area of the loop structure

**ANSWER D:** To make a loop antenna more powerful, either the number of wire turns in the loop are increased, or the loop is made physically larger.

---

**E4E09 Why is an antenna with a cardioid pattern desirable for a direction-finding system?**

A. The broad-side responses of the cardioid pattern can be aimed at the desired station
B. The deep null of the cardioid pattern can pinpoint the direction of the desired station
C. The sharp peak response of the cardioid pattern can pinpoint the direction of the desired station
D. The high-radiation angle of the cardioid pattern is useful for short-distance direction finding

**ANSWER B:** A cardioid pattern is shaped like a heart. There is a broad reception pattern in one direction and a deep null in the other. When the button is pushed to "sense" the direction of the incoming signal on a direction-finding system that has a cardioid antenna pattern, the deep null in the pattern is normally 180 degrees away from the incoming signal. It's easier to use the null point to sense the signal direction of the incoming signal than it is the broad pattern – look for the minimum signal, not the maximum.

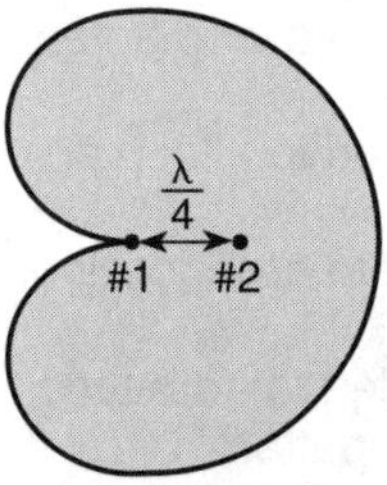

$\frac{\lambda}{4}$

#1 #2

**Two 1/4 wave vertical antennas fed 90-degrees out of phase will have a cardoid radiation pattern. The deep null will be 180-degrees from the transmitter, and is very helpful in direction-finding.**

---

**E4E10 What type of terrain can cause errors in direction finding?**

A. Homogeneous terrain
B. Smooth grassy terrain
C. Varied terrain
D. Terrain with no buildings or mountains

**ANSWER C:** Hills and valleys, plus tall buildings, can create errors in determining the exact direction of the incoming signal. Find yourself some experts in "T-hunting" and let them show you the ropes on how to track down those elusive transmitters, or if you are the transmitter, to hide yourself as a "hidden T." Transmitter hunting is a fun way to improve your skills in understanding how VHF and UHF radio waves propagate over different terrains.

---

**E4E11 What is the amateur station activity known as fox hunting?**

A. Attempting to locate a hidden transmitter by using receivers and direction-finding techniques

B. Attempting to locate a hidden receiver by using receivers and direction-finding techniques

C. Assisting government agents with tracking transmitter collars worn by foxes

D. Assembling stations using generators and portable antennas to test emergency communications skills

**ANSWER A:** Amateur operators will use receivers and directional antennas to help locate hidden transmitters. This is called a "fox hunt."

**T-hunting in action!**

## Subelement E5 — Electrical Principles [9 Exam Questions — 9 Groups]

***E5A Characteristics of resonant circuits: Series resonance (capacitor and inductor to resonate at a specific frequency); Parallel resonance (capacitor and inductor to resonate at a specific frequency); half-power bandwidth***

**E5A01 What can cause the voltage across reactances in series to be larger than the voltage applied to them?**

A. Resonance
B. Capacitance
C. Conductance
D. Resistance

**ANSWER A:** Coils (having inductance) and capacitors (having capacitance) will be at resonance when the capacitive reactance equals the inductive reactance. At resonance, large voltages greater than the applied voltage can be present across these components in the circuit. You can see this with a mobile whip antenna if something were to touch the whip in transmit. You may see quite a spark!

**Equation for Resonant Frequency**

$X_L$ = Inductive Reactance

$X_L = 2\pi fL$

$X_C$ = Capacitive Reactance

$$X_C = \frac{1}{2\pi fC}$$

L = Inductance in henrys

C = Capacitance in farads

f = Frequency in hertz

At Resonance $X_L = X_C$

$$2\pi f_r L = \frac{1}{2\pi f_r C}$$

Solving for $f_r$

$$f_r^2 = \frac{1}{(2\pi)^2 LC}$$

The Resonant Frequency, $f_r$, is:

$$f_r = \frac{1}{2\pi\sqrt{LC}}$$

**E5A02 What is resonance in an electrical circuit?**

A. The highest frequency that will pass current
B. The lowest frequency that will pass current
C. The frequency at which capacitive reactance equals inductive reactance
D. The frequency at which power factor is at a minimum

**ANSWER C:** At resonance, $X_L$ (inductive reactance) equals $X_C$ (capacitive reactance). In a mobile series resonant whip antenna, there will be maximum current through the loading coil when the whip is tuned to resonance. On your base station vertical trap antenna or trap beam antenna, parallel resonant circuits offer high impedance, trapping out a specific length of the antenna for a certain frequency. At resonance, the resonant parallel circuit looks like a high impedance to any of your power getting through to the longer length of the antenna system.

**E5A03 What are the conditions for resonance to occur in an electrical circuit?**

A. The power factor is at a minimum
B. Inductive and capacitive reactances are equal
C. The square root of the sum of the capacitive and inductive reactance is equal to the resonant frequency
D. The square root of the product of the capacitive and inductive reactance is equal to the resonant frequency

**ANSWER B:** When a coil's inductive reactance equals a capacitor's capacitive reactance, resonance occurs. On antenna systems, you fine-tune your elements for resonance at a certain frequency.

---

**E5A04 When the inductive reactance of an electrical circuit equals its capacitive reactance, what is this condition called?**

A. Reactive quiescence
B. High Q
C. Reactive equilibrium
D. Resonance

**ANSWER D:** Resonance in a circuit occurs at the frequency where inductive reactance $X_L$ equals capacitive reactance $X_C$.

---

**E5A05 What is the magnitude of the impedance of a series R-L-C circuit at resonance?**

A. High, as compared to the circuit resistance
B. Approximately equal to capacitive reactance
C. Approximately equal to inductive reactance
D. Approximately equal to circuit resistance

**ANSWER D:** Did you ever wonder why the serious mobile ham always uses a gigantic loading coil for his series resonant antenna? The bigger the loading coil, the less resistance. At resonance, the magnitude of the impedance is equal to the *circuit resistance* (key words).

**A large loading coil on a mobil whip helps antennas achieve high Q resonance.**

---

**E5A06 What is the magnitude of the impedance of a circuit with a resistor, an inductor and a capacitor all in parallel, at resonance?**

A. Approximately equal to circuit resistance
B. Approximately equal to inductive reactance
C. Low, as compared to the circuit resistance
D. Approximately equal to capacitive reactance

**ANSWER A:** Whether the system is parallel or series R-L-C resonant, the impedance will always be equal to the *circuit resistance* (key words).

---

**E5A07 What is the magnitude of the current at the input of a series R-L-C circuit at resonance?**

A. It is at a minimum
B. It is at a maximum
C. It is DC
D. It is zero

**ANSWER B:** A mobile whip antenna will have maximum current in a series R-L-C circuit at resonance. One way you can tell whether or not a mobile whip is working properly is to feel the coil after the transmitter has been shut down. If it's warm, there has been current through the coil and up to the whip tip stinger.

**E5A08 What is the magnitude of the circulating current within the components of a parallel L-C circuit at resonance?**

A. It is at a minimum
B. It is at a maximum
C. It is DC
D. It is zero

**ANSWER B:** A parallel circuit at resonance is similar to a tuned trap in a multi-band trap antenna. At resonance, the trap keeps power from going any further to the longer antenna elements. But be assured there is maximum circulating current within the parallel R-L-C circuit. Now don't get confused with this question – a parallel circuit has maximum current within it, and minimum current going on to the next stage. That's why they call those parallel resonant circuits "traps."

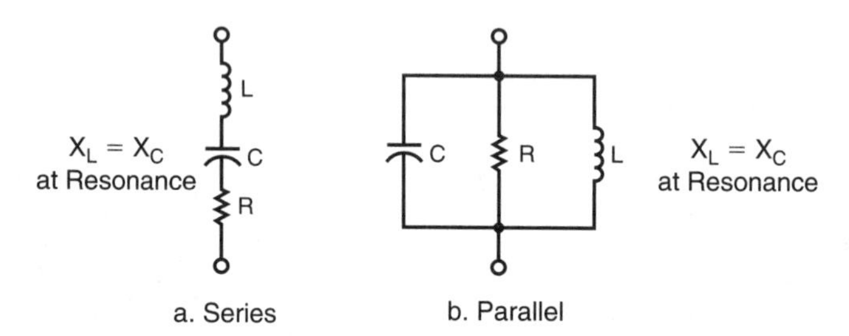


**Series and Parallel Resonant Circuits**

**E5A09 What is the magnitude of the current at the input of a parallel R-L-C circuit at resonance?**

A. It is at a minimum
B. It is at a maximum
C. It is DC
D. It is zero

**ANSWER A:** The total current into and out of a parallel RLC circuit at resonance is at a minimum. On old radios, we would tune a plate control to see the current dip at resonance.

**E5A10 What is the relationship between the current through a resonant circuit and the voltage across the circuit?**

A. The voltage leads the current by 90 degrees
B. The current leads the voltage by 90 degrees
C. The voltage and current are in phase
D. The voltage and current are 180 degrees out of phase

**ANSWER C:** When a circuit is at resonance, voltage and current are in phase. If you remember "ELI the ICE man," it will help you visualize the relationship of voltage and current when they are NOT in resonance.

ELI = voltage (E) leads current (I) in an inductive (L) circuit.

ICE = current (I) leads voltage (E) in a capacitive (C) circuit.

At *Resonance*: $X_L = X_C$ = voltage and current in phase (neither is leading the other).

**E5A11 What is the relationship between the current into (or out of) a parallel resonant circuit and the voltage across the circuit?**

A. The voltage leads the current by 90 degrees
B. The current leads the voltage by 90 degrees

C. The voltage and current are in phase
D. The voltage and current are 180 degrees out of phase

**ANSWER C:** Whether the circuit is series or parallel resonant, at resonance voltage and current are always in phase. When a mobile whip antenna is tuned for resonance, voltage and current are in phase, and your antenna stainless steel stinger receives all of the power from your transceiver, minus the slight resistance in the network. On that big 3-element trap beam you put up for tri-band operation, the traps are actually parallel resonant circuits, and current WITHIN the trap is at maximum, and only that portion of the antenna BEFORE the trap is radiating. Just remember that voltage and current are in phase at resonance.

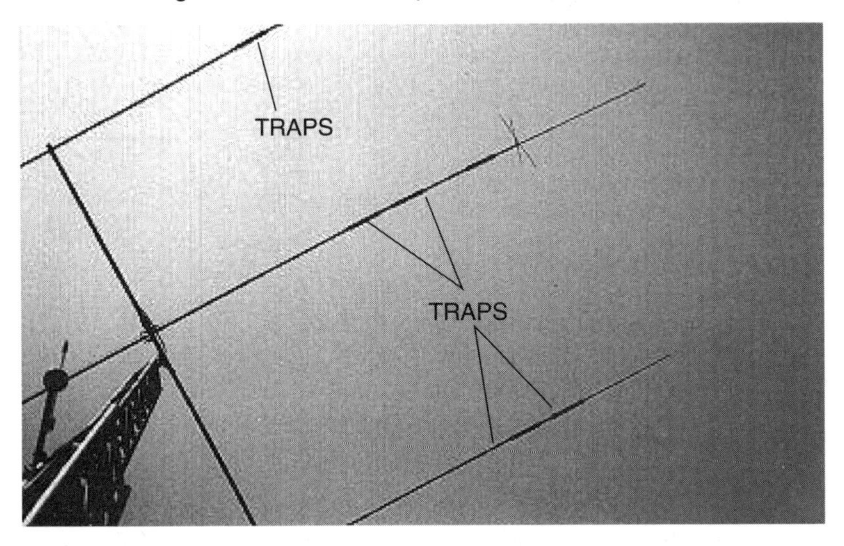


**Yagi Antenna with Traps**

**E5A12 What is the half-power bandwidth of a parallel resonant circuit that has a resonant frequency of 1.8 MHz and a Q of 95?**

A. 18.9 kHz
B. 1.89 kHz
C. 189 Hz
D. 58.7 kHz

**ANSWER A:** Here is a series of questions on the half-power bandwidth of a parallel resonant circuit. There is a good chance one of these questions will be asked on your actual Extra Class examination.
Here is the formula:

Half-power bandwidth $= \frac{f_r}{Q}$; where: $f_r$ is the resonant frequency in **kHz**
Q is a **quality factor** of the circuit

This is a handy formula to allow you to look at a circuit design and see what frequencies are covered in its half-power bandwidth. Since the answers come out in kHz, you will need to take their stated resonant frequency in MHz and convert it to kHz by moving the decimal point three places to the right before you solve the problem. Here's how you do it on your calculator:
Clear, 1800 ÷ 95 = 18.94 kHz, so the answer is 18.9 kHz. Yikes, take a look at the possible answers and see if you goofed up on moving the decimal point!

**E5A13 What is the half-power bandwidth of a parallel resonant circuit that has a resonant frequency of 7.1 MHz and a Q of 150?**

A. 211 kHz
B. 16.5 kHz
C. 47.3 kHz
D. 21.1 kHz

**ANSWER C:** 7.1 MHz is 7100 kHz, divided by 150. Remember for kHz from MHz, move three decimal places to the right. Clear, 7100 ÷ 150 = 47.33.

**Half-Power Bandwidth**

An amplifier's voltage gain will vary with frequency as shown. At the cutoff frequencies, the voltage gain has dropped to 0.707 of what it is in the mid-band. These frequencies, $f_1$ and $f_2$, are sometimes called the half-power frequencies.

If the output voltage is 10 volts across a 100-ohm load when the gain is A at the mid-band, then the power output, $P_O$, at mid-band is:

$$P_O = \frac{V^2}{R} = \frac{10^2}{100} = \frac{100}{100} = 1 \text{ Watt}$$

At the cutoff frequency, the output voltage will be 0.707 of what it is at the mid-band; therefore, 7.07 volts. The power output is:

$$P_{0.707} = \frac{(7.07)^2}{100} = \frac{50}{100} = 0.5 \text{ Watt}$$

The power output at the cutoff frequency points is one-half the mid-band power. The half-power bandwidth is between the frequencies $f_1$ and $f_2$.

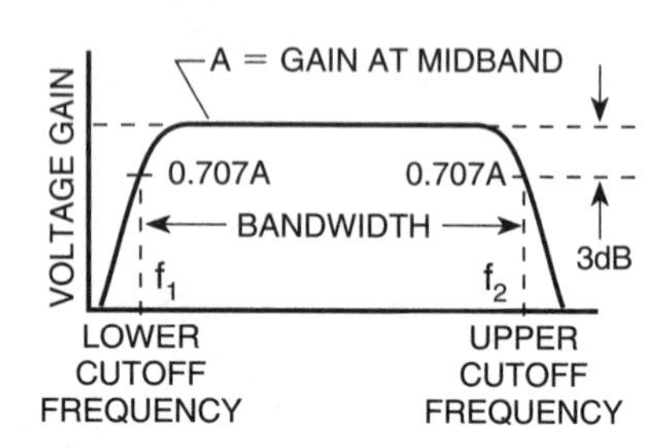


**Power Ratio in dB**

$$P_{dB} = 10 \log_{10} \frac{P_O}{P_{0.707}}$$

$$= 10 \log_{10} \frac{1}{0.5}$$

$$= 10 \log_{10} 2$$

$$= 10 \times 0.301$$

$$= 3 \text{ db}$$

The power output at the 0.707 frequencies is 3 db down from the mid-band power.

---

**E5A14 What is the half-power bandwidth of a parallel resonant circuit that has a resonant frequency of 14.25 MHz and a Q of 150?**

A. 95 kHz
B. 10.5 kHz
C. 10.5 MHz
D. 17 kHz

**ANSWER A:** Same routine, 14.25 MHz moved three decimal points to the right is 14,250 kHz. It divided by a Q of 150 is 95. Easy!

---

**E5A15 What is the half-power bandwidth of a parallel resonant circuit that has a resonant frequency of 21.15 MHz and a Q of 95?**

A. 4.49 kHz
B. 44.9 kHz
C. 22.3 kHz
D. 222.6 kHz

**ANSWER D:** Again, 21.15 MHz is 21,150 kHz divided by a Q of 95 is 222.63. No sweat!

---

**E5A16 What is the half-power bandwidth of a parallel resonant circuit that has a resonant frequency of 3.7 MHz and a Q of 118?**

A. 22.3 kHz
B. 76.2 kHz
C. 31.4 kHz
D. 10.8 kHz

**ANSWER C:** With this many half-power bandwidth questions in the pool, you can count on at least one being on your upcoming Extra Class Element 4 exam. With this question, go from MHz to kHz, and then divide it by the Q. Clear, enter 3700 ÷ 118 = 31.36. Round the answer to 31.4.

---

**E5A17 What is the half-power bandwidth of a parallel resonant circuit that has a resonant frequency of 14.25 MHz and a Q of 187?**

A. 22.3 kHz
B. 10.8 kHz
C. 76.2 kHz
D. 13.1 kHz

**ANSWER C:** Here is your last half-power bandwidth problem. Hopefully by now you have memorized the formula of E5A12. Remember the conversion to kHz so the general steps are: MHz to kHz, enter the number in the calculator and divide it by the Q. There is your answer, in kHz.

---

**E5A18 What is the resonant frequency of a series RLC circuit if R is 47 ohms, L is 50 microhenrys and C is 40 picofarads?**

A. 79.6 MHz
B. 1.78 MHz
C. 3.56 MHz
D. 7.96 MHz

**ANSWER C:** The schematic diagram shown is a simple series resonant circuit. Don't panic. The following resonance formula will carry you through many questions on series and parallel circuits – one of which may be on your exam. You can throw away the resistance found in the parallel or series diagrams. You won't need it for the resonance formula.

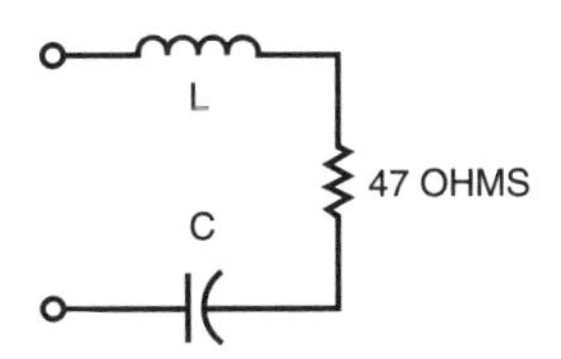


**Series Resonant Circuit**

Here is the resonance formula, which you should memorize:

$$f_r = \frac{10^6}{2\pi\sqrt{LC}}$$

where: $f_r$ is resonant frequency in **kHz**
$2\pi = 6.28$
L is inductance in **microhenrys**
C is capacitance in **picofarads**

Your first step is multiplying L × C. For this question, that is 50 × 40 = 2000. Now we need the square root of this, since L and C are under the square root sign. With 2000 showing on your calculator, simply press the square root key. Presto, you should have 44.72. Now multiply by 2π, which is 6.28 × 44.72 = 280.84. This leaves us with 1,000,000 ÷ 280.84. Press your clear button, and enter 1,000,000 ÷ 280.84 = 3560.7463 kHz.

Your answer came out in kHz, but they are looking for MHz. No problem here – simply move the decimal point three places to the *left*, and you get answer C. That's all there is to it!

**E5A19 What is the resonant frequency of a series RLC circuit if R is 47 ohms, L is 40 microhenrys and C is 200 picofarads?**

A. 1.99 kHz
B. 1.78 MHz
C. 1.99 MHz
D. 1.78 kHz

**ANSWER B:** Use the formula at E5A18. Solve the denominator first, then divide it into $10^6$. Here are the calculator keystrokes using a calculator that has a square root ($\sqrt{\ }$) key: Clear, 40 × 200 = $\sqrt{\ }$ × 6.28 = 561.7. Remember 561.7. Clear, 1,000,000 ÷ 561.7 = 1780 kHz = 1.78 MHz. Be careful about answer D, it's in kHz.

Here is the resonance formula again, which you should memorize:

$$f_r = \frac{10^6}{2\pi\sqrt{LC}}$$

where: $f_r$ is resonant frequency in **kHz**
$2\pi = 6.28$
L is inductance in **microhenrys**
C is capacitance in **picofarads**

---

**E5A20 What is the resonant frequency of a series RLC circuit if R is 47 ohms, L is 50 microhenrys and C is 10 picofarads?**

A. 3.18 MHz
B. 3.18 kHz
C. 7.12 kHz
D. 7.12 MHz

**ANSWER D:** Use the formula at E5A18. First multiply microhenrys and picofarads, take their square root, multiply by 6.28, and then divide your answer into 1,000,000. Finally, change kHz to MHz. Easy! Here are the keystrokes: Clear, 50 × 10 = $\sqrt{\ }$ × 6.28 = 140.4. Remember 140.4. Clear, 1,000,000 ÷ 140.4 = 7122 kHz which is 7.12 MHz, answer D. C is **not** correct, it's in kHz.

---

**E5A21 What is the resonant frequency of a series RLC circuit if R is 47 ohms, L is 25 microhenrys and C is 10 picofarads?**

A. 10.1 MHz
B. 63.7 MHz
C. 10.1 kHz
D. 63.7 kHz

**ANSWER A:** Use the formula at E5A18. L × C, $\sqrt{\ }$, multiply by $2\pi$, and then divide your answer into 1,000,000. Remember to change kHz to MHz. The keystrokes are: Clear, 25 × 10 = $\sqrt{\ }$ × 6.28 = 99.29. Clear, 1,000,000 ÷ 99.3 = 10070.5 kHz which is 10.1 MHz. Answer C is **not** correct.

---

**E5A22 What is the resonant frequency of a series RLC circuit if R is 47 ohms, L is 3 microhenrys and C is 40 picofarads?**

A. 13.1 MHz
B. 14.5 MHz
C. 14.5 kHz
D. 13.1 kHz

**ANSWER B:** Use the formula at E5A18. The general calculator keystrokes are L × C, =, $\sqrt{\ }$, times 6.28, =. Divide the answer into 1,000,000, and change kHz to MHz. Presto! By golly, I think you have it.

---

**E5A23 What is the resonant frequency of a series RLC circuit if R is 47 ohms, L is 4 microhenrys and C is 20 picofarads?**

A. 19.9 kHz
B. 17.8 kHz
C. 19.9 MHz
D. 17.8 MHz

**ANSWER D:** Use the formula at E5A18. Press L × C, =, $\sqrt{\ }$, times 6.28, =. Divide that into 1,000 and guess what; you save a step and the answer comes out in MHz! The same general keystrokes, but 1,000,000 is changed to 1000.

**E5A24 What is the resonant frequency of a series RLC circuit if R is 47 ohms, L is 8 microhenrys and C is 7 picofarads?**

A. 2.84 MHz
B. 28.4 MHz
C. 21.3 MHz
D. 2.13 MHz

**ANSWER C:** Use formula at E5A18, same general keystrokes: L × C, =, $\sqrt{\ }$, times 6.28, =. However, now divide the answer into $10^3$ rather than $10^6$, and it comes out MHz!

---

**E5A25 What is the resonant frequency of a series RLC circuit if R is 47 ohms, L is 3 microhenrys and C is 15 picofarads?**

A. 23.7 MHz
B. 23.7 kHz
C. 35.4 kHz
D. 35.4 MHz

**ANSWER A:** Use formula at E5A18 and general keystrokes at E5A22. Don't forget to take the square root of 3 × 15 before you multiply by 6.28! If you don't have a calculator with a square root key, approximate the square root. You know that a square root multiplied by itself gives you the number you want. You want the $\sqrt{45}$. You know that 6 × 6 = 36 and 7 × 7 = 49; therefore, the square root of 45 is between 6 and 7, but closer to 7. 6.8 × 6.8 = 46.2, so that's too large. 6.7 × 6.7 = 44.9 so that's close enough. Use 6.7. Multiply it by 6.28 and divide into 1000. The answer is 23.74, which is answer A.

---

***E5B Exponential charge/discharge curves (time constants): definition; time constants in RL and RC circuits***

---

**E5B01 What is the term for the time required for the capacitor in an RC circuit to be charged to 63.2% of the supply voltage?**

A. An exponential rate of one
B. One time constant
C. One exponential period
D. A time factor of one

**ANSWER B:** In an RC circuit, assuming there is no initial charge on the capacitor, it takes one time constant to charge a capacitor to 63.2 percent of its final supply voltage value.

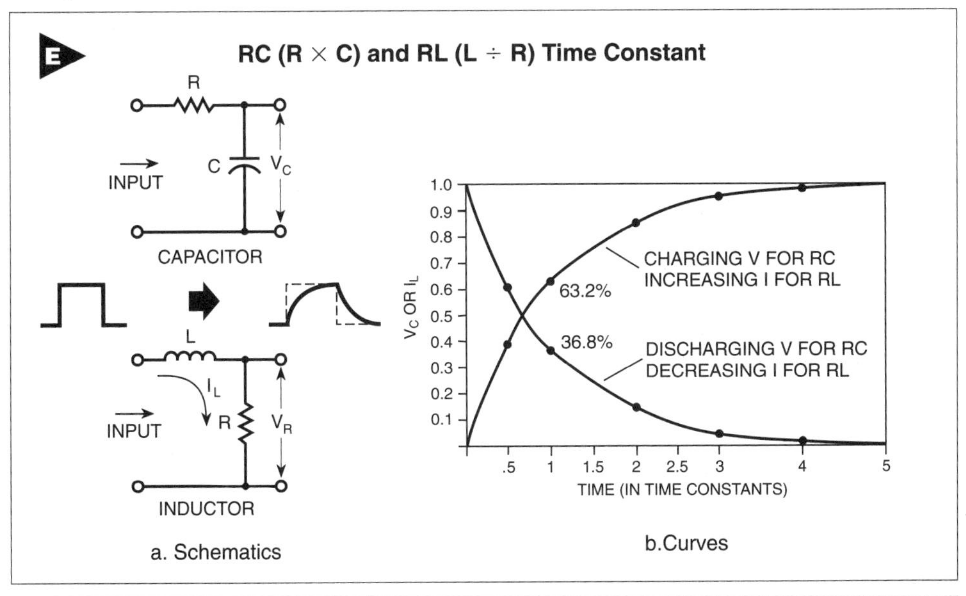

a. Schematics
b.Curves

**E5B02 What is the term for the time required for the current in an RL circuit to build up to 63.2% of the maximum value?**

A. One time constant
B. An exponential period of one
C. A time factor of one
D. One exponential rate

**ANSWER A:** In an RL circuit, assuming there is no initial current through the inductance, it takes one time constant for the current in the circuit to build up to 63.2 percent of its final maximum value. Measuring $V_R$ tracks $I_L$.

---

**E5B03 What is the term for the time it takes for a charged capacitor in an RC circuit to discharge to 36.8% of its initial value of stored charge?**

A. One discharge period
B. An exponential discharge rate of one
C. A discharge factor of one
D. One time constant

**ANSWER D:** Consider the filter capacitors in a power supply, where the capacitors discharge when the input power is removed. The time it takes a capacitor in an RC circuit to discharge to 36.8 percent of its initial value of stored charge is one time constant.

---

**E5B04 The capacitor in an RC circuit is charged to what percentage of the supply voltage after two time constants?**

A. 36.8%
B. 63.2%
C. 86.5%
D. 95%

**ANSWER C:** To calculate the percentage of the supply voltage after two time constants in an RC circuit, write down the percent after a single time constant as 63.2 percent. The remaining percent of charge that the capacitor must charge is 36.8 percent. It is found by deducting 63.2 percent from 100 percent. Since 36.8 percent is the final value to which the capacitor will charge after it has charged to 63.2 percent, in the next time constant the capacitor will charge to 63.2 percent of the 36.8 percent. In other words, another percent charge is added to the capacitor in the second time constant equal to:

$$36.8\% \times 63.2\% = 23.26\% \text{ (rounded to 23.3\%)}$$

This really needs to be calculated in decimal format as:

$$0.368 \times 0.632 = 0.2326$$

Percent values are converted to decimals by moving the decimal point two places to the left. Decimal values are converted to percent values by multiplying by 100 (moving decimal point to right two places.)

Since the final percent of charge in two time constants is desired, it is only necessary to add the two percent charges together:

| | Percent | Decimal |
|---|---|---|
| Charge in first time constant = | 63.2% | 0.632 |
| Charge in second time constant = | 23.3% | 0.233 |
| Total charge after two time constants | 86.5% | 0.865 |

The process can be continued for the third time constant. Since the capacitor has charged to 86.5 percent in two time constants, if you start from that point the final value of charge would be another 13.5 percent. Therefore, in the next time constant, the third, the capacitor would charge another:

$$13.5\% \times 63.2\% = 8.53\% \text{ (rounded to 8.5\%)}$$

After the third time constant the total charge is:

$$86.5\% + 8.5\% = 95\%$$

If the process is continued through five time constants, the capacitor charge is 99.3 percent. Therefore, in electronic calculations, the capacitor is considered to be fully charged (or discharged) after five time constants

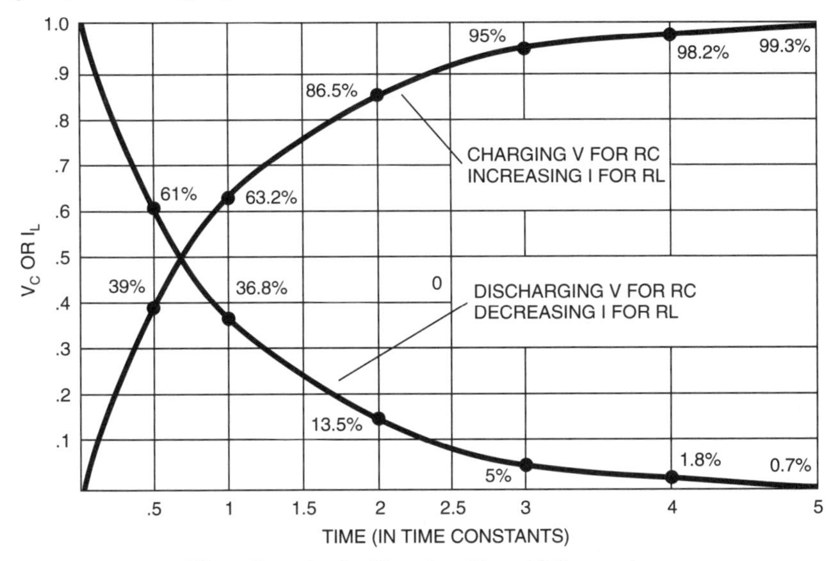


**Time Constants Showing V and I Percentages**

**E5B05 The capacitor in an RC circuit is discharged to what percentage of the starting voltage after two time constants?**

A. 86.5%
B. 63.2%
C. 36.8%
D. 13.5%

**ANSWER D:** This question is the reverse of the previous question. It is asking to what percent the capacitor has *discharged.* The capacitor discharges the same percent in one time constant as it charges (for the same RC circuit values of course.) However, the question is asking for the amount of charge remaining. At one time constant, the capacitor discharges 63.2% (in the previous question, it charged 63.2% in one time constant); therefore, the percent charge remaining on the capacitor is 36.8% (100%−63.2%). From the previous question, the capacitor charges another 23.3% in the second time constant. Thus, the capacitor on discharge, discharges another 23.3% in the second time constant. After two time constants the capacitor has discharged by 86.5%; therefore, the remaining charge is 13.5% (100%−86.5%). In five time constants the capacitor would have discharged 99.3% of its charge, so 0.7% of charge remains – it is considered to be fully discharged.

**E5B06 What is the time constant of a circuit having two 100-microfarad capacitors and two 470-kilohm resistors all in series?**

A. 47 seconds
B. 101.1 seconds
C. 103 seconds
D. 220 seconds

**ANSWER A:** For an RC circuit $\tau = RC$, but first we must combine resistor and capacitor values to come up with a single value for each. Since this is a series circuit, the resistor values add ($R_T = R_1 + R_2$) to arrive at a total resistance, but for the capacitance total you have to solve the formula $C_T = (C_1 \times C_2) \div (C_1 + C_2)$. Therefore,

Powers of 10:

$$C_T = \frac{(1 \times 10^{+2} \times 10^{-6} \times 1 \times 10^{+2} \times 10^{-6})}{(100 \times 10^{-6} + 100 \times 10^{-6})}$$

$$C_T = \frac{(1 \times 10^{-8})}{(2 \times 10^{+2} \times 10^{-6})}$$

$$C_T = \frac{(1 \times 10^{-8})}{(2 \times 10^{-4})} = 0.5 \times 10^{-4} = 50 \times 10^{-6}$$

$$C_T = 0.00005 \text{ farads or } 50 \text{ microfarads}$$

For the resistance:

$$R_T = 470{,}000 + 470{,}000 = 940{,}000$$
$$R_T = 4.7 \times 10^{+5} + 4.7 \times 10^{+5} = 9.4 \times 10^{+5}$$

Now that we have calculated the total series capacitance and the total resistance, we simply multiply $0.00005 \times 940{,}000 = 47$ seconds. In powers of ten this is $\tau = 0.5 \times 10^{-4} \times 9.4 \times 10^{+5} = 4.7 \times 10^{+1} = 47$.

Of course, the old timers who recognize that two equal capacitors in series equals a single capacitor at half the value would immediately jump to $\tau = 50$ microfarads $\times$ 940 kilohms, pull out the calculator and press the following:

Clear, $.00005 \times 940000 = 47$. The time constant is in seconds. Notice if you miss a single decimal point, they have an incorrect answer for you. Watch out!

**Remember:**

| Capacitors in Series: | Resistors in Series: |
|---|---|
| $C_T = \dfrac{C_1 \times C_2}{C_1 + C_2}$ | $R_T = R_1 + R_2$ |
| Capacitors in Parallel: | Resistors in Parallel: |
| $C_T = C_1 + C_2$ | $R_T = \dfrac{R_1 \times R_2}{R_1 + R_2}$ |
| C = Capacitance in **farads** | R = Resistance in **ohms** |

**Capacitors, Resistors in Series and Parallel**

**E5B07 What is the time constant of a circuit having two 220-microfarad capacitors and two 1-megohm resistors all in parallel?**

A. 47 seconds
B. 101.1 seconds
C. 103 seconds
D. 220 seconds

**ANSWER D:** Another RC circuit with all components in parallel. It will be easy if you remember that two equal capacitors in parallel have a total capacitance of twice the value of one of the capacitors, and that two equal value resistors in parallel have a total resistance equal to one-half the value of one of the resistors. And more good news – "megs" are being multiplied by "micros" so the powers of ten after the numerical value cancel. Two 220 microfarad capacitors in parallel is a larger 440 microfarad capacitor. Two one-megohm resistors in parallel is a resistance of 0.5 megohms. Solve for the time constant by multiplying $.5 \times 440$ ("megs" times "micros") and you end up with 220 seconds as the correct answer. The calculator keystrokes are: Clear, $.5 \times 440 = 220$.

---

**E5B08 What is the time constant of a circuit having a 220-microfarad capacitor in series with a 470-kilohm resistor?**

A. 47 seconds
B. 80 seconds
C. 103 seconds
D. 220 seconds

**ANSWER C:** Use "megs" and "micros" and convert the resistance value to megohms. 470,000 ohms is 0.47 megohms. The time constant $\tau = 0.47 \times 220 = 103.4$ seconds, rounded off to 103 seconds.

---

**E5B09 How long does it take for an initial charge of 20 V DC to decrease to 7.36 V DC in a 0.01-microfarad capacitor when a 2-megohm resistor is connected across it?**

A. 0.02 seconds
B. 0.08 seconds
C. 450 seconds
D. 1350 seconds

**ANSWER A:** Note from the time-constant chart at E5B04 that a capacitor discharges (or charges) 63.2% in one time constant. Thus, after discharging for one time constant, 36.8% of the charge remains on the capacitor. Therefore, in one time constant the 20 volts will drop down to 36.8% or 7.36 volts ($.368 \times 20 = 7.36$). Write down one time constant = 7.36 volts for later reference. Since 7.36 volts matches the voltage referred to in the question, the time asked for is one time constant. Use the "megs" times "micros" again to calculate the time constant as $2 \times 0.01 = 0.02$ seconds. That's the correct answer.

---

**E5B10 How long does it take for an initial charge of 20 V DC to decrease to 0.37 V DC in a 0.01-microfarad capacitor when a 2-megohm resistor is connected across it?**

A. 0.02 seconds
B. 0.08 seconds
C. 450 seconds
D. 1350 seconds

**ANSWER B:** The discharge continues from question E5B09. Refer to the time-constant chart at question E5B04. Now the voltage asked for is down to 0.37 volts. Continuing the same thought process, a fourth time constant discharge reduces the voltage across the capacitor to $1\ V \times 0.368 = 0.368$, rounded to 0.37 volts. Write down 0.37 volts = four time constants. Four time constants is $4 \times 0.02 = 0.08$ seconds, the correct answer. Remember, one time constant for this circuit (question E5B09) is $2 \times 0.01 = 0.02$ seconds; $4 \times 0.02 = 0.08$.

**E5B11 How long does it take for an initial charge of 800 V DC to decrease to 294 V DC in a 450-microfarad capacitor when a 1-megohm resistor is connected across it?**

A. 0.02 seconds
B. 0.08 seconds
C. 450 seconds
D. 1350 seconds

**ANSWER C**: Since we're having so much fun, let's go for some new numbers, such as 800 volts DC and a 450 microfarad capacitor with a 1 megohm resistor connected across it. Let's continue the same steps as in the previous questions. Since the voltage across a capacitor drops to 36.8% of its initial value, what will the voltage be across the capacitor in one time constant? 800 × 0.368 = 294.4 volts. Write down 294 volts = one time constant. That's the voltage for this question so we know the time wanted is one time constant. What's the time constant?

Using "megs" × "micros", the time constant = 1 × 450 = 450 seconds.

***E5C Impedance diagrams: Basic principles of Smith charts; impedance of RLC networks at specified frequencies; PC based impedance analysis (including Smith Charts)***

**E5C01 What type of graph can be used to calculate impedance along transmission lines?**

A. A Smith chart
B. A logarithmic chart
C. A Jones chart
D. A radiation pattern chart

**ANSWER A:** The Smith Chart (see chart at page 113) is an invaluable graph for calculating the impedance along Amateur Radio transmission lines. Many times you will see a completed Smith Chart, generated by a computer, illustrating the resonance of a commercial-quality VHF or UHF collinear base station and repeater antenna.

**E5C02 What type of coordinate system is used in a Smith chart?**

A. Voltage circles and current arcs
B. Resistance circles and reactance arcs
C. Voltage lines and current chords
D. Resistance lines and reactance chords

**ANSWER B:** If you look carefully at a Smith Chart, you will see two circles – resistance circles centered on a resistance axis – all tangent to an open circuit point (infinity resistance and infinity reactance), and portions of reactance circles (arcs) fanning out from the resistance axis and passing through the same open circuit point. Inductive reactance curves are plotted positively from the resistance axis, and capacitive reactance curves negatively.

**E5C03 What type of calculations can be performed using a Smith chart?**

A. Beam headings and radiation patterns
B. Satellite azimuth and elevation bearings
C. Impedance and SWR values in transmission lines
D. Circuit gain calculations

**ANSWER C:** Commercial-quality coaxial cable, coiled on a 500-foot spool, may also contain a Smith Chart that has been computer-generated to show impedance and SWR value.

## Smith Chart—Finding Antenna Impedance

**Finding Antenna Impedance, $Z_A$, when Fed with Known Length of Transmission Line (λ in Wavelengths) of Given Impedance, $Z_o$.**
(Impedance and wavelength of transmission line are at the frequency of transmission)

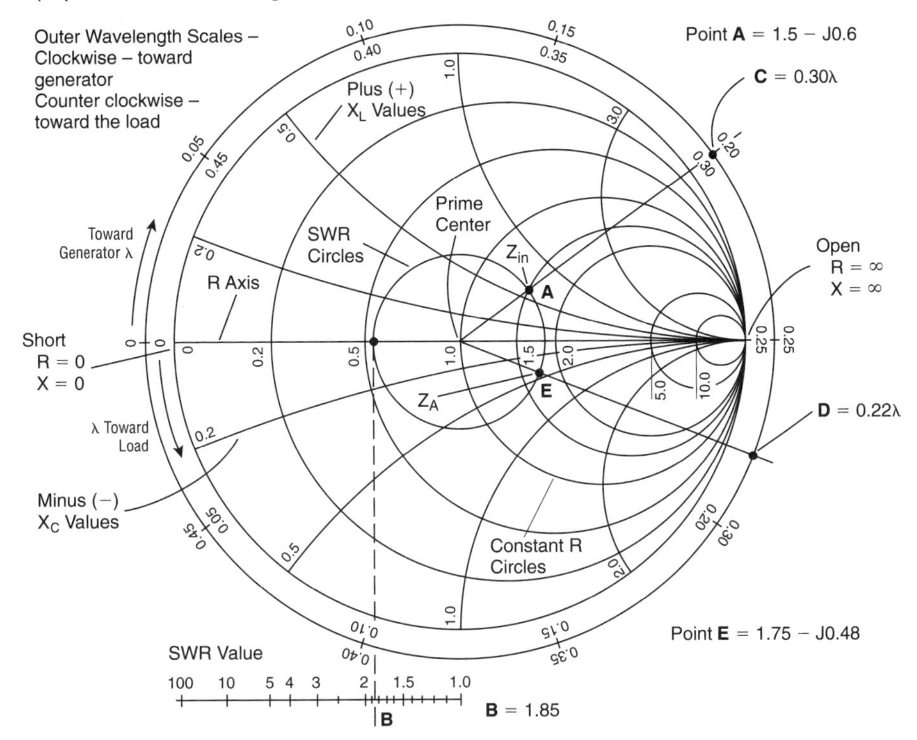


Known Values: $Z_o = 50\Omega$, $\lambda_Z = 0.42\lambda$

Steps

1. Connect transmission line to antenna.
2. Measure the input impedance $Z_{in}$ to transmission line. (Example 75 + J30)
3. Normalize $Z_{in}$ by dividing by $Z_o$.

$$Z_{in} = \frac{75}{50} + J\frac{30}{50} = 1.5 + J0.6$$

4. Plot 1.5 + J0.6 on chart (point A is at intersection of a resistance circle of 1.5 and a positive reactance curve of 0.6)
5. Draw SWR circle with prime center as center and radius through A. Project tangent to SWR circle at point where circle intersects the resistance axis down to linear scale and read value of SWR (Point B = 1.85).
6. Draw a line (radial line) from prime center through $Z_{in}$ at A and project to wavelength scale. Use "toward load" scale because you are going from $Z_{in}$ at generator to the load presented by Antenna. (Point C = 0.30λ)
7. To find the load impedance line on the wavelength (towards load) scale, add the transmission line length to Point C value. (Point D = 0.42 + .30 = 0.72λ)
8. Since the wavelength scales are only plotted for 0.5λ, the values repeat every 0.5λ. Therefore, subtract 0.5 for 0.72 to arrive at 0.22. (Point D = 0.22 toward load)
9. Draw another radial line from prime center to Point D. The intersection with the SWR circle is the load impedance of the antenna. (Point E = 1.75 – J0.48). You may have to interprolate between lines.
10. This is still a normalized value so multiply by $Z_o = 50\Omega$ to obtain true value or antenna impedance. $Z_A = 50$ (1.75 – J0.48) = 87.5 – J24. Which is an impedance of 87.5 ohms of resistance and 24 ohms of capacitive reactance.

**E5C04 What are the two families of circles that make up a Smith chart?**

A. Resistance and voltage
B. Reactance and voltage
C. Resistance and reactance
D. Voltage and impedance

**ANSWER C:** Resistance and reactance values comprise the "two families of circles" on the Smith Chart. Circles for positive inductive reactance are to the right of the resistance axis in Figure E5-1; circles for negative capacitance reactance are to the left of the resistance axis.

---

**E5C05 What type of chart is shown in Figure E5-1?**

A. Smith chart
B. Free-space radiation directivity chart
C. Vertical-space radiation pattern chart
D. Horizontal-space radiation pattern chart

**ANSWER A:** The Smith chart is an excellent way of visualizing the contours of resistance or reactance. The chart is used for circuit analysis and provides a graphic presentation for this purpose.

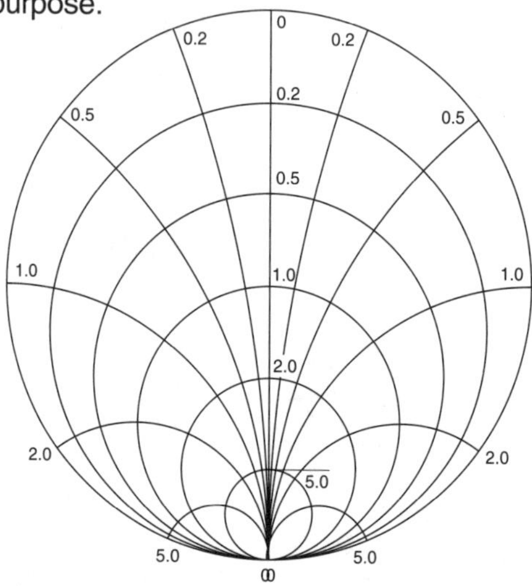


**Figure E5-1**

---

**E5C06 On the Smith chart shown in Figure E5-1, what is the name for the large outer circle bounding the coordinate portion of the chart?**

A. Prime axis
B. Reactance axis
C. Impedance axis
D. Polar axis

**ANSWER B:** The large outer circle is the *reactance axis*. Within the reactance axis are curved reactance circles that are tangent to the resistance axis. Where the reactance circle intersects with the reactance axis, you will see an assigned value of reactance. Inductive reactance is to the right, and capacitive reactance is to the left, in Figure E5-1.

---

**E5C07 On the Smith chart shown in Figure E5-1, what is the only straight line shown?**

A. The reactance axis
B. The current axis
C. The voltage axis
D. The resistance axis

**ANSWER D:** The straight line on a Smith chart is always the *resistance axis*.

**E5C08 What is the process of normalizing with regard to a Smith chart?**

A. Reassigning resistance values with regard to the reactance axis
B. Reassigning reactance values with regard to the resistance axis
C. Reassigning impedance values with regard to the prime center
D. Reassigning prime center with regard to the reactance axis

**ANSWER C:** In order that the Smith chart coordinate system apply to all types of transmission lines, the coordinates are "customized" to a particular transmission line by dividing test measurements by the characteristic impedance of the transmission line. This process of reassigning resistance and reactance values with regard to the prime center is called "normalizing".

---

**E5C09 What is the third family of circles, which are added to a Smith chart during the process of solving problems?**

A. Standing-wave ratio circles
B. Antenna-length circles
C. Coaxial-length circles
D. Radiation-pattern circles

**ANSWER A:** When a computer analyzes an antenna design, it will draw a third family of circles on the Smith chart for standing-wave ratios.

---

**E5C10 In rectangular coordinates, what is the impedance of a network comprised of a 10-microhenry inductor in series with a 40-ohm resistor at 500 MHz?**

A. 40 + j31,400
B. 40 – j31,400
C. 31,400 + j40
D. 31,400 – j40

**ANSWER A:** Don't panic, I promise I won't send you back to high school for algebra, trigonometry, and calculus to solve this problem. We have a circuit containing both reactance and resistance, giving us impedance, represented by the letter Z. Impedance has both a resistive (R) and reactive part ($X_C$ or $X_L$). The parts are represented by vectors that are at right angles (perpendicular) to each other. As a result, $Z = \sqrt{R^2 + X^2}$. Each part forms a leg of a right triangle and contributes to the magnitude of the impedance, which is the hypotenuse of the triangle. You determine the magnitude of the impedance by calculating each part, R and X, and then Z. Let's first calculate the inductive reactance $X_L$ of this question's circuit with the formula for inductive reactance:

$X_L = 2\pi fL$ Where: L is inductance in **henrys**
f is frequency in **hertz**
$\pi$ is equal to 3.14

At 500 MHz,

$$X_L = 2 \times 3.14 \times 500 \times 10^{+6} \times 10 \times 10^{-6}$$

$$X_L = 31{,}400$$

The resistance, R, is given as 40 ohms.
Therefore,

$$Z = \sqrt{40^2 + 31{,}400^2} = \sqrt{1600 + 985{,}960{,}000} = \sqrt{985{,}961{,}600}$$

$$Z = 31{,}400.2 \text{ rounded to } 31{,}400$$

The examination question asks for the answer in rectangular coordinates. This is great news because it allows us to simplify the problem-solving – even to the point of doing the problem in our head. Here is why – rectangular coordinates consist of the 2 parts of the impedance discussed above – the first part is the resistance, and the second part is the reactance. The reactance is preceded by a (+j) or a (−j) which indicates whether the circuit's reactance is inductive with a (+j), or capacitive with a (−j). The resistance is placed first, followed by the reactance: therefore, the impedance of an inductive circuit is $Z = R + jX_L$, and for a capacitive circuit it is $Z = R - jX_C$. The problem has an inductor in series with a 40-ohm resistor. Since it is an inductor, the reactance is inductive and a (+j) precedes it. Therefore, for this major problem, the easy answer is 40+j. Now that wasn't so hard, was it?

### Representing an AC Quantity

#### Rectangular Coordinates

A quantity can be represented by points that are measured on an X-Y plane with an X and Y axis perpendicular to each other. The point where the axes cross is the origin, the points on the X and Y axis are called coordinates, and the magnitude is represented by the length of a vector from the origin to the unique point defined by the *rectangular coordinates.*

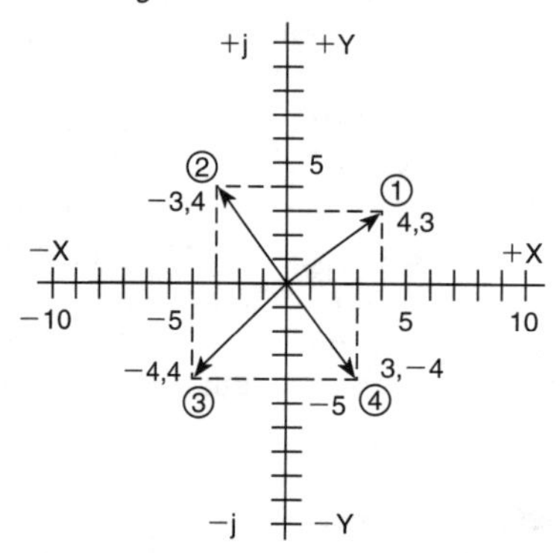


To be able to identify ac quantities and their phase angle, the X axis is called the real axis and the Y axis is called the imaginary axis. There are positive and negative real axis coordinates and positive and negative imaginary axis coordinates. The imaginary axis coordinates have a **j** operator to identify that they are imaginary coordinates.

In rectangular coordinates:

Vector 1 is 4 + j3
Vector 2 is −3 + j4
Vector 3 is −4 − j4
Vector 4 is 3 − j4

#### Polar Coordinates

A quantity can be represented by a vector starting at an origin, the length of which is the magnitude of the quantity, and rotated from a zero axis through 360°.

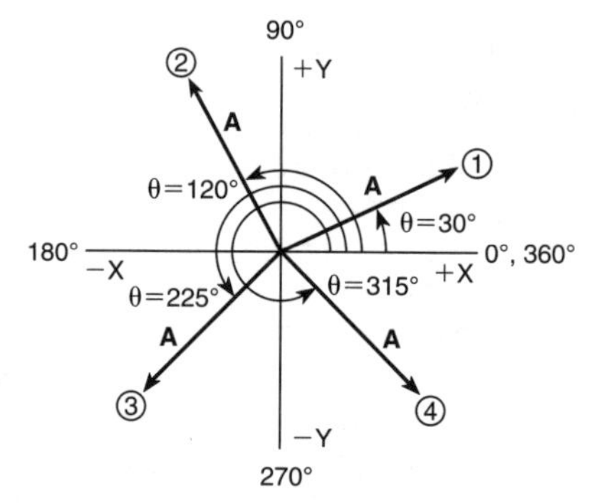


The position of the Vector A can be defined by the angle through which it is rotated. When we do so we are defining the ac quantity in *polar coordinates.*

In polar coordinates:

Vector 1 is $A\angle 30°$
Vector 2 is $A\angle 120°$
Vector 3 is $A\angle 225°$
Vector 4 is $A\angle 315°$

**E5C11 In polar coordinates, what is the impedance of a network comprised of a 100-picofarad capacitor in parallel with a 4,000-ohm resistor at 500 kHz?**

A. 2490 ohms, $\angle 51.5$ degrees
B. 4000 ohms, $\angle 38.5$ degrees
C. 2490 ohms, $\angle -51.5$ degrees
D. 5112 ohms, $\angle -38.5$ degrees

**ANSWER C:** Shift gears now and be alert because this test question requires your answer in POLAR COORDINATES, not the easy rectangular coordinates as the previous Q & A. Disconnect yourself from the thought process used for rectangular coordinates in the previous question. Just because the resistance is given as 4,000, do not look at the 4,000 ohm resistor as your best choice for the correct answer. Polar coordinate values are calculated differently. In polar coordinates, the ac quantity – let's say impedance Z in this case – is described as a vector at a particular phase angle. You solve for the vector magnitude Z separately and for the phase angle separately. The vector Z (which is the hypotenuse of a right triangle) is defined by a resistance vector (R) and a reactance vector (X) at right angles to each other. If the reactance vector is inductive, it is a $+X_L$; if it is capacitive the reactance is $-X_C$. Positive inductive reactances produce positive angles and negative capacitive reactances produce negative angles. Let's begin the necessary calculations by finding the value of the capacitive reactance with the formula:

$$X_C = \frac{1}{2\pi fC}$$

Where: C is capacitance in **farads**
f is frequency in **hertz**
$\pi = 3.14$

$$X_C = \frac{1}{2 \times 3.14 \times 0.5 \times 10^{+6} \times 100 \times 10^{-12}}$$

Clearing the powers of ten:

$$X_C = \frac{10^{+6}}{6.28 \times 0.5 \times 100}$$

**Note:**
This equation can be used directly when:
f is frequency in **MHz**
C is in **picofarads**

$$X_C = \frac{1{,}000{,}000}{314}$$

$$X_C = 3184.7 \text{ ohms}$$

The resistance is already given at 4,000 ohms, so R = 4000.

You now need to find the magnitude of the impedance, Z in ohms, of the *parallel circuit.* This must be done with vector (phasor) algebra, as follows:

$$Z = \frac{Z_1 \times Z_2}{Z_1 + Z_2}$$

**Special Case:**
When $Z_1 = R \angle 0^\circ$ and $Z_2 = X_C \angle -90^\circ$

$$Z = \frac{RX_C \angle -90^\circ}{\sqrt{R^2 + X_C^2} \angle \arctan \frac{-X_C}{R}}$$

$$Z = \frac{R \angle 0^\circ \times X_C \angle -90^\circ}{(R + j0) + (0 - jX_C)}$$

$$Z = \frac{4000 \angle 0^\circ \times 3185 \angle -90^\circ}{(4000 - j3185)} = \frac{4000 \times 3185 \angle -90^\circ}{\sqrt{4000^2 + 3185^2} \angle \arctan -3185/4000}$$

$$Z = \frac{12.74 \times 10^{+6} \angle -90°}{\sqrt{16 \times 10^{+6} + 10.144 \times 10^{+6}} \angle \arctan -0.7963}$$

$$Z = \frac{12.74 \times 10^{+6} \angle -90°}{5.113 \times 10^{3} \angle -38.5°}$$

$$Z = 2.493 \times 10^{3} \angle -90° + 38.5° = 2493 \angle -51.5° \text{ ohms}$$

Answer C is the correct answer. You will need a scientific calculator or trig table to look up the tangent of the phase angles. The solution is not easy. You must keep your wits about you because you cannot add and subtract vectors directly. You must deal with the magnitude and the angle in polar coordinates and the real (resistance) and imaginary (reactance) parts in rectangular coordinates. You can further double-check you answer by recognizing that there must be a minus sign in front of the phase angle because the circuit has capacitive reactance.

---

**E5C12 Which point on Figure E5-2 best represents the impedance of a series circuit consisting of a 300-ohm resistor, a 0.64-microhenry inductor and a 85-picofarad capacitor at 24.900 MHz?**

A. Point 1　　　　C. Point 5
B. Point 3　　　　D. Point 8

**ANSWER D:** Let's now look at Figure E5-2, and notice it is a rectangular-coordinate system with axes at right angles to each other. The indicated points on the plane show X and Y coordinate values that are possible answers to this question. In this problem we must calculate both inductive and capacitive reactance.

$X_L = 2\pi \times 24.9 \times 10^6 \times 0.64 \times 10^{-6} = 100$ ohms.

$$X_C = \frac{1}{2\pi \times 24.9 \times 10^6 \times 85 \times 10^{-12}} = \frac{1}{1.33 \times 10^{-2}} = 75 \text{ ohms.}$$

The inductive reactance (100 ohms) minus the capacitive reactance (75 ohms) results in 25 ohms final reactance for this series circuit, which is slightly inductive at 24.9 MHz. Move out to the right to the 300-ohm resistance line on the +X axis. Move up slightly on the +Y axis to 25 ohms and see that they intersect at Point 8, which is the correct answer D.

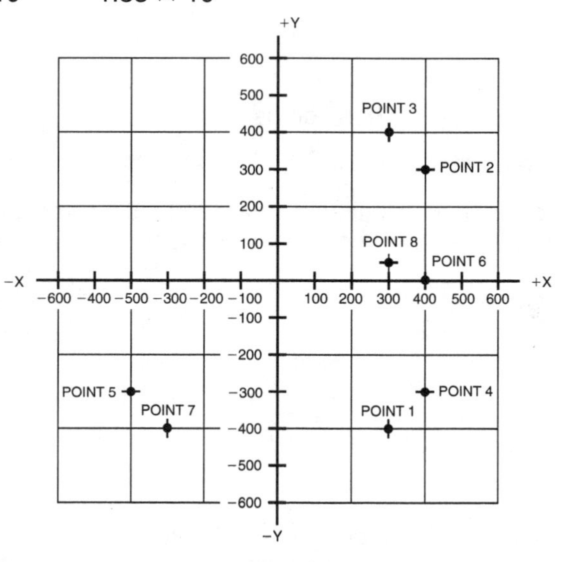


Figure E5-2

**E5C13 What are the curved lines on a Smith chart?**

A. Portions of current circles
B. Portions of voltage circles
C. Portions of resistance circles
D. Portions of reactance circles

**ANSWER D:** The curved lines on a Smith chart are portions of reactance circles. Remember that the large outer circle bounding the coordinate portion of the chart is the reactance axis.

---

**E5C14 How are the wavelength scales on a Smith chart calibrated?**

A. In portions of transmission line electrical frequency
B. In portions of transmission line electrical wavelength
C. In portions of antenna electrical wavelength
D. In portions of antenna electrical frequency

**ANSWER B:** The wavelength scales on the Smith chart are calibrated in portions of transmission line electrical wavelength.

---

***E5D Phase angle between voltage and current; impedances and phase angles of series and parallel circuits***

---

**E5D01 What is the phase angle between the voltage across and the current through a series R-L-C circuit if $X_C$ is 25 ohms, R is 100 ohms, and $X_L$ is 100 ohms?**

A. 36.9 degrees with the voltage leading the current
B. 53.1 degrees with the voltage lagging the current
C. 36.9 degrees with the voltage lagging the current
D. 53.1 degrees with the voltage leading the current

**ANSWER A:** Here's another series of questions on the phase angle between voltage and current in a series RLC circuit. When an applied voltage produces a current in a series R-L-C circuit, the voltage and current will be out of phase due to the reactance ($X_L$ and $X_C$) in the circuit. Refer to the schematic below.

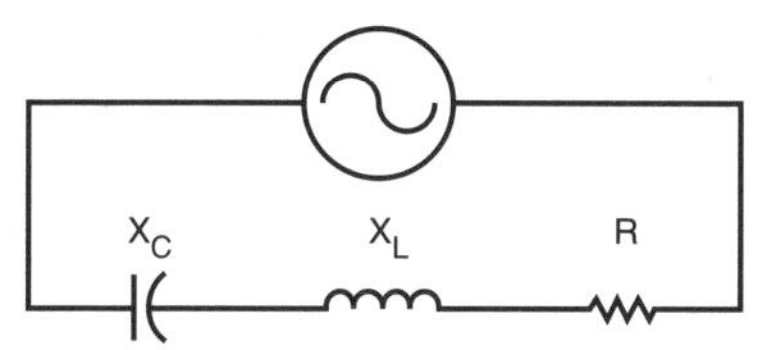


**Series RLC Circuit Schematic for Phase Angle Calculations**

Here is the formula, memorize it:

$\text{Tan } \phi = \frac{X}{R}$ where: $\phi$ is the phase angle in **degrees**
$X = (X_L - X_C)$ is total reactance in the circuit in **ohms**
R is circuit resistance in **ohms**

Don't panic, it gets easier. Since $X_L$ and $X_C$ are in opposition, we simply subtract one from the other. Now recall ELI the ICE man. Remember him? ELI the ICE man! Say it out loud, "ELI the ICE man." In an inductive (L) circuit, voltage (E) leads current (I) just like in the word "ELI" where E is before I. In a capacitive (C) circuit, current (I) leads voltage (E), just like the word "ICE" where I comes before E.

In the first problem, $X_C$ is 25 ohms, and $X_L$ is 100 ohms. Since 100 is greater than 25, it's going to be an inductive circuit with voltage leading current (Mr. ELI). If we subtract 25 from 100, that leaves us with 75. But don't stop now – to find the tangent of $\phi$, we must divide 100 *into* 75, and this ends up 0.75. If you had a trig table – or if your calculator can do trigonometry – you would find the tangent of 37 degrees equals 0.75. Since all the problems on your test would either end up 0.75 or 0.25, just keep in mind that 0.75 equals 37 degrees. This means your answer is 36.9 degrees with voltage leading the current. Watch out! Answer C has the right numbers but has voltage *lagging* the current.

**E5D02 What is the phase angle between the voltage across and the current through a series R-L-C circuit if $X_C$ is 500 ohms, R is 1 kilohm, and $X_L$ is 250 ohms?**

A. 68.2 degrees with the voltage leading the current
B. 14.0 degrees with the voltage leading the current
C. 14.0 degrees with the voltage lagging the current
D. 68.2 degrees with the voltage lagging the current

**ANSWER C:** This is an ICE circuit, because $X_C$ is 500 ohms and $X_L$ is only 250 ohms. The total reactance (X) is 500 − 250 = 250. By dividing 1000 (the resistance) into 250 the tangent of $\phi$ comes out 0.25. We know that a tangent of 0.25 is a phase angle of 14 degrees, and since the circuit is ICE, voltage lags the current. (Another way of saying current leads the voltage!) See below for more detail.

### Vector Addition

Here is more detail. In an ac circuit, when calculating the impedance of the circuit, the reactance and resistance must be added vectorially rather than algebraically. This vector addition can be understood best by looking at the following diagram:

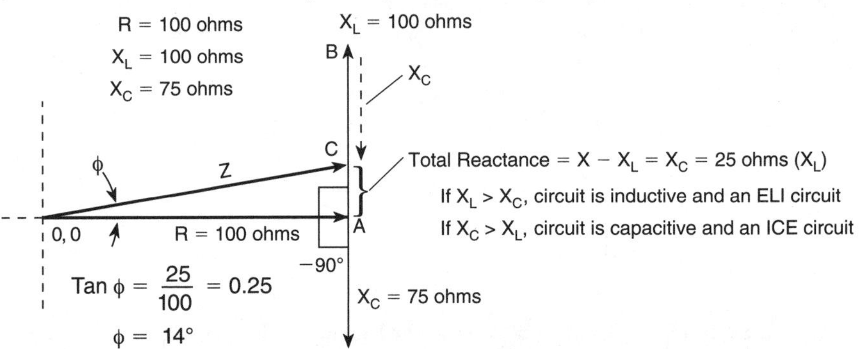


The resistance vector (of a size proportional to its value) is plotted horizontally (along x axis). Any reactance vector (using same scale as for resistance value) is plotted at right angles (90°) to the resistance vector at the tip of the resistance vector (Point A). Inductive reactance ($X_L$) would be plotted at +90°, while capacitive reactance ($X_C$) would be plotted at −90°. The total reactance is determined by plotting the $X_C$ vector from the tip (Point B) of the $X_L$ vector. The result is Point C on the $X_L$ vector. The total impedance vector, Z, is drawn from the origin (tail of R vector) to point C. The triangle formed is a right triangle; therefore, by trigonometry and the Pythagorean theorem:

$$Z = \sqrt{R^2 + (X_L - X_C)^2} = \sqrt{100^2 + 25^2}$$
$$= \sqrt{10625} = 103.08 \text{ ohms}$$

and:

$$\text{Tan } \phi = \frac{X}{R} = \frac{25}{100} = 0.25 \qquad \phi = 14°$$

The tangent of the angle $\phi$ is found by dividing the value of the side opposite (X) vector by the value of the side adjacent (R) vector. $\phi$ is the phase angle of the impedance.

**E5D03 What is the phase angle between the voltage across and the current through a series R-L-C circuit if $X_C$ is 50 ohms, R is 100 ohms, and $X_L$ is 25 ohms?**

A. 76 degrees with the voltage lagging the current
B. 14 degrees with the voltage leading the current
C. 76 degrees with the voltage leading the current
D. 14 degrees with the voltage lagging the current

**ANSWER D:** This is an ICE circuit because $X_C$ is greater than $X_L$, and the total X is 50 − 25 = 25. Tangent of $\phi$ is 0.25, and $\phi$ is 14 degrees. Since this is an ICE circuit, current leads voltage, but be careful, the answer is stated correctly as voltage LAGGING the current.

---

**E5D04 What is the phase angle between the voltage across and the current through a series R-L-C circuit if $X_C$ is 100 ohms, R is 100 ohms, and $X_L$ is 75 ohms?**

A. 14 degrees with the voltage lagging the current
B. 14 degrees with the voltage leading the current
C. 76 degrees with the voltage leading the current
D. 76 degrees with the voltage lagging the current

**ANSWER A:** This is more ICE than ELI because $X_C$ is greater than $X_L$, and the tangent of $\phi$ of 0.25 gives us a 14-degree phase angle. Since the circuit is ICE, voltage (E) is lagging current (I).

---

**E5D05 What is the phase angle between the voltage across and the current through a series R-L-C circuit if $X_C$ is 50 ohms, R is 100 ohms, and $X_L$ is 75 ohms?**

A. 76 degrees with the voltage leading the current
B. 76 degrees with the voltage lagging the current
C. 14 degrees with the voltage lagging the current
D. 14 degrees with the voltage leading the current

**ANSWER D:** Here again we are with a Mr. ELI circuit since $X_L$ is larger. The tangent of $\phi$ is $(75 - 50) \div 100 = 0.25$, and gives us an angle of 14 degrees with voltage leading the current.

---

**E5D06 What is the relationship between the current through and the voltage across a capacitor?**

A. Voltage and current are in phase
B. Voltage and current are 180 degrees out of phase
C. Voltage leads current by 90 degrees
D. Current leads voltage by 90 degrees

**ANSWER D:** A capacitor is an "ICE" circuit where current leads the voltage, as you see in "ICE" where the letter "I" is leading the letter "E".

---

**E5D07 What is the relationship between the current through an inductor and the voltage across an inductor?**

A. Voltage leads current by 90 degrees
B. Current leads voltage by 90 degrees
C. Voltage and current are 180 degrees out of phase
D. Voltage and current are in phase

**ANSWER A:** This is an "ELI" condition where voltage leads the current as we see in "ELI" where the letter "E" is ahead of the letter "I".

**E5D08 What is the phase angle between the voltage across and the current through a series RLC circuit if $X_C$ is 25 ohms, R is 100 ohms, and $X_L$ is 50 ohms?**

A. 14 degrees with the voltage lagging the current
B. 14 degrees with the voltage leading the current
C. 76 degrees with the voltage lagging the current
D. 76 degrees with the voltage leading the current

**ANSWER B:** Look again at the formula at E5D01. When an applied voltage produces a current in a series R-L-C circuit, the voltage and current will be out of phase due to the reactance ($X_L$ and $X_C$) in the circuit. Right off the bat, you can zero in on 2 out of 4 answers by looking to see whether voltage is lagging the current, or voltage is leading the current. Voltage lagging is an ICE, more capacitive circuit. Voltage leading the current, ELI, is more inductive.

Now let's take a look at this problem where $X_C$ is 25 ohms, and $X_L$ is 50 ohms, and R is 100 ohms. Let's first work on the reactance (X) which is 50 − 25 = 25. Simple subtraction. Easy stuff. We find the tangent by dividing 100 (the resistance) into 25 = 0.25. From our trig table stored in our brain, the tangent of 14 degrees is 0.25. This gives us a memorized phase angle of 14 degrees, and voltage is leading the current (ELI). Keep in mind that .75 = 37 degrees, and **.25** = **14 degrees**. It looks to me that these series R-L-C circuits always end up at 14 degrees with you calculating ELI or ICE. Now that wasn't so tough, was it?

---

**E5D09 What is the phase angle between the voltage across and the current through a series RLC circuit if $X_C$ is 75 ohms, R is 100 ohms, and $X_L$ is 100 ohms?**

A. 76 degrees with the voltage leading the current
B. 14 degrees with the voltage leading the current
C. 14 degrees with the voltage lagging the current
D. 76 degrees with the voltage lagging the current

**ANSWER B:** This is an ELI circuit, right? $X_L$ is greater than $X_C$. X equals 100 − 75 and divided by 100 comes out 0.25, and that's 14 degrees ELI.

---

**E5D10 What is the phase angle between the voltage across and the current through a series RLC circuit if $X_C$ is 75 ohms, R is 100 ohms, and $X_L$ is 50 ohms?**

A. 76 degrees with the voltage lagging the current
B. 14 degrees with the voltage leading the current
C. 14 degrees with the voltage lagging the current
D. 76 degrees with the voltage leading the current

**ANSWER C:** Here is another ICE circuit, coming out at 14 degrees. Voltage is lagging current (ICE). X is 75 − 50 = 25, and tangent of $\phi$ = 0.25, 14 degrees ICE.

---

**E5D11 What is the phase angle between the voltage across and the current through a series RLC circuit if $X_C$ is 250 ohms, R is 1 kilohm, and $X_L$ is 500 ohms?**

A. 81.47 degrees with the voltage lagging the current
B. 81.47 degrees with the voltage leading the current
C. 14.04 degrees with the voltage lagging the current
D. 14.04 degrees with the voltage leading the current

**ANSWER D:** Here is a Mr. ELI circuit, coming out at 14 degrees, where voltage leads the current. X is 500 − 250, and the tangent $\phi$ is 250 ÷ 1000 = 0.25, 14 degrees ELI.

***E5E Algebraic operations using complex numbers: rectangular coordinates (real and imaginary parts); polar coordinates (magnitude and angle)***

**E5E01 In polar coordinates, what is the impedance of a network comprised of a 100-ohm-reactance inductor in series with a 100-ohm resistor?**

A. 121 ohms, /35 degrees
B. 141 ohms, /45 degrees
C. 161 ohms, /55 degrees
D. 181 ohms, /65 degrees

**ANSWER B:** This one is relatively easy – impedance equals the square root of the sum of the resistance squared and the inductive reactance squared. Z in polar coordinates is:

$$Z = \sqrt{100^2 + 100^2} \ \underline{/\arctan 100/100}$$

$$Z = \sqrt{20000} \ \underline{/\arctan 1}$$

$$Z = 141.4 \ \underline{/+45°}$$

You might recognize the square root easier if you work in powers of 10. The magnitude of $Z = \sqrt{(1 \times 10^2)^2 + (1 \times 10^2)^2} = \sqrt{2 \times 10^4} = \sqrt{2} \times 10^2 = 1.414 \times 10^2$. You may recognize the $\sqrt{2}$ as 1.414, so the answer is 141.4. From just the magnitude, you could have selected the correct answer, or you might have used the phase angle. A right triangle with equal R and X has a phase angle of 45° because, as you probably recall, the tangent of 45° is 1.

---

**E5E02 In polar coordinates, what is the impedance of a network comprised of a 100-ohm-reactance inductor, a 100-ohm-reactance capacitor, and a 100-ohm resistor all connected in series?**

A. 100 ohms, /90 degrees
B. 10 ohms, /0 degrees
C. 10 ohms, /100 degrees
D. 100 ohms, /0 degrees

**ANSWER D:** Easy one again – since inductive reactance cancels capacitive reactance, in this circuit with both reactances equal, the final reactance is zero, and all we have left is 100 ohms resistance. Since there is no reactance vector, there is no phase difference, so the phase angle is zero degrees. The resistance vector lies on the 0° axis. One of the easiest ways to check this is to put the impedance in rectangular coordinates: $Z = 100 + j100 - j100 = 100 + j0$, just a resistance of 100 ohms lying on the zero axis.

---

**E5E03 In polar coordinates, what is the impedance of a network comprised of a 300-ohm-reactance capacitor, a 600-ohm-reactance inductor, and a 400-ohm resistor, all connected in series?**

A. 500 ohms, /37 degrees
B. 400 ohms, /27 degrees
C. 300 ohms, /17 degrees
D. 200 ohms, /10 degrees

**ANSWER A:** Even though the answer is requested in polar coordinates, it is best to start out using rectangular coordinates. $Z = (400 + j600 - j300) = (400 + j300)$. Remember, in a right triangle with a 3:4:5 relationship, it is easy to determine the value of the hypotenuse. With sides of 300 and 400, the hypotenuse will be 500, with a phase angle whose tangent is 3/4 or 4/3, depending upon the relationship of the sides. With the resistance side of a right triangle at 400 ohms and the reactance side at 300 ohms, you know that the hypotenuse magnitude of the impedance is $Z = 500$. The angle is a positive one with an arctan of (an angle whose tangent is) 300/400, or 0.75. Grab a trig table or scientific calculator for an angle of 36.8° rounded to 37°.

---

### Complex Numbers (Real and Imaginary) and Operator j

Recall from algebra that $+2 \times +2 = +4$ and $-2 \times -2 = +4$ $\therefore \sqrt{4}$ has two roots $+2$ or $-2$. In ac circuit analysis there is a need to take the square root of a negative number. For example, $\sqrt{-4}$ has a root $\sqrt{-1} \times \sqrt{4} = j2$ and $-j2$. The root of a negative number is called an *imaginary number* and this is indicated by writing the operator j in front of the root. j is equal to $\sqrt{-1}$ and is the basic imaginary quantity. Since $j = \sqrt{-1}$, it is interesting to note the following because they will be encountered in ac circuit analysis:

$$j^2 = j \times j = \sqrt{-1} \times \sqrt{-1} = -1$$

$$j^3 = j \times j \times j = j^2 \times j = -1 \times j = -j$$

$$j^4 = j \times j \times j \times j = j^2 \times j^2 = -1 \times -1 = +1$$

$$\frac{1}{j} = \frac{1}{j} \times \frac{j}{j} = \frac{j}{j^2} = \frac{j}{-1} = -j$$

**Summary**

$j = \sqrt{-1}$

$j^2 = -1$

$j^3 = -j$

$j^4 = +1$

$\frac{1}{j} = -j$

**j Operator as Vector Rotator**

Note that an imaginary quantity can be considered as a vector being rotated by the operator j.

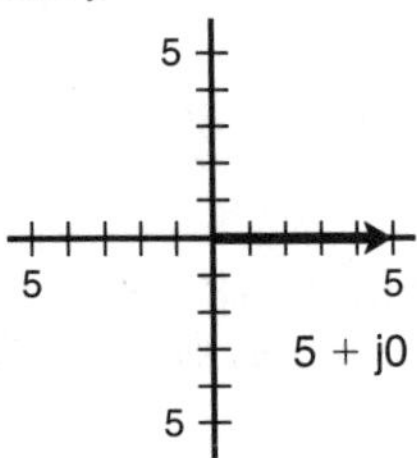


Vector starts at 0°

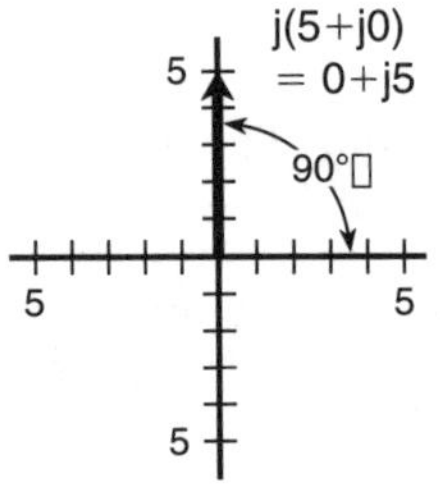


The j operator, when it multiplies a vector, rotates the vector by 90° counter clockwise.

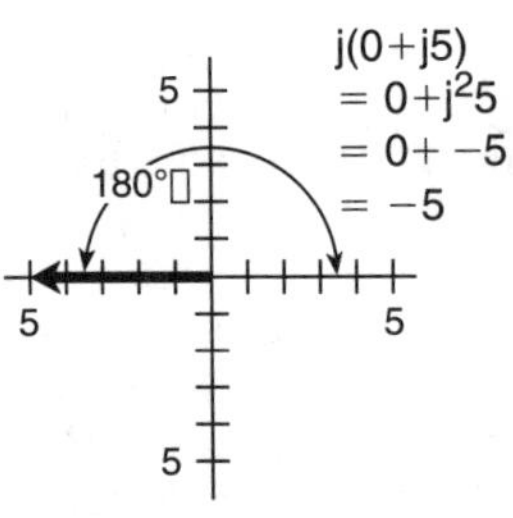


A second 90° rotation puts the vector at 180°

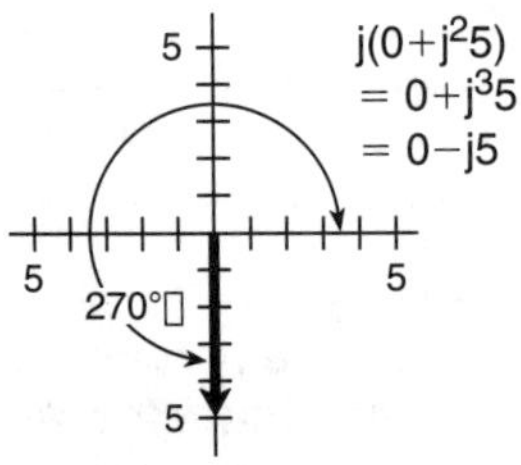
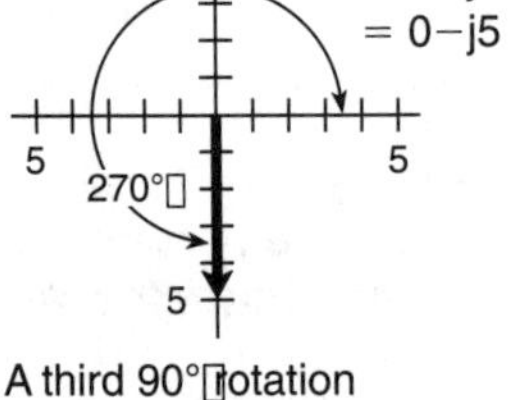


A third 90° rotation puts the vector at 270°

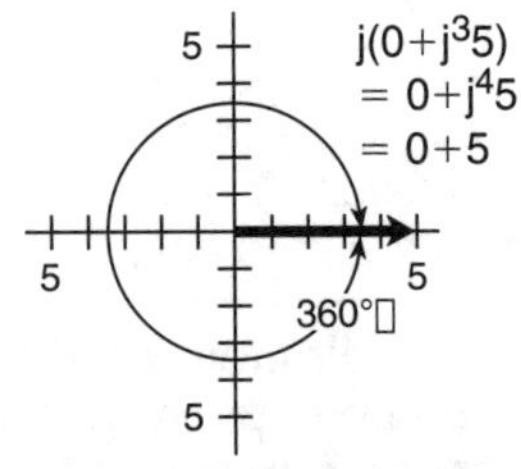


A fourth 90° rotation puts the vector back at 0°

**E5E04 In polar coordinates, what is the impedance of a network comprised of a 400-ohm-reactance capacitor in series with a 300-ohm resistor?**

A. 240 ohms, ∠36.9 degrees
B. 240 ohms, ∠−36.9 degrees
C. 500 ohms, ∠53.1 degrees
D. 500 ohms, ∠−53.1 degrees

**ANSWER D:** Here is a series problem which is easily solved by first calculating the magnitude of the impedance which equals the square root of resistance squared plus reactance squared. The square root of 300 squared plus 400 squared is the square root of 250,000 or 500 ohms. Since the circuit is capacitive, the phase angle will have a minus sign in front of it. Remember again, a right triangle with a 3:4:5 relationship of the sides and hypotenuse is an easy one to determine the hypotenuse value. With sides of 300 and 400, the hypotenuse will be 500, with a phase angle whose tangent is 3/4 or 4/3 depending on the relationship of the sides. In this case it's 4/3 and the angle is −53.1 degrees.

---

**E5E05 In polar coordinates, what is the impedance of a network comprised of a 400-ohm-reactance inductor in parallel with a 300-ohm resistor?**

A. 240 ohms, $\underline{/36.9}$ degrees
B. 240 ohms, $\underline{/-36.9}$ degrees
C. 500 ohms, $\underline{/53.1}$ degrees
D. 500 ohms, $\underline{/-53.1}$ degrees

**ANSWER A:** We know right off the bat that the answer is going to be either A or C because an inductive reactance will have a positive polar coordinate angle. To come up with 240 ohms at an angle of +36.9° as the correct impedance, let's start out by defining the two impedances that are in parallel:

$Z_1 = 300 \underline{/0^\circ}$ since it is a resistance in **ohms**
$Z_2 = 400 \underline{/+90^\circ}$ since it is an inductive reactance in **ohms**

$$Z = \frac{Z_1 \times Z_2}{Z_1 + Z_2} \quad \text{or Special Case: } Z = \frac{RX_L \underline{/+90^\circ}}{\sqrt{R^2 + X_L^2}\ \underline{/\arctan \frac{X_L}{R}}}$$

Additions and subtractions are easier in rectangular coordinates, and multiplication and division are easier in polar coordinates.

$$Z = \frac{300 \times 400 \underline{/0^\circ + 90^\circ}}{(300 + j0) + (0 + j400)} = \frac{120{,}000\underline{/+90^\circ}}{500 \underline{/\arctan 400/300}}$$

$$Z = 240 \underline{/+90^\circ - 53.1^\circ}$$

$$Z = 240 \underline{/36.9^\circ} \text{ ohms}$$

The correct answer is A. We used the 3:4:5 ratio, and both rectangular and polar coordinates.

---

**E5E06 In polar coordinates, what is the impedance of a network comprised of a 100-ohm-reactance capacitor in series with a 100-ohm resistor?**

A. 121 ohms, $\underline{/-25}$ degrees
B. 191 ohms, $\underline{/-85}$ degrees
C. 161 ohms, $\underline{/-65}$ degrees
D. 141 ohms, $\underline{/-45}$ degrees

**ANSWER D:** R = 100 ohms and $-X_C$ = 100 ohms. Calculate $Z = \sqrt{100^2 + 100^2} = \sqrt{2 \times 10^4}$. This works out to be the square root of 20,000, which is 141 ohms. The phase angle will be negative so the correct answer D can be selected without going further. However, the arctan is 100/100 or 1, and from this we know that the angle is 45°.

---

**E5E07 In polar coordinates, what is the impedance of a network comprised of a 100-ohm-reactance capacitor in parallel with a 100-ohm resistor?**

A. 31 ohms, $\underline{/-15}$ degrees
B. 51 ohms, $\underline{/-25}$ degrees
C. 71 ohms, $\underline{/-45}$ degrees
D. 91 ohms, $\underline{/-65}$ degrees

**ANSWER C:** Polar coordinates again, but you are given $X_C$ so you don't have to calculate it. First calculate impedance:

$$Z = \frac{100 \underline{/0^\circ} \times 100 \underline{/-90^\circ}}{\sqrt{100^2 + 100^2}\ \underline{/\arctan -1}} = \frac{10000 \underline{/-90^\circ}}{141.4 \underline{/-45^\circ}}$$

$$Z = 70.7 \underline{/-45^\circ}$$

You can determine the correct answer as C just from the impedance magnitude, but since the arctan is 1, you know the phase angle is 45°.

---

**E5E08 In polar coordinates, what is the impedance of a network comprised of a 300-ohm-reactance inductor in series with a 400-ohm resistor?**

A. 400 ohms, /27 degrees
B. 500 ohms, /37 degrees
C. 500 ohms, /47 degrees
D. 700 ohms, /57 degrees

**ANSWER B:** Easy one here – let's first find the magnitude of the impedance. The phase angle will be positive because the circuit is inductive. You don't really have to calculate if you recognize that the right triangle sides have a ratio of 3:4:5. Since resistance is 400 ohms, inductive reactance 300 ohms, Z will have a magnitude of 500 ohms. Now for the phase angle. The phase angle in a series circuit:

$$\text{Tangent } \Theta = \frac{\text{X Reactance}}{\text{R Resistance}} = \frac{300}{400} = 0.75$$

Trig table shows the tan 0.75 = 37 degrees

**E5E09 When using rectangular coordinates to graph the impedance of a circuit, what does the horizontal axis represent?**

A. The voltage or current associated with the resistive component
B. The voltage or current associated with the reactive component
C. The sum of the reactive and resistive components
D. The difference between the resistive and reactive components

**ANSWER A:** The horizontal axis represents the *resistive component* of the impedance. When the impedance is in a circuit and voltage is applied, there will be a current, and the voltage drop across the resistance is plotted on the horizontal axis.

**E5E10 When using rectangular coordinates to graph the impedance of a circuit, what does the vertical axis represent?**

A. The voltage or current associated with the resistive component
B. The voltage or current associated with the reactive component
C. The sum of the reactive and resistive components
D. The difference between the resistive and reactive components

**ANSWER B:** The vertical axis, when using rectangular coordinates, represents the *reactive component* of the impedance. When the impedance is in a circuit and voltage is applied, there will be a current, and the voltage across the reactance is plotted perpendicular to the horizontal axis and parallel to the vertical axis.

Remember – horizontal is *resistive*, and vertical is *reactive*.

**E5E11 What do the two numbers represent that are used to define a point on a graph using rectangular coordinates?**

A. The horizontal and inverted axes
B. The vertical and inverted axes
C. The coordinate values along the horizontal and vertical axes
D. The phase angle with respect to its prime center

**ANSWER C:** When we see the two numbers defining a point on a rectangular coordinate graph, these two numbers are the coordinate values along the horizontal and vertical axes, respectively. The horizontal coordinate is usually given first.

**E5E12 If you plot the impedance of a circuit using the rectangular coordinate system and find the impedance point falls on the right side of the graph on the horizontal line, what do you know about the circuit?**

A. It has to be a direct current circuit
B. It contains resistance and capacitive reactance

C. It contains resistance and inductive reactance
D. It is equivalent to a pure resistance

**ANSWER D:** If you are computing the impedance using rectangular coordinates, and the impedance point falls to the right side and directly on the horizontal line, the impedance is equivalent to a pure resistance.

---

**E5E13 Why would you plot the impedance of a circuit using the polar coordinate system?**

A. To display the data on an XY chart
B. To give a visual representation of the phase angle
C. To graphically represent the DC component
D. To show the reactance which is present

**ANSWER B:** Using polar coordinates, we will see a visual representation of the value of the impedance and its phase angle.

---

**E5E14 What coordinate system can be used to display the resistive, inductive, and/or capacitive reactance components of an impedance?**

A. Maidenhead grid
B. National Bureau of Standards
C. Faraday
D. Rectangular

**ANSWER D:** The coordinate system to display resistive, inductive, and/or capacitive reactance components of an impedance is rectangular coordinates. It represents each component reactance or resistance as a vector whose length represents the magnitude of the reactance or resistance.

---

**E5E15 What coordinate system can be used to display the phase angle of a circuit containing resistance, inductive and/or capacitive reactance?**

A. Maidenhead grid
B. National Bureau of Standards
C. Faraday
D. Polar

**ANSWER D:** To display the impedance of a circuit with its phase angle, you would use a polar coordinate system.

---

**E5E16 In polar coordinates, what is the impedance of a circuit of 100 −j100 ohms impedance?**

A. 141 ohms, $\angle -45$ degrees
B. 100 ohms, $\angle 45$ degrees
C. 100 ohms, $\angle -45$ degrees
D. 141 ohms, $\angle 45$ degrees

**ANSWER A:** Look at the vector diagram below:

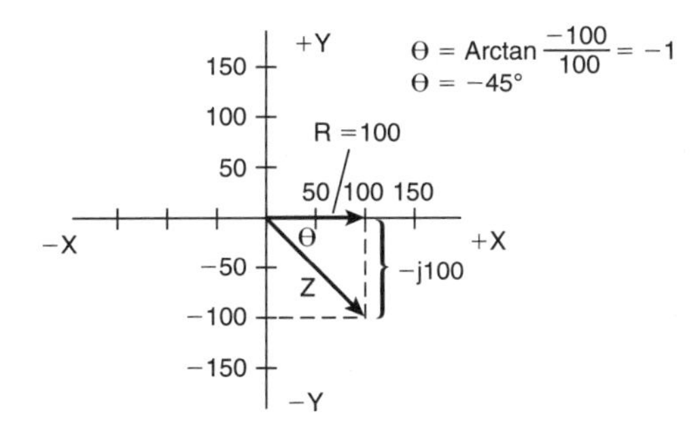


**Vector Diagram for E5E16**

$R = 100$ ohms and $X_l = -j100$ ohms. Calculate
$Z = \sqrt{100^2 + 100^2} = \sqrt{2 \times 10^4} = \sqrt{20{,}000} = \sqrt{2} \times 10^2 = 141$.
The circuit is capacitive because of $-j100$. Since the resistance is 100 ohms and the capacitive reactance is 100 ohms, the angle ($\Theta$) has a tangent of $-1$; therefore, the angle is a negative 45°. The impedance in polar coordinates is $141\angle{-45°}$, answer A.

---

**E5E17 In polar coordinates, what is the impedance of a circuit that has an admittance of 7.09 millisiemens at 45 degrees?**

A. $5.03 \times 10(-5)$ ohms, $\angle 45$ degrees
B. 141 ohms, $\angle{-45}$ degrees
C. 19,900 ohms, $\angle{-45}$ degrees
D. 141 ohms, $\angle 45$ degrees

**ANSWER B:** Admittance is the reciprocal of impedance. Admittance is 1/Z; therefore, $Z = 1/A = 1/7.09 \times 10^{-3}\angle 45° = 141\angle{-45°}$. Remember, the +45° angle in the denominator becomes minus 45° when divided into 1.

---

**E5E18 In rectangular coordinates, what is the impedance of a circuit that has an admittance of 5 millisiemens at −30 degrees?**

A. $173 - j100$ ohms
B. $200 + j100$ ohms
C. $173 + j100$ ohms
D. $200 - j100$ ohms

**ANSWER C:** Following the procedure of the question E5E17,
$Z = 1/5 \times 10^{-3}\angle{-30°} = 0.2 \times 10^3\angle 30° = 200\angle 30°$.
Converting polar coordinates to rectangular gives $R = 200 \cos 30° = 173$ ohms, and $jX = 200 \sin 30° = 100$ ohms. The impedance in rectangular coordinates is $173 + j100$, answer C. Be careful of answer A. You will be tempted to put a $-j$ in front of the reactance because the admittance is given initially with a −30° angle.

---

**E5E19 In rectangular coordinates, what is the admittance of a circuit that has an impedance of 240 ohms at 36.9 degrees?**

A. $3.33 \times 10(-3) - j2.50 \times 10(-3)$ siemens
B. $3.33 \times 10(-3) + j2.50 \times 10(-3)$ siemens
C. $192 + j144$ siemens
D. $3.33 - j2.50$ siemens

**ANSWER A:** Remember, admittance is the reciprocal of impedance, therefore:
$A = 1/240\angle 36.9° = 4.2 \times 10^{-3}\angle{-36.9°}$. Converting polar coordinate impedance $4.2 \times 10^{-3}\angle{-36.9°}$ into rectangular coordinates gives conductance =
$4.2 \times 10^{-3} \cos 36.9° = 4.2 \times 10^{-3} \times 0.8 = 3.36 \times 10^{-3}$ siemens, and susceptance as $4.2 \times 10^3 \sin 36.9° = 4.2 \times 10^{-3} \times 0.6 = 2.52 \times 10^{-3}$ siemens. The susceptance is $-j$ because of the −36.9°. The admittance in rectangular coordinates is $3.36 \times 10^{-3} - j2.52 \times 10^{-3}$ siemens. The closest answer is A.

---

**E5E20 In polar coordinates, what is the impedance of a series circuit consisting of a resistance of 4 ohms, an inductive reactance of 4 ohms, and a capacitive reactance of 1 ohm?**

A. 6.4 ohms, $\angle 53$ degrees
B. 5 ohms, $\angle 37$ degrees
C. 5 ohms, $\angle 45$ degrees
D. 10 ohms, $\angle{-51}$ degrees

**ANSWER B:** Now it's back to polar coordinates. Start your calculation by using rectangular coordinates. $Z = (4 + j4 - j1) = (4 + j3)$. Since the circuit is more inductive than capacitive, the answer will have positive degrees with nothing in front of the angle, as opposed to a negative sign in front of the angle. And if you remember trigonometry, you will know that a right triangle with a 3:4:5 relationship of the sides and hypotenuse is an easy one to determine the hypotenuse value. With sides 3 and 4, the hypotenuse will be 5, with a phase angle whose tangent is 3/4 or 4/3 depending on the relationship of the sides. You have the 3 and 4, and answer B has the 5 with no negative sign in front of 37 degrees. The vector diagram is shown below. Watch out for incorrect answer C.

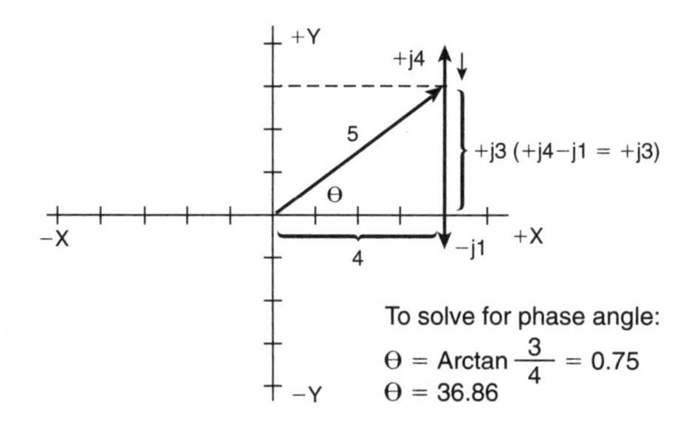


**Vector Diagram for E5E20**

**E5E21 Which point on Figure E5-2 best represents the impedance of a series circuit consisting of a 400 ohm resistor and a 38 picofarad capacitor at 14 MHz?**

A. Point 2
B. Point 4
C. Point 5
D. Point 6

**ANSWER B:** Look at Figure E5-2 on the next page, and notice it is a rectangular-coordinate system with axes at right angles to each other. The indicated points on the plane indicate X and Y coordinate values. In this question, the resistance of 400 ohms can be found along the +X axis; and since the circuit is capacitive, we go down to 300 on the −Y axis and project a line parallel to the X axis to intersect at point 4 a projection from 400 on the +X axis. You can work the problem all the way out, but in this question you can visually spot it at a glance. To check yourself, solve the formula $X_C = 1/2\pi fC$ as follows:

$$X_C = \frac{1}{2\pi fC} = \frac{1}{6.28 \times 14 \times 10^6 \times 38 \times 10^{-12}}$$

$$= \frac{1}{3.341 \times 10^3 \times 10^{-6}} = 0.2993 \times 10^3 = 299.3 \text{ ohms or } 300 \text{ ohms}$$

**E5E22 Which point in Figure E5-2 best represents the impedance of a series circuit consisting of a 300 ohm resistor and an 18 microhenry inductor at 3.505 MHz?**

A. Point 1
B. Point 3
C. Point 7
D. Point 8

**ANSWER B:** The inductive reactance at 3.505 MHz is $X_L = 2\pi \times 3.505 \times 10^6 \times 18 \times 10^{-6} = 396.2$ or approximately 400 ohms of inductive reactance. Follow the +X axis out to 300 for the resistance value. Move up on a line parallel to the +Y axis until you intersect the 400 line. You need to go in a positive Y direction because the circuit is inductive. Point 3 is the answer.

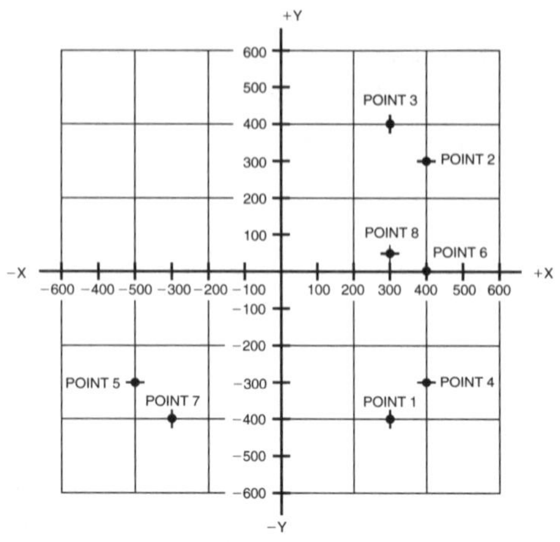


**Figure E5-2**

**E5E23 Which point on Figure E5-2 best represents the impedance of a series circuit consisting of a 300 ohm resistor and a 19 picofarad capacitor at 21.200 MHz?**

A. Point 1
B. Point 3
C. Point 7
D. Point 8

**ANSWER A:** The capacitive reactance is $X_C = 1/2\pi fC = 1/6.28 \times 21.2 \times 10^6 \times 19 \times 10^{-12} = 1/2.53 \times 10^{-3} = 395$ ohms. We first move to the right along the +X axis to the value of the 300-ohm resistor, and move DOWN on a line perpendicular to the X axis (parallel to the −Y axis) to Point 1 at the intersection with the minus 400 line. 395 ohms is considered as 400 ohms. (If we went up, rather than down, we would be incorrect. Going up the +Y axis would mean the circuit is inductive. 19 picofarads makes the circuit capacitive, so we go down.)

***E5F Skin effect; electrostatic and electromagnetic fields***

**E5F01 What is the result of skin effect?**

A. As frequency increases, RF current flows in a thinner layer of the conductor, closer to the surface
B. As frequency decreases, RF current flows in a thinner layer of the conductor, closer to the surface
C. Thermal effects on the surface of the conductor increase the impedance
D. Thermal effects on the surface of the conductor decrease the impedance

**ANSWER A:** If you ever had a chance to inspect an old wireless station, you would have seen that the antenna "plumbing" leading out of the transmitter was usually constructed of hollow copper tubing. Radio frequency current always travels along the thinner outside layer of a conductor, and as frequency increases, there is almost no current in the center of the conductor. The higher the frequency, the greater the skin effect. This is why we use wide copper ground foil to minimize the resistance to AC current that we need to pass to ground.

**E5F02 What effect causes most of an RF current to flow along the surface of a conductor?**

A. Layer effect
B. Seeburg effect
C. Skin effect
D. Resonance effect

**ANSWER C:** Most vertical antennas and beam antennas for home installation use hollow tubing as their radiators. This is because most of the RF current travels along the surface of the conductor. The phenomenon is called *skin effect.*

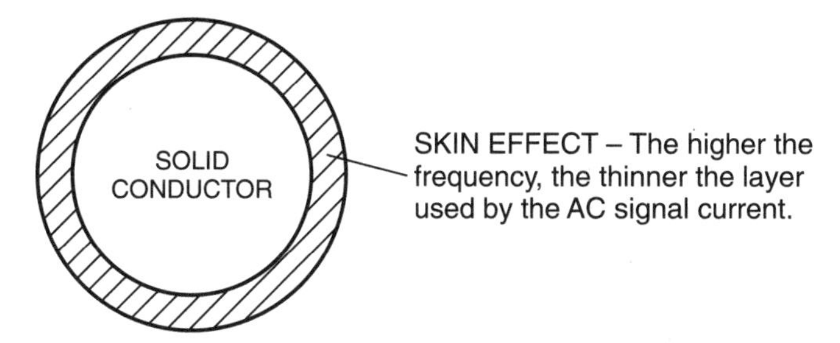


**Cross-Section of Wire Conductor Showing Skin Effect**

---

**E5F03 Where does almost all RF current flow in a conductor?**
A. Along the surface of the conductor
B. In the center of the conductor
C. In a magnetic field around the conductor
D. In a magnetic field in the center of the conductor

**ANSWER A:** As frequency increases, RF currents travel in the thin surface layers of conductors, not within the body. This is known as skin effect. Some hams will shine up their VHF and UHF antennas before a big contest to improve the skin effect of their radiators.

---

**E5F04 Why does most of an RF current flow near the surface of a conductor?**
A. Because a conductor has AC resistance due to self-inductance
B. Because the RF resistance of a conductor is much less than the DC resistance
C. Because of the heating of the conductor's interior
D. Because of skin effect

**ANSWER D:** It is this skin effect that keeps the RF current traveling within a few thousandths of an inch on a conductor's surface.

---

**E5F05 Why is the resistance of a conductor different for RF currents than for direct currents?**
A. Because the insulation conducts current at high frequencies
B. Because of the Heisenburg Effect
C. Because of skin effect
D. Because conductors are non-linear devices

**ANSWER C:** The higher the frequency of an alternating current (AC) signal, the greater the skin effect. For DC, the internal cross-sectional area of a conductor is very important, but for RF signals on the amateur bands, it's not important.

---

**E5F06 What device is used to store electrical energy in an electrostatic field?**
A. A battery
B. A transformer
C. A capacitor
D. An inductor

**ANSWER C:** A capacitor is made up of parallel plates separated by a dielectric (non-conductor). When a voltage is placed across a capacitor, energy is stored in the electrostatic field developed between the capacitor plates. As a reminder, notice the letters "AC" in the word capacitor, and the letters "ATIC" in the word electrostatic. It should lead you to the correct answer.

---

**E5F07 What unit measures electrical energy stored in an electrostatic field?**

A. Coulomb
B. Joule
C. Watt
D. Volt

**ANSWER B:** The joule is a measure of energy, or the capacity to do work. The amount of electrical energy stored in an electrostatic field associated with capacitors is expressed in joules. If a capacitor is charged with one coulomb of electrons in one second by a voltage of one volt, one joule of energy is stored in the capacitor's electrostatic field. Using energy at one joule per second is one watt of electrical power.

---

**E5F08 What is a magnetic field?**

A. Current through the space around a permanent magnet
B. The space through which a magnetic force acts
C. The space between the plates of a charged capacitor, through which a magnetic force acts
D. The force that drives current through a resistor

**ANSWER B:** In marine ham radio installations, we always route our power cables well away from the ship's compass. Do you know why? When you pull current through a conductor, the current produces a *magnetic field* as a force around the conductor – the more current; the stronger the field. The force field can affect the reading of a magnetic compass.

---

**E5F09 In what direction is the magnetic field oriented about a conductor in relation to the direction of electron flow?**

A. In the same direction as the current
B. In a direction opposite to the current
C. In all directions; omnidirectional
D. In a direction determined by the left-hand rule

**ANSWER D:** If you grab a big straight cable with your left hand, with your thumb pointing in the direction the electrons are flowing in the conductor (opposite from conventional current), your fingers will naturally point in the direction of the magnetic lines of force. This is known as the *left-hand rule.* Have you ever been to a junk yard and watched the electromagnet coil magically lift a car off the ground and drop it into a nearby box car? When the coil is energized, it becomes a magnet. As quickly as the energy is cut off, magnetism stops, and the car drops off.

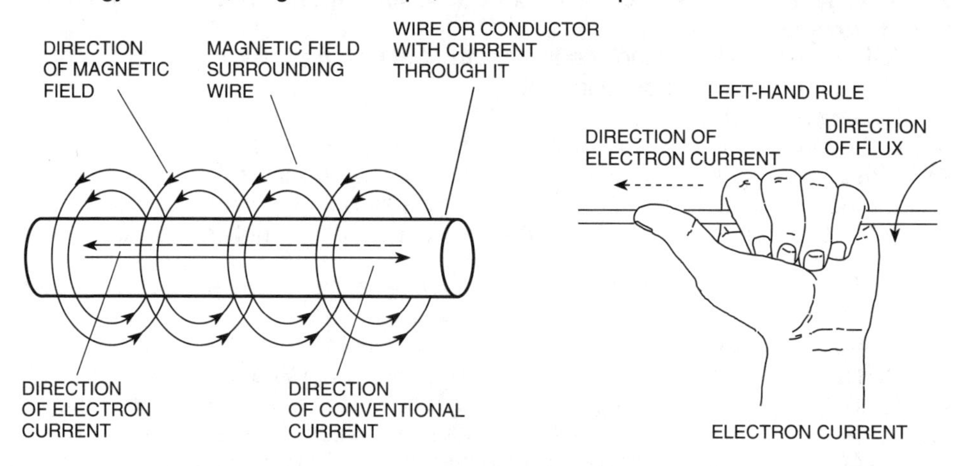


**Left-Hand Rule**

**E5F10 What determines the strength of a magnetic field around a conductor?**

A. The resistance divided by the current
B. The ratio of the current to the resistance
C. The diameter of the conductor
D. The amount of current

**ANSWER D:** As increasing current passes through a conductor, the strength of the magnetic field will continue to build.

---

**E5F11 What is the term for energy that is stored in an electromagnetic or electrostatic field?**

A. Amperes-joules
B. Potential energy
C. Joules-coulombs
D. Kinetic energy

**ANSWER B:** Energy stored in an electromagnetic field in an inductor, or an electrostatic field in a capacitor, is called *potential energy*.

---

***E5G Circuit Q; reactive power; power factor***

---

**E5G01 What is the Q of a parallel R-L-C circuit if the resonant frequency is 14.128 MHz, L is 2.7 microhenrys and R is 18 kilohms?**

A. 75.1
B. 7.51
C. 71.5
D. 0.013

**ANSWER A:** Here come a series of questions about parallel resonant circuits, one of which will probably appear on your upcoming Extra Class exam.

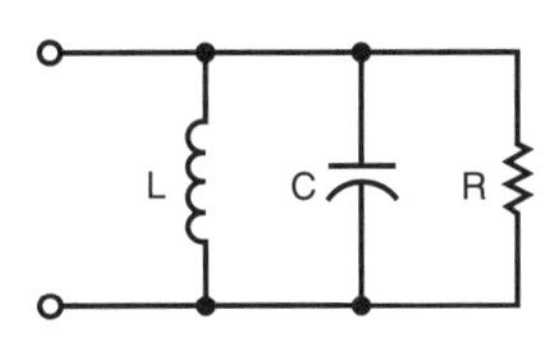


**Parallel Resonant Circuit**

Memorize this formula:

$$Q = \frac{R}{X_L}$$

where: Q is a **quality factor** of this circuit
R is the resistance in **ohms**
$X_L = 2\pi f_r L$
$f_r$ is the resonant frequency in **Hz**
L is the inductance in **henrys**
$2\pi$ is 6.28

We are looking for the Q (quality of a circuit) where the resonant frequency is known, the inductance is known, and the resistance is known. In parallel resonant circuits, as resistance increases the Q decreases and the bandwidth increases. As resistance decreases, the Q increases and bandwidth becomes sharper, which is usually what we want in tuned circuits. Remember, this is a parallel resonant circuit. But before we divide the inductive reactance ($X_L$) into resistance to find out the Q, we first must calculate the inductive reactance. This is easy, as long as you keep your decimal point positions on target. We'll show you a shortcut!

The frequency is given in MHz, and the inductance is given in microhenrys. Since we're multiplying these together, mega ($10^6$) and micro ($10^{-6}$) cancel, so its a simple multiplication of 6.28 × the frequency of 14.128 (in MHz) × the inductance of 2.7 (in microhenrys) = 239.55. Clear your calculator, and 18,000 ÷ 239.55 = 75.14, and there's your answer!

The shortcut is simply leaving MHz and microhenrys at their same decimal point position, and multiplying the two together times 6.28. The result is divided into the resistance in ohms, and there is your answer.

**E5G02 What is the Q of a parallel R-L-C circuit if the resonant frequency is 4.468 MHz, L is 47 microhenrys and R is 180 ohms?**

A. 0.00735 C. 0.136
B. 7.35 D. 13.3

**ANSWER C:** Remember $X_L = 2\pi fL$ is calculated first and then divided into the resistance to arrive at the value for Q. Here are the steps: 6.28 × 4.468 × 47 = 1318.8, divided into 180, and your answer comes out 0.136.

**Q and Half-Power Bandwidth**

**For tuned circuits the quality factor, Q, is:**

Series Resonant Circuit

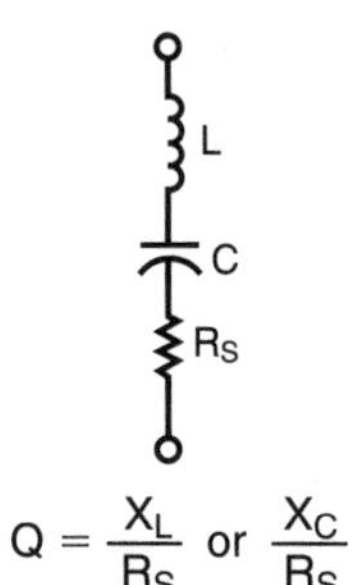


$X_L = 2\pi f_r L$

$X_C = \frac{1}{2\pi f_r C}$

$X_L = X_C$ at $f_r$, the resonant frequency

$Q = \frac{X_L}{R_S}$ or $\frac{X_C}{R_S}$

Parallel Resonant Circuit

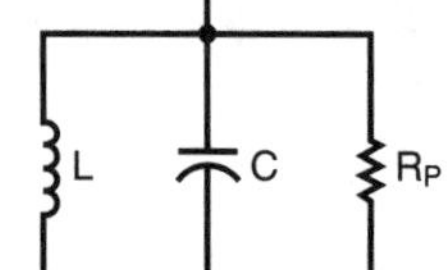


$Q = \frac{R_P}{X_L}$ or $\frac{R_P}{X_C}$

**For tuned circuits with Q greater than 10:**

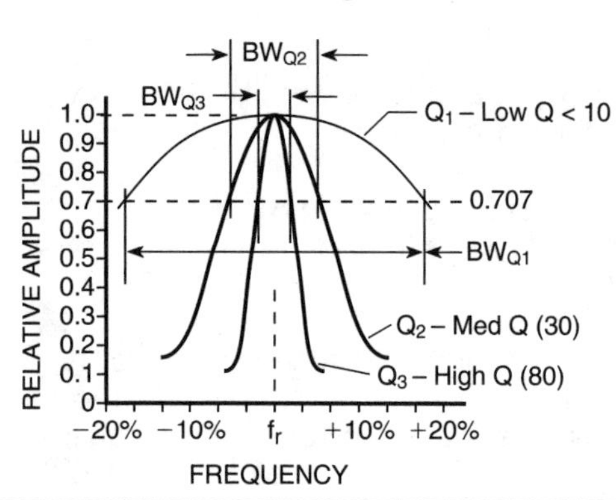


$f_r$ = Resonant frequency in Hz

BW = Half-power Bandwidth (−3db)

$Q = \frac{f_r}{BW}$

$\therefore BW = \frac{f_r}{Q}$

**E5G03 What is the Q of a parallel R-L-C circuit if the resonant frequency is 7.125 MHz, L is 8.2 microhenrys and R is 1 kilohm?**

A. 36.8 C. 0.368
B. 0.273 D. 2.72

**ANSWER D:** A simple formula $Q = R \div X_L$, but you must calculate $X_L$ first. Remember mega and micro cancel each other. Enter $6.28 \times 7.125 \times 8.2$, and divide this into a resistance of 1000 ohms, and there's your answer, 2.72. Simple!

---

**E5G04 What is the Q of a parallel R-L-C circuit if the resonant frequency is 7.125 MHz, L is 12.6 microhenrys and R is 22 kilohms?**

A. 22.1
B. 39
C. 25.6
D. 0.0256

**ANSWER B:** You don't even need to look at the formula at E5G01. All you need to do is to look at the numbers given in the examination question and do the calculations. $6.28 \times$ MHz $\times$ microhenrys and divide into 22,000 equals 39.02.

---

**E5G05 What is the Q of a parallel R-L-C circuit if the resonant frequency is 3.625 MHz, L is 42 microhenrys and R is 220 ohms?**

A. 23
B. 0.00435
C. 4.35
D. 0.23

**ANSWER D:** You should have it memorized by now: Clear, $6.28 \times 3.625 \times 42 = 956.13$. Enter $220 \div 956.13 = 0.23$. Oh! Oh! Another answer less than 1. Good work on your calculator! Your kids ARE impressed.

---

**E5G06 Why is a resistor often included in a parallel resonant circuit?**

A. To increase the Q and decrease the skin effect
B. To decrease the Q and increase the resonant frequency
C. To decrease the Q and increase the bandwidth
D. To increase the Q and decrease the bandwidth

**ANSWER C:** Resistance in a parallel resonant circuit will *decrease* the circuit Q and *increase* the bandwidth. Antennas using large inductive and capacitive values will have increased Q, but tighter bandwidth. With smaller inductance and capacitance in antenna systems, the Q decreases, but resonance broadens.

---

**E5G07 What is the term for an out-of-phase, nonproductive power associated with inductors and capacitors?**

A. Effective power
B. True power
C. Peak envelope power
D. Reactive power

**ANSWER D:** If it is non-productive, it is reactive power in an AC circuit. The letters AC are found in the word reactive (meaning out-of-phase). It indicates *non-productive power* (key words) produced in circuits containing inductors and capacitors. Since out-of-phase power in inductors will tend to cancel out-of-phase power in capacitors, reactive power is wattless. It is not converted to heat and dissipated. It is energy being stored temporarily in a field and then returned to the circuit. It circulates back and forth in a coil's magnetic field and/or a capacitor's electrostatic field.

---

**E5G08 In a circuit that has both inductors and capacitors, what happens to reactive power?**

A. It is dissipated as heat in the circuit
B. It goes back and forth between magnetic and electric fields, but is not dissipated
C. It is dissipated as kinetic energy in the circuit
D. It is dissipated in the formation of inductive and capacitive fields

**ANSWER B:** In a circuit that has both capacitors and inductors, the reactive power oscillates back and forth between magnetic and electric fields, and does not get dissipated.

---

**E5G09 In a circuit where the AC voltage and current are out of phase, how can the true power be determined?**

A. By multiplying the apparent power times the power factor
B. By subtracting the apparent power from the power factor
C. By dividing the apparent power by the power factor
D. By multiplying the RMS voltage times the RMS current

**ANSWER A:** If we multiply apparent power times the power factor of an AC voltage and current out of phase, we can determine true power. See E5G10 figure.

---

**E5G10 What is the power factor of an R-L circuit having a 60 degree phase angle between the voltage and the current?**

A. 1.414 C. 0.5
B. 0.866 D. 1.73

**ANSWER C:** If you are using a scientific calculator, you will find the cosine of 60 degrees, given in the problem, comes out 0.5, answer C. The power factor is cos $\phi$. If you don't have a scientific calculator, you might memorize this table:

| $\phi$ | cos $\phi$ |
|---|---|
| 30° | 0.866 |
| 45° | 0.707 |
| 60° | 0.5 |

**Apparent and True Power**

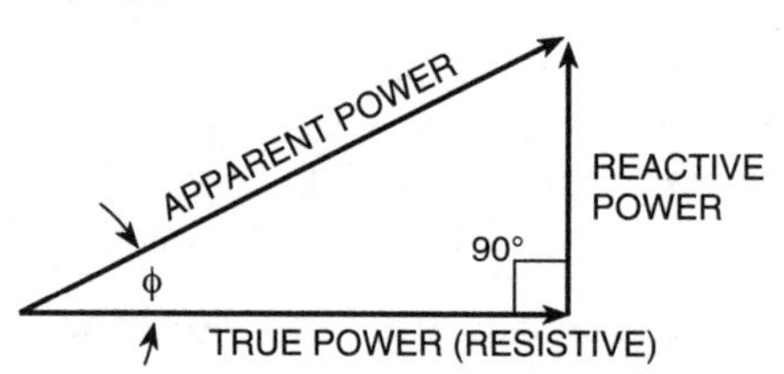


True power is power dissipated as heat, while reactive power of a circuit is stored in inductances or capacitances and then returned to the circuit.

Apparent power is voltage times current without taking into account the phase angle between them. According to right angle trigonometry,

$$\text{cosine } \phi = \frac{\text{side adjacent}}{\text{hypotenuse}}$$

therefore,

$$\text{cosine } \phi = \frac{\text{True Power}}{\text{Apparent Power}}$$

therefore,

True Power = Apparent Power × Cosine $\phi$

Cosine $\phi$ is called the *power factor* (PF) of a circuit.

Since apparent power is E × I,

True Power = E × I × Cos $\phi$

or

True Power = E × I × PF

**E5G11 How many watts are consumed in a circuit having a power factor of 0.2 if the input is 100-V AC at 4 amperes?**

A. 400 watts
B. 80 watts
C. 2000 watts
D. 50 watts

**ANSWER B:** This problem is based on the formula:

True power = E (volts) × I (amps) × PF (power factor).

True power works out to be 100 volts × 4 amps × 0.2, giving us 80 watts. Easy enough, right?

---

**E5G12 Why would the power used in a circuit be less than the product of the magnitudes of the AC voltage and current?**

A. Because there is a phase angle greater than zero between the current and voltage
B. Because there are only resistances in the circuit
C. Because there are no reactances in the circuit
D. Because there is a phase angle equal to zero between the current and voltage

**ANSWER A:** When the *phase angle is greater than zero* (key words) between voltage and current, the rate at which electrical energy is used in a circuit is going to be less than the product of the magnitudes of the AC voltage and current. Since the voltage and current aren't in phase, the true power will be less because there is some reactive power present. See E5G10.

---

**E5G13 What is the Q of a parallel RLC circuit if the resonant frequency is 14.128 MHz, L is 4.7 microhenrys and R is 18 kilohms?**

A. 4.31
B. 43.1
C. 13.3
D. 0.023

**ANSWER B:** Here are some more relatively easy questions on calculating the Q of a parallel R-L-C circuit. Look again at the formula at E5G01. The frequency is given in megahertz, and the inductance is given in microhenries. Since we are multiplying these together, mega ($10^6$) and micro ($10^{-6}$) CANCEL, so it's a simple multiplication of 6.28 × 14.128 × 4.7 = 417, which divided into 18,000 gives a Q = 43.1. Just remember that mega and micro cancel.

---

**E5G14 What is the Q of a parallel RLC circuit if the resonant frequency is 14.225 MHz, L is 3.5 microhenrys and R is 10 kilohms?**

A. 7.35
B. 0.0319
C. 71.5
D. 31.9

**ANSWER D:** $2\pi$ × frequency × inductance to find $X_L$ . The result is divided into the resistance to get Q. 6.28 × 14.225 × 3.5 = 312.67. Enter 10,000 ÷ 312.67 = 31.98. Got it? Remember mega ($10^6$) canceled micro ($10^{-6}$).

---

**E5G15 What is the Q of a parallel RLC circuit if the resonant frequency is 7.125 MHz, L is 10.1 microhenrys and R is 100 ohms?**

A. 0.221
B. 4.52
C. 0.00452
D. 22.1

**ANSWER A:** Short and sweet! The steps are easy. $2\pi$ × MHz × microhenrys, divided into resistance. 6.28 × 7.125 × 10.1 = 451.92. 100 ÷ 451.92 = 0.221.

**E5G16 What is the Q of a parallel RLC circuit if the resonant frequency is 3.625 MHz, L is 3 microhenrys and R is 2.2 kilohms?**

A. 0.031
B. 32.2
C. 31.1
D. 25.6

**ANSWER B:** Same as before. Keep in mind that micro and mega cancel, so don't change the decimal points. Calculator keystrokes: Clear, 6.28 × 3.625 × 3 = 68.29. Enter 2200 ÷ 68.29 = 32.22.

***E5H Effective radiated power; system gains and losses***

**E5H01 What is the effective radiated power of a repeater station with 50 watts transmitter power output, 4-dB feed line loss, 2-dB duplexer loss, 1-dB circulator loss and 6-dBd antenna gain?**

A. 199 watts
B. 39.7 watts
C. 45 watts
D. 62.9 watts

**ANSWER B:** You may well have one question out of the following 10 questions based on simple dB calculations at a repeater site. Don't worry about logarithms, you can do these in your head, and all you must remember is: 3 dB = a 2X increase; to twice the value; and −3 dB = a 2X loss to one-half the value.

To solve the problems, just add and subtract all the dB's, and if it ends up about +3 dB, then double your transmitter power output to end up with effective radiated power. If your answer ends up around −3 dB gain, cut your transmitter power in half for effective radiated power. In this question, we have −7 dB loss, offset by 6 dB gain for 1 dB net loss. If we started out with 50 watts, our effective radiated power (ERP) will be slightly less. 39.7 is calculated by multiplying 50W × 0.794 (−1 dB),the only answer in the ballpark.

**dB Power Gain or Loss**

| ± dB | Decimal Gain Value | Approx. ±Times Power Change | Decimal Loss Value |
|---|---|---|---|
| 1 dB | 1.259 | 1¼ X | 0.794 |
| 2 dB | 1.585 | 1½ X | 0.633 |
| 3 dB | 1.995 | 2 X | 0.501 |
| 4 dB | 2.511 | 2½ X | 0.398 |
| 5 dB | 3.162 | 3 X | 0.316 |
| 6 dB | 3.981 | 4 X | 0.251 |
| 7 dB | 5.011 | 5 X | 0.200 |
| 8 dB | 6.310 | 6¼ X | 0.158 |
| 9 dB | 7.943 | 8 X | 0.126 |
| 10 dB | 10.0 | 10 X | 0.1 |

**E5H02 What is the effective radiated power of a repeater station with 50 watts transmitter power output, 5-dB feed line loss, 3-dB duplexer loss, 1-dB circulator loss and 7-dBd antenna gain?**

A. 79.2 watts
B. 315 watts
C. 31.5 watts
D. 40.5 watts

**ANSWER C:** Most of these 10 ERP Extra questions come out close enough that you can work them in your head. Our chart on decimal gain values and decimal loss values will help, but the chart is not permitted during the exam, so you should remember decimal loss values as follows: 1 dB loss = 0.794; 2 dB loss = 0.633. In this problem, we have a 2 dB loss, so multiple 0.633 × 50, and you end up with an answer close to 31.5 watts.

---

**E5H03 What is the effective radiated power of a station with 75 watts transmitter power output, 4-dB feed line loss and 10-dBd antenna gain?**

A. 600 watts
B. 75 watts
C. 150 watts
D. 299 watts

**ANSWER D:** We have 4 dB loss, and a great antenna system giving us 10 dB gain. This gives us a +6 dB, which is 4 times our power output. There it is, only one answer near 300, 299 watts! 4 × 75 = 300 watts.

---

**E5H04 What is the effective radiated power of a repeater station with 75 watts transmitter power output, 5-dB feed line loss, 3-dB duplexer loss, 1-dB circulator loss and 6-dBd antenna gain?**

A. 37.6 watts
B. 237 watts
C. 150 watts
D. 23.7 watts

**ANSWER A:** 9 dB loss, offset by 6 dB antenna gain gives a 3 dB net loss, or half power. Half of 75 watts is answer A, 37.6 watts.

**Decibels**

It is important to know about decibels because they are used extensively in electronics. Look at the derivation for decibels and note it is a measure of the ratio of two powers, $P_1$ and $P_2$. Remember the power changes for different dB values. Also, since 6 dB is a four times change, 16 dB (6 dB + 10 dB) will be a 4 × 10 = 40 times change, and 26 dB (6 dB + 20 dB) will be a 4 × 100 = 400 times change. Thus, any dB value from 10 dB and above can be evaluted by using the power change of 1 dB to 9 dB and multiplying it by the 10 dB, 20 dB, 30 dB, 40 dB, 50 dB, 60 dB, etc. change. Thus, 57 dB (7 dB + 50 dB) will be a 5 × 100,000 = 500,000 times change.

| dB | Power Change $\frac{P_1}{P_2}$ | |
|---|---|---|
| 1 dB | 1¼X | Power change |
| 2 dB | 1½X | Power change |
| 3 dB | 2X | Power change |
| 4 dB | 2½X | Power change |
| 5 dB | 3X | Power change |
| 6 dB | 4X | Power change |
| 7 dB | 5X | Power change |

| dB | Power Change $\frac{P_1}{P_2}$ | |
|---|---|---|
| 8 dB | 6¼X | Power change |
| 9 dB | 8X | Power change |
| 10 dB | 10X | Power change |
| 20 dB | 100X | Power change |
| 30 dB | 1000X | Power change |
| 40 dB | 10,000X | Power change |
| 50 dB | 100,000X | Power change |
| 60 dB | 1,000,000X | Power change |

**Derivation:**

If $\text{dB} = 10 \log_{10} \frac{P_1}{P_2}$

then what power ratio is 60 dB?

$$60 = 10 \log_{10} \frac{P_1}{P_2}$$

$$\frac{60}{10} = \log_{10} \frac{P_1}{P_2}$$

$$6 = \log_{10} \frac{P_1}{P_2}$$

Remember: logarithm of a number is the exponent to which the base must be raised to get the number.

$$\therefore 10^6 = \frac{P_1}{P_2}$$

$$1{,}000{,}000 = \frac{P_1}{P_2}$$

Or $P_1 = 1{,}000{,}000\, P_2$

60 dB means $P_1$ is one million times $P_2$

**E5H05 What is the effective radiated power of a station with 100 watts transmitter power output, 1-dB feed line loss and 6-dBd antenna gain?**

A. 350 watts
B. 500 watts
C. 20 watts
D. 316 watts

**ANSWER D:** Here we have a nice system with only 1 dB of feed line loss and a powerful 6 dB gain antenna. A 6 dB antenna yields a 4 times increase in power to 400 watts, but 1 dB of feed line loss will cut our effective radiated power down to 316 watts (400 × 0.794).

---

**E5H06 What is the effective radiated power of a repeater station with 100 watts transmitter power output, 5-dB feed line loss, 3-dB duplexer loss, 1-dB circulator loss and 10-dBd antenna gain?**

A. 794 watts
B. 126 watts
C. 79.4 watts
D. 1260 watts

**ANSWER B:** A total of 9 dB loss, offset by 10 dB gain, net +1 dB. Since we had 100 watts to start with and the net gain is 1 dB, the only close answer is answer B, 126 watts.

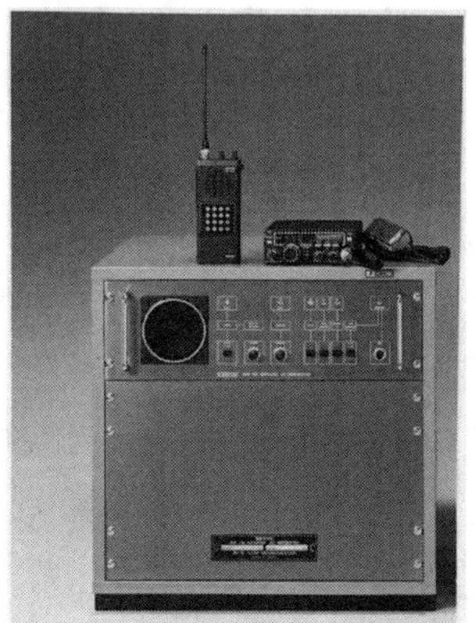

**A Repeater**

---

**E5H07 What is the effective radiated power of a repeater station with 120 watts transmitter power output, 5-dB feed line loss, 3-dB duplexer loss, 1-dB circulator loss and 6-dBd antenna gain?**

A. 601 watts
B. 240 watts
C. 60 watts
D. 79 watts

**ANSWER C:** Here we have 9 dB loss with 6 dB gain. That's a net loss of 3 dB. It cuts our 120-watt transmitter to an effective radiated power of only 60 watts. Quit buying junk coax!

---

**E5H08 What is the effective radiated power of a repeater station with 150 watts transmitter power output, 2-dB feed line loss, 2.2-dB duplexer loss and 7-dBd antenna gain?**

A. 1977 watts
B. 78.7 watts
C. 420 watts
D. 286 watts

**ANSWER D:** Our total loss is about 4.2 dB, and our total gain is 7 dB. This is an approximate 3 dB overall gain, and this almost doubles your 150 watts to an approximate 286 watts.

---

**E5H09 What is the effective radiated power of a repeater station with 200 watts transmitter power output, 4-dB feed line loss, 3.2-dB duplexer loss, 0.8-dB circulator loss and 10-dBd antenna gain?**

A. 317 watts
B. 2000 watts
C. 126 watts
D. 300 watts

**ANSWER A:** Our total loss is 8 dB, and our total gain is 10 dB gain. This gives us a net gain of 2 dB; therefore, from the chart at E5H01, we have a gain of 1.585 times. Therefore, with 200 watts input, the ERP is 317 watts (200 × 1.585).

To check your answer, remember

$$\text{dB} = 10 \log_{10} \frac{P_1}{P_2}$$

where: $P_1$ = Power Output
$P_2$ = Power Input

Therefore,

$$\text{dB} = 10 \log_{10} \frac{317}{200}$$

$$\text{dB} = 10 \log_{10} 1.585$$ The $\log_{10}$ of 1.585 is 0.2

$$\text{dB} = 10 \times 0.2 = 2$$

Since the output power is 317 watts and the input power is 200 watts, there is a net gain of 2 dB.

---

**E5H10 What is the effective radiated power of a repeater station with 200 watts transmitter power output, 2-dB feed line loss, 2.8-dB duplexer loss, 1.2-dB circulator loss and 7-dBd antenna gain?**

A. 159 watts
B. 252 watts
C. 632 watts
D. 63.2 watts

**ANSWER B:** Here is our last calculation with dB values. Our total loss adds up to a 6 dB loss, and our total gain is 7 dB, for an actual gain of 1 dB. This would be a slight increase to our 200 watts output in effective radiated power, so ERP is 252 watts.

---

**E5H11 What term describes station output (including the transmitter, antenna and everything in between), when considering transmitter power and system gains and losses?**

A. Power factor
B. Half-power bandwidth
C. Effective radiated power
D. Apparent power

**ANSWER C:** It's important to always calculate effective radiated power when you set up your ham station. This table shows you the dB's you could lose with cheap coax cable, and the valuable dB's you can gain by using a quality collinear VHF or UHF antenna.

**Coax Cable Type, Size and Loss per 100 Feet**

| Coax Type | Size | Loss at HF 100 MHz | Loss at UHF 400 MHz |
|---|---|---|---|
| RG-6 | Large | 2.3 dB | 4.7 dB |
| RG-59 | Medium | 2.9 dB | 5.9 dB |
| RG-58U | Small | 4.3 dB | 9.4 dB |
| RG-8X | Medium | 3.7 dB | 8.0 dB |
| RG-8U | Large | 1.9 dB | 4.1 dB |
| RG-213 | Large | 1.9 dB | 4.5 dB |
| Hardline | Large, Rigid | 0.5 dB | 1.5 dB |

**E5H12 What is reactive power?**

A. Wattless, nonproductive power
B. Power consumed in wire resistance in an inductor
C. Power lost because of capacitor leakage
D. Power consumed in circuit Q

**ANSWER A:** Reactive power is *non-productive power* (key words) produced in circuits containing inductors and capacitors. Reactive power is wattless. It is not converted to heat and dissipated. It is energy being stored temporarily in a field and then returned to the circuit.

---

**E5H13 What is the power factor of an RL circuit having a 45 degree phase angle between the voltage and the current?**

A. 0.866
B. 1.0
C. 0.5
D. 0.707

**ANSWER D:** Using your scientific calculator, or from memory, remember that the cosine of 45 degrees is 0.707. Think of an airplane – the very popular 707 jet aircraft – taking off at an angle of 45 degrees.

---

**E5H14 What is the power factor of an RL circuit having a 30 degree phase angle between the voltage and the current?**

A. 1.73
B. 0.5
C. 0.866
D. 0.577

**ANSWER C:** At 30 degrees, the power factor is 0.866. If you don't pass your Extra Class exam, you could be 86'd out of the room!

---

**E5H15 How many watts are consumed in a circuit having a power factor of 0.6 if the input is 200V AC at 5 amperes?**

A. 200 watts
B. 1000 watts
C. 1600 watts
D. 600 watts

**ANSWER D:** Here are two simple problems that are based on the formula:

True power = E (volts) × I (amps) × PF (power factor).

Therefore, 200 volts × 5 amps × 0.6 = 600 watts. Easy, huh?

---

**E5H16 How many watts are consumed in a circuit having a power factor of 0.71 if the apparent power is 500 watts?**

A. 704 W
B. 355 W
C. 252 W
D. 1.42 mW

**ANSWER B:** Here they have already calculated apparent power at 500 watts. Multiply apparent power by power factor 0.71, and you end up with 355 watts consumed in the circuit.

---

### *E5I Photoconductive principles and effects*

---

**E5I01 What is photoconductivity?**

A. The conversion of photon energy to electromotive energy
B. The increased conductivity of an illuminated semiconductor junction
C. The conversion of electromotive energy to photon energy
D. The decreased conductivity of an illuminated semiconductor junction

**ANSWER B:** The photocell has high resistance when no light shines on it, and a varying resistance when light shines on it. It can be used in an ON or OFF state as a simple beam alarm across a doorway, or may be found in almost all modern SLR cameras as a sensitive light meter to judge the amount of light present.

**E5I02 What happens to the conductivity of a photoconductive material when light shines on it?**

A. It increases
B. It decreases
C. It stays the same
D. It becomes unstable

**ANSWER A:** In the dark, a photocell has high resistance. When you shine light on it, the conductivity increases as the resistance decreases.

**E5I03 What happens to the resistance of a photoconductive material when light shines on it?**

A. It increases
B. It becomes unstable
C. It stays the same
D. It decreases

**ANSWER D:** More light, less resistance.

**Resistance Variation of a Photodiode**

| Light Incidence | Current I | Resistance | Test Conditions |
|---|---|---|---|
| Dark | 5 nA | 2000 MΩ | $V_R = 10$ V<br>*$E_e = 0$ |
| Light | 15 μA | 666.7 kΩ | $V_R = 10$ V<br>$E_e = 250$ μW/cm$^2$ at 940 nm |

*Irradiance ($E_e$) is the radiant power per unit area incident on surface

**E5I04 What happens to the conductivity of a semiconductor junction when light shines on it?**

A. It stays the same
B. It becomes unstable
C. It increases
D. It decreases

**ANSWER C:** Conductivity increases when light shines on a semiconductor junction – just like the photocell.

**E5I05 What is an optocoupler?**

A. A resistor and a capacitor
B. A frequency modulated helium-neon laser
C. An amplitude modulated helium-neon laser
D. An LED and a phototransistor

**ANSWER D:** Optocouplers are found in many modern, high-frequency transceivers. When you spin the dial to tune, you are not turning a giant capacitor, but rather interrupting a light beam on a phototransistor.

**E5I06 What is an optoisolator?**

A. An LED and a phototransistor
B. A P-N junction that develops an excess positive charge when exposed to light
C. An LED and a capacitor
D. An LED and a solar cell

**ANSWER A:** Another name for an optocoupler is an optoisolator. An LED shines through a window at a phototransistor in an optoisolator.

**E5I07 What is an optical shaft encoder?**

A. An array of neon or LED indicators whose light transmission path is controlled by a rotating wheel

B. An array of optocouplers whose light transmission path is controlled by a rotating wheel

C. An array of neon or LED indicators mounted on a rotating wheel in a coded pattern

D. An array of optocouplers mounted on a rotating wheel in a coded pattern

**ANSWER B:** When you spin the big knob of your new high-frequency transceiver it is connected to an optical shaft encoder that determines the tuned frequency.

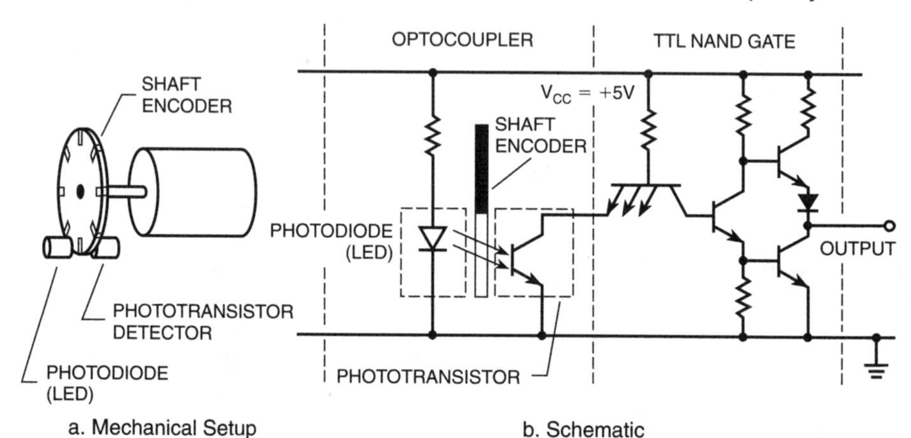


**Optocoupler Used for Shaft Encoder**

**E5I08 What characteristic of a crystalline solid will photoconductivity change?**

A. The capacitance

B. The inductance

C. The specific gravity

D. The resistance

**ANSWER D:** The resistance of the crystalline solid varies when light shines on it because of the photoconductive effect.

**E5I09 Which material will exhibit the greatest photoconductive effect when visible light shines on it?**

A. Potassium nitrate

B. Lead sulfide

C. Cadmium sulfide

D. Sodium chloride

**ANSWER C:** Cadmium sulfide offers the greatest photoconductive effect when light shines on it. Photocells used in amateur radio optical encoders may contain cadmium sulfide.

**E5I10 Which material will exhibit the greatest photoconductive effect when infrared light shines on it?**

A. Potassium nitrate

B. Lead sulfide

C. Cadmium sulfide

D. Sodium chloride

**ANSWER B:** Lead sulfide is used for infrared light systems. Remember this – lead for infrared, and cadmium for visible.

**E5I11 Which material is affected the most by photoconductivity?**

A. A crystalline semiconductor
B. An ordinary metal
C. A heavy metal
D. A liquid semiconductor

**ANSWER A**: The crystalline semiconductor is found inside the optical shaft encoder mechanism, and might also be found outside your ham shack as part of your solar-charging system. The light-emitting diode is normally the light source inside your tuning dial optocoupler.

---

**E5I12 What characteristic of optoisolators is often used in power supplies?**

A. They have low impedance between the light source and the phototransistor
B. They have very high impedance between the light source and the phototransistor
C. They have low impedance between the light source and the LED
D. They have very high impedance between the light source and the LED

**ANSWER B:** The optoisolator offers isolation between the light-emitting diode and phototransistor. The light-emitting diode is in the secondary, connected to the output terminals, and the phototransistor might be in a circuit at AC line potential.

---

**E5I13 What characteristic of optoisolators makes them suitable for use with a triac to form the solid-state equivalent of a mechanical relay for a 120 V AC household circuit?**

A. Optoisolators provide a low impedance link between a control circuit and a power circuit
B. Optoisolators provide impedance matching between the control circuit and power circuit
C. Optoisolators provide a very high degree of electrical isolation between a control circuit and a power circuit
D. Optoisolators eliminate (isolate) the effects of reflected light in the control circuit

**ANSWER C:** The chief advantage of the optoisolator at household voltages is to provide a very high degree of electrical isolation between the control circuit and the house power circuit. Look for the answer "isolation" to agree with the optoisolator device.

## Subelement E6 — Circuit Components [5 Exam Questions — 5 Groups]

***E6A Semiconductor material: Germanium, Silicon, P-type, N-type; Transistor types: NPN, PNP, junction, power; field-effect transistors (FETs): enhancement mode; depletion mode; MOS; CMOS; N-channel; P-channel***

**E6A01 In what application is gallium arsenide used as a semiconductor material in preference to germanium or silicon?**

A. In high-current rectifier circuits
B. In high-power audio circuits
C. At microwave-frequency frequencies
D. At very low frequency RF circuits

**ANSWER C:** Here's a new word for your amateur radio vocabulary: GaAsFET. The GaAs stands for gallium arsenide, and FET stands for field effect transistor. We use the GaAsFET transistor in VHF, UHF, and microwave receivers. This transistor offers excellent signal gain with an extremely-low noise figure.

---

**E6A02 What type of semiconductor material contains more free electrons than pure germanium or silicon crystals?**

A. N-type
B. P-type
C. Bipolar
D. Insulated gate

**ANSWER A:** If there are MORE free electrons orbiting the nucleus, it is an N-type material.

---

**E6A03 What are the majority charge carriers in P-type semiconductor material?**

A. Free neutrons
B. Free protons
C. Holes
D. Free electrons

**ANSWER C:** The charge carriers are in abundance in P-type semiconductor materials, the holes.

---

**E6A04 What is the name given to an impurity atom that adds holes to a semiconductor crystal structure?**

A. Insulator impurity
B. N-type impurity
C. Acceptor impurity
D. Donor impurity

**ANSWER C:** The impurity atom that adds holes to a semiconductor crystal structure is called an acceptor impurity.

There is a good chance that you will have one question from this group about semiconductor materials. Remember this phrase, "ANFREE" for an N-type material where antimony atoms or arsenic atoms have been added, and there are more free electrons in an N-type semiconductor. "PIUMH" (pronounced like sneezing) is a P-type material produced by adding either gallIUM or indIUM, producing more "Potholes" than free electrons. Also, anything that PROVIDES is a donor, and an Acceptor impurity will Add holes. Got it? Good reading about semiconductors is found in the latest edition of the American Radio Relay League Handbook for Radio Amateurs.

**E6A05 What is the alpha of a bipolar transistor?**

A. The change of collector current with respect to base current
B. The change of base current with respect to collector current
C. The change of collector current with respect to emitter current
D. The change of collector current with respect to gate current

**ANSWER C:** In bipolar transistors, the term "alpha" is the variation of *COLLECTOR current* with respect to *EMITTER current* (CCEC, key words). This is an important consideration when designing a circuit that will utilize a bipolar transistor.

---

**E6A06 In Figure E6-1, what is the schematic symbol for a PNP transistor?**

A. 1 C. 4
B. 2 D. 5

**ANSWER A:** Remember this on your Technician Class examination? Which way is the arrow pointing? If it's pointing in, it is PNP – "P in."

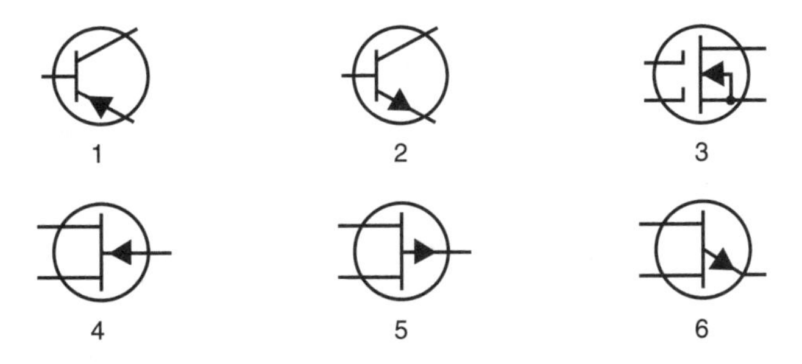


Figure E6-1

---

**E6A07 What term indicates the frequency at which a transistor grounded base current gain has decreased to 0.7 of the gain obtainable at 1 kHz?**

A. Corner frequency
B. Alpha rejection frequency
C. Beta cutoff frequency
D. Alpha cutoff frequency

**ANSWER D:** Grounded base is the same as a common base configuration. Common base means alpha. Since 1 kHz is mentioned in this question, they are referring to the alpha cutoff frequency in a transistor.

---

**E6A08 What is a depletion-mode FET?**

A. An FET that has a channel with no gate voltage applied; a current flows with zero gate voltage
B. An FET that has a channel that blocks current when the gate voltage is zero
C. An FET without a channel; no current flows with zero gate voltage
D. An FET without a channel to hinder current through the gate

**ANSWER A:** Your handheld battery may become depleted if depletion-mode FETs are used in many of the circuits. Even though there is no gate voltage applied, current still continues to flow.

## Bipolar Transistor Basics – PNP and NPN

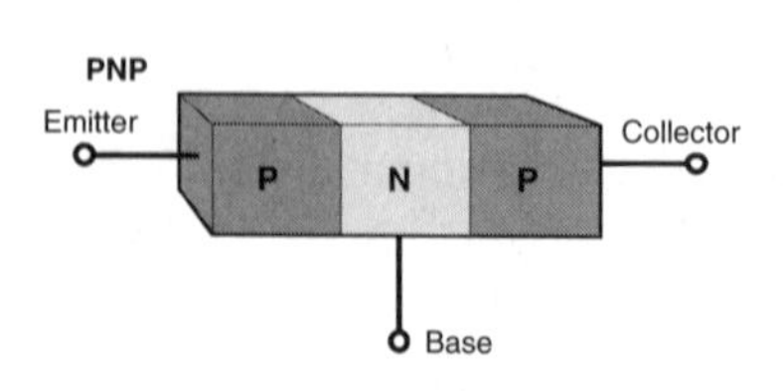


a. Junction Structure

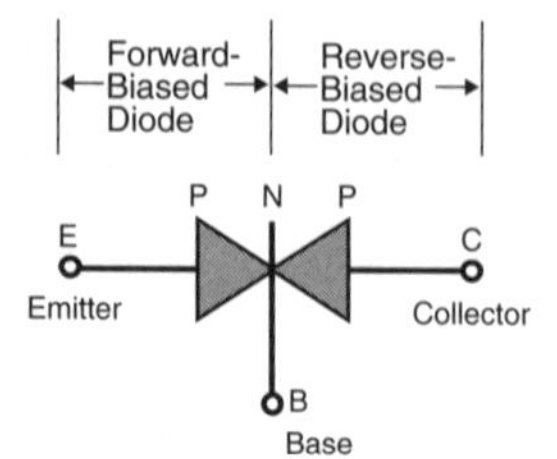


b. Diode Juction Equivalent

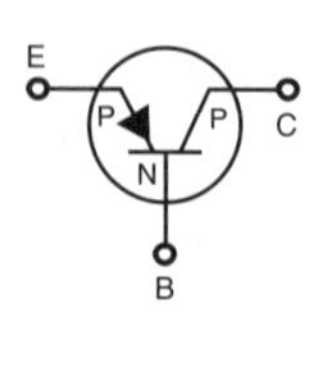


c. PNP Symbol

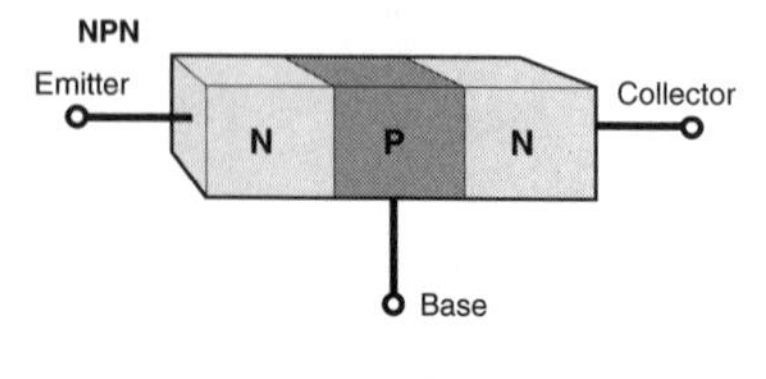


d. Junction Structure

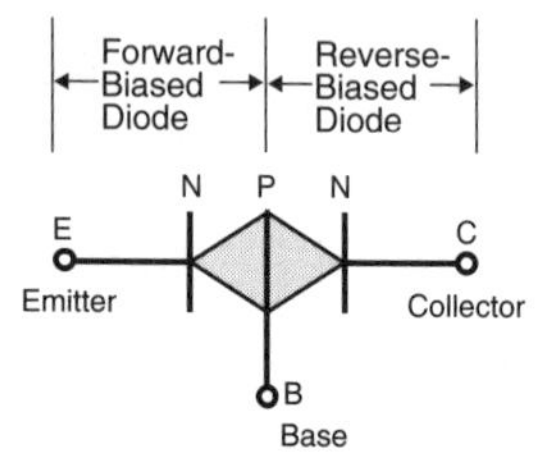


e. Diode Junction Equivalent

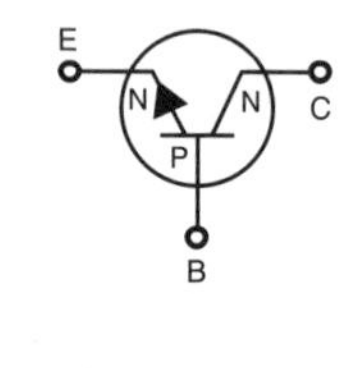


f. NPN Symbol

There are two types of transistors, bipolar and field-effect. A bipolar transistor is a combination of two junctions of semiconductor material built into a semiconductor chip (usually silicon). There are two types – PNP and NPN. Their junction structures are shown in Figure a and d, respectively. For a transistor that produces gain, the emitter-base junction is a forward-biased diode, and the collector-base junction is a reverse-biased diode. The diode equivalents of PNP and NPN transistors are shown in Figure b and e, respectively. However, unlike a reverse-biased diode that does not conduct current (except for a small leakage current), the reverse-biased collector-base junction of a bipolar transistor conducts collector current that is controlled by the current into the base at the base-emitter junction. The normal active

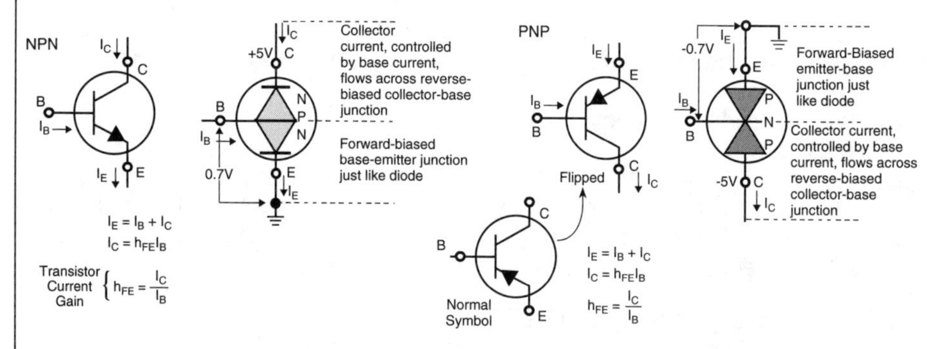


g. NPN Operation

h. PNP Operation

An NPN silicon transistor P base is 0.7 V more positive that its N emitter, and the N collector is several more volts more positive than the emitter. The emitter current is the sum of the base and collector current. The current gain under any DC operating condition is hFE, the ratio of $I_C$ to $I_B$, and current gains of 50 to 200 are common in modern-day silicon transistors. hFE is actually called "the common-emitter" current gain because the emitter is common in the circuit.

A PNP silicon transistor N base is 0.7 V negative with respect to its emitter in order to have the P emitter more positive than the N base. The P collector is several volts negative from the emitter to keep the collector-base junction reverse biased. As shown in Figure h, the same current equations apply, and the current gain, hFE, is the same. The major difference is in the polarity of the voltages for operation. For NPN common-emitter operation the base and collector voltages are positive with respect to the emitter; while for the PNP the voltages are negative.

Source: *Basic Communications Electronics*, Hudson & Luecke, © 1999 Master Publishing, Inc., Lincolnwood, IL

**E6A09 In Figure E6-2, what is the schematic symbol for an N-channel dual-gate MOSFET?**

A. 2 C. 5
B. 4 D. 6

**ANSWER B:** Note that the arrow is pointing iN, and there are now two gates.

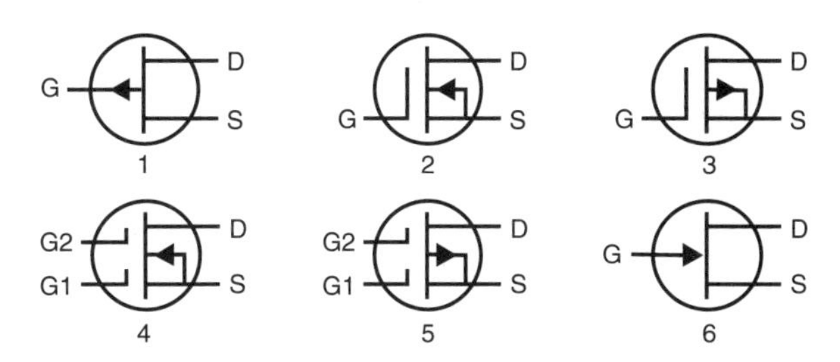


**Figure E6-2**

**E6A10 In Figure E6-2, what is the schematic symbol for a P-channel junction FET?**

A. 1 C. 3
B. 2 D. 6

**ANSWER A:** In the P-channel junction FET, the arrow is Pointing out. For an N-channel junction FET, the arrow is always pointing iN.

**E6A11 Why do many MOSFET devices have built-in gate-protective Zener diodes?**

A. To provide a voltage reference for the correct amount of reverse-bias gate voltage
B. To protect the substrate from excessive voltages
C. To keep the gate voltage within specifications and prevent the device from overheating
D. To prevent the gate insulation from being punctured by small static charges or excessive voltages

**ANSWER D:** The "front end" transistors of the modern ham transceiver must sometimes sustain major amounts of incoming signals from nearby transmitters, or maybe even a nearby lightning strike. Lightning static could destroy a MOSFET if it weren't for the built-in, gate-protective Zener diodes.

**E6A12 What do the initials CMOS stand for?**

A. Common mode oscillating system
B. Complementary mica-oxide silicon
C. Complementary metal-oxide semiconductor
D. Complementary metal-oxide substrate

**ANSWER C:** When we talk about transistorized devices, they are in fact semiconductors. A CMOS is made from layers of metal-oxide semiconductor material.

**E6A13 How does DC input impedance on the gate of a field-effect transistor compare with the DC input impedance of a bipolar transistor?**

A. They cannot be compared without first knowing the supply voltage
B. An FET has low input impedance; a bipolar transistor has high input impedance

C. An FET has high input impedance; a bipolar transistor has low input impedance
D. The input impedance of FETs and bipolar transistors is the same

**ANSWER C:** An FET is much easier to work with in circuits than the simple bipolar transistor. Its higher input impedance is less likely to load down a circuit that it is attached to, and the FET can also handle big swings in signal levels.

**Field-Effect Transistor (FET) Basics**

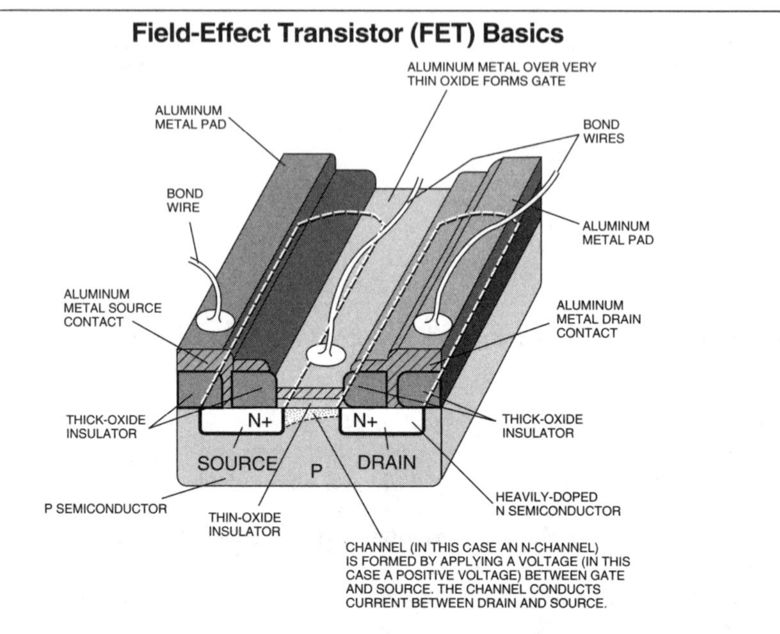


**Pictorial of FET Construction (N-Channel Enhancement)**

Unlike bipolar transistors that depend on current into the base to control collector current, field-effect transistor current between source and drain is controlled by a voltage on a gate. The basic structure of an N-channel MOSFET is shown here. Heavily-doped N semiconductor material forms source and drain regions in a P semiconductor material substrate. The region between the source and drain is the gate region, where a thin layer of oxide insulates the P semiconductor substrate underneath from a metal plate that is deposited over the thin oxide. A thick oxide layer over the source and drain regions insulates metal connection pads from the substrate. Holes in this thick oxide layer allow the metal pads to contact the source and drain. There are four common types of MOSFETs: P-channel depletion and enhancement mode devices; and N-channel depletion and enhancement mode devices.

Source: *Basic Communications Electronics*, Hudson & Luecke, © 1999 Master Publishing, Inc., Lincolnwood, IL

**E6A14 What two elements widely used in semiconductor devices exhibit both metallic and nonmetallic characteristics?**

A. Silicon and gold
B. Silicon and germanium
C. Galena and germanium
D. Galena and bismuth

**ANSWER B:** Both silicon and germanium exhibit a combination of metallic and non-metallic properties. Think of silicon rectifiers and germanium diodes.

**E6A15 What type of semiconductor material contains fewer free electrons than pure germanium or silicon crystals?**

A. N-type
B. P-type
C. Superconductor-type
D. Bipolar-type

**ANSWER B:** If there are more holes than electrons, it is a P-type semiconductor.

**E6A16 What are the majority charge carriers in N-type semiconductor material?**

A. Holes
B. Free electrons
C. Free protons
D. Free neutrons

**ANSWER B:** In an N-type semiconductor material, there are more free electrons than holes.

**E6A17 What are the three terminals of a field-effect transistor?**

A. Gate 1, gate 2, drain
B. Emitter, base, collector
C. Emitter, base 1, base 2
D. Gate, drain, source

**ANSWER D:** Gosh darn semiconductors! Gate, drain, source. Gosh darn field effect transistor, a semiconductor!

***E6B Diodes: Zener, tunnel, varactor, hot-carrier, junction, point contact, PIN and light emitting; operational amplifiers (inverting amplifiers, noninverting amplifiers, voltage gain, frequency response, FET amplifier circuits, single-stage amplifier applications); phase-locked loops***

**E6B01 What is the principal characteristic of a Zener diode?**

A. A constant current under conditions of varying voltage
B. A constant voltage under conditions of varying current
C. A negative resistance region
D. An internal capacitance that varies with the applied voltage

**ANSWER B:** Always remember that a Zener diode is for voltage regulation, for constant voltage even though current changes.

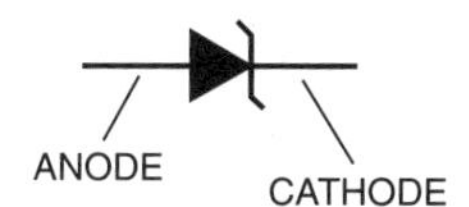


Here is the schematic symbol of a Zener diode. Since a diode only passes energy in one direction, look for that one-way arrow, plus a "Z" indicating it is a Zener diode. Doesn't that vertical line look like a tiny "Z".

**Zener Diode**

**E6B02 What is the principal characteristic of a tunnel diode?**

A. A high forward resistance
B. A very high PIV
C. A negative resistance region
D. A high forward current rating

**ANSWER C:** As a tunnel diode is conducting current, there is a spot where current increases as the voltage drop across the diode decreases. They call this a negative resistance region.

**E6B03 What special type of diode is capable of both amplification and oscillation?**

A. Point contact
B. Zener
C. Tunnel
D. Junction

**ANSWER C:** The tunnel diode is commonly used in both amplifier and oscillator circuits. The negative resistance region is particularly useful in oscillators.

**E6B04 What type of semiconductor diode varies its internal capacitance as the voltage applied to its terminals varies?**

A. Varactor
B. Tunnel
C. Silicon-controlled rectifier
D. Zener

**ANSWER A:** We use the varactor diode to tune VHF and UHF circuits by varying the voltage applied to the varactor diode.

---

**E6B05 In Figure E6-3, what is the schematic symbol for a varactor diode?**

A. 8
B. 6
C. 2
D. 1

**ANSWER D:** The internal capacitance of a varactor diode varies as the voltage applied to its terminals changes. Your author builds microwave equipment with varactor diodes, so be sure to identify the proper symbol. The symbol is essentially a capacitor and a diode combined.

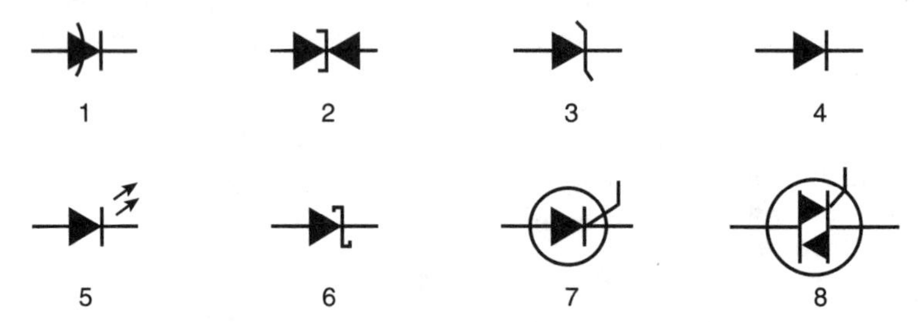


Figure E6-3

---

**E6B06 What is a common use of a hot-carrier diode?**

A. As balanced mixers in FM generation
B. As a variable capacitance in an automatic frequency control circuit
C. As a constant voltage reference in a power supply
D. As VHF and UHF mixers and detectors

**ANSWER D**: When you get involved with VHF and UHF equipment, you will find some common uses of the hot-carrier diode because of its low noise figure characteristics.

---

**E6B07 What limits the maximum forward current rating in a junction diode?**

A. Peak inverse voltage
B. Junction temperature
C. Forward voltage
D. Back EMF

**ANSWER B:** Guess what will kill most electronic components? High temperature! The limit of the maximum forward current in a junction diode is the junction temperature. This is why you will find these diodes mounted to the chassis of your equipment. The chassis acts as a heat sink to keep the junction temperature from exceeding its maximum limit.

---

**E6B08 Structurally, what are the two main categories of semiconductor diodes?**

A. PN junction and metal-semiconductor junction
B. Electrolytic and PN junction
C. CMOS-field effect and metal-semiconductor junction
D. Vacuum and point contact

**ANSWER A:** The two main categories of semiconductor diodes are PN junction and metal-semiconductor junction. Look at the incorrect answers and you will see they deal with capacitors, transistors, and maybe even a tube!

**E6B09 What is a common use for point contact diodes?**

A. As a constant current source
B. As a constant voltage source
C. As an RF detector
D. As a high voltage rectifier

**ANSWER C:** The little point contact diode, since it only conducts in one direction, may be used as an RF detector in some VHF and UHF equipment.

**E6B10 In Figure E6-3, what is the schematic symbol for a light-emitting diode?**

A. 1
B. 5
C. 6
D. 7

**ANSWER B:** The light-emitting diode (LED) is a giveaway on the schematic diagram! Look at the two arrows indicating an illuminating device.

**E6B11 What voltage gain can be expected from the circuit in Figure E6-4 when $R_1$ is 10 ohms and $R_F$ is 470 ohms?**

A. 0.21
B. 94
C. 47
D. 24

**ANSWER C:** Take a look at the operational amplifier basics chart, and remember the gain formula of an inverting IC op amp:

$G = -R_F \div R_1 = R_F/R_1.$

The minus sign means the output is out of phase with the input. $R_F = 470$ ohms; R1 = 10 ohms. Divide 10 into 470, coming out 47. On a calculator, the keystrokes are: Clear, 470 ÷ 10 = 47. Not really a brain-buster, was it?

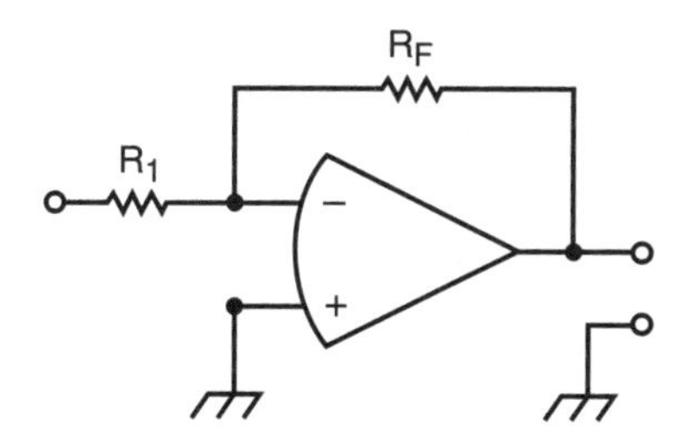


**Figure E6-4**

**E6B12 How does the gain of a theoretically ideal operational amplifier vary with frequency?**

A. It increases linearly with increasing frequency
B. It decreases linearly with increasing frequency
C. It decreases logarithmically with increasing frequency
D. It does not vary with frequency

**ANSWER D:** The gain on an ideal op amp should not vary with frequency.

## Operational Amplifier Basics

**Ideal Operational Amplifier**

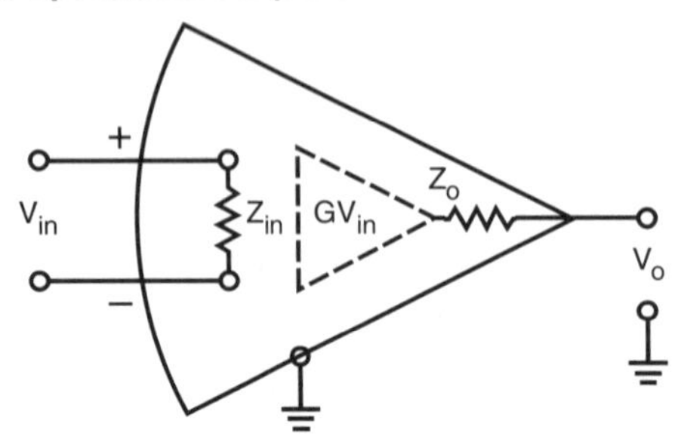


$Z_{in}$ = Infinity

G = Gain = Infinity

$Z_o$ = Zero

Bandwith = Infinity

$V_o$ has no offset ($V_o$=0 when $V_{in}$=0)

**IC Operational Ampliifer**

Even though IC operational amplifiers do not meet all the ideal specifications, G and $Z_{in}$ are very large. Because G is very large, any small input voltage would drive the output into saturation (for practical supply voltages); therefore, normal operational amplifier operation is with feedback to set the gain. Here's an example for an inverting amplifier:

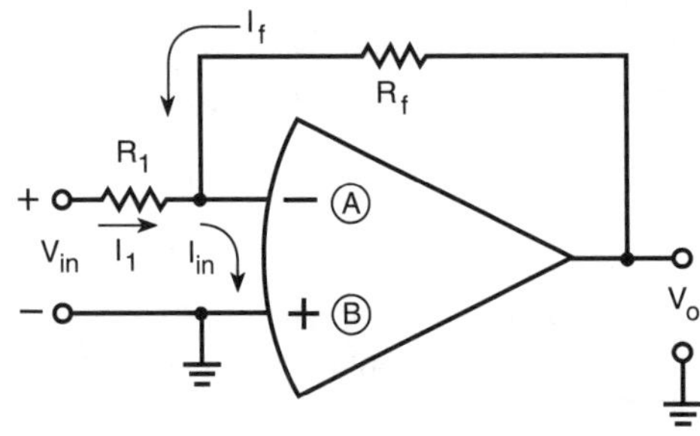


Since $Z_{in}$ is very large, $I_{in} = 0$

$$\therefore I_1 + I_f = 0 \qquad I_1 = \frac{V_{in}}{R_1} \qquad I_f = \frac{V_o}{R_f}$$

$$\therefore \frac{V_{in}}{R_1} + \frac{V_o}{R_f} = 0 \qquad \frac{V_o}{R_f} = -\frac{V_{in}}{R_1} \qquad \therefore \frac{V_o}{V_{in}} = -\frac{R_f}{R_1}$$

Ⓐ Inverting Input — An input voltage that is more positive on this input will cause the output voltage to be less positive.

Ⓑ Non-Inverting Input — An input voltage that is more positive on this input will cause the output voltage to be more positive.

The gain of an *inverting IC operational amplifier* is:

**E** $$G = -\frac{R_f}{R_1}$$

The minus sign means the output is out of phase with the input.

**E6B13 What essentially determines the output impedance of a FET common-source amplifier?**

A. The drain resistor
B. The input impedance of the FET
C. The drain supply voltage
D. The gate supply voltage

**ANSWER A:** The resistance connected to the drain (drain resistor) determines the output impedance of a FET common-source amplifier.

**E6B14 What will be the voltage of the circuit shown in Figure E6-4 if $R_1$ is 1000 ohms and RF is 10,000 ohms and 0.23 volts is applied to the input?**

A. 0.23 volts
B. 2.3 volts
C. −0.23 volts
D. −2.3 volts

**ANSWER D:** This time we work the formula: $V_{OUT}/V_{IN} = -R_F/R_1$, when R1 = 1000, and $R_F$ = 10,000, the gain of the op amp is $V_{OUT}/V_{IN} = -10$. As a result, the output = −10 × VIN = −10 × 0.23 = −2.3 volts.

---

**E6B15 What voltage gain can be expected from the circuit in Figure E6-4 when R1 is 1800 ohms and RF is 68 kilohms?**

A. 1
B. 0.03
C. 38
D. 76

**ANSWER C:** Now we're back to the gain of an inverting IC operational amplifier. Simply divide 1,800 ohms into 68,000 ohms. Your calculator keystrokes are: clear, 68000 ÷ 1800 = 37.777, with your correct answer of 38 rounded off. Simple, huh?

---

**E6B16 What voltage gain can be expected from the circuit in Figure E6-4 when R1 is 3300 ohms and RF is 47 kilohms?**

A. 28
B. 14
C. 7
D. 0.07

**ANSWER B:** One more time, an easy one - divide the small number (3,300) into the large number (47,000), and you end up with 14.24, with the correct answer of 14.

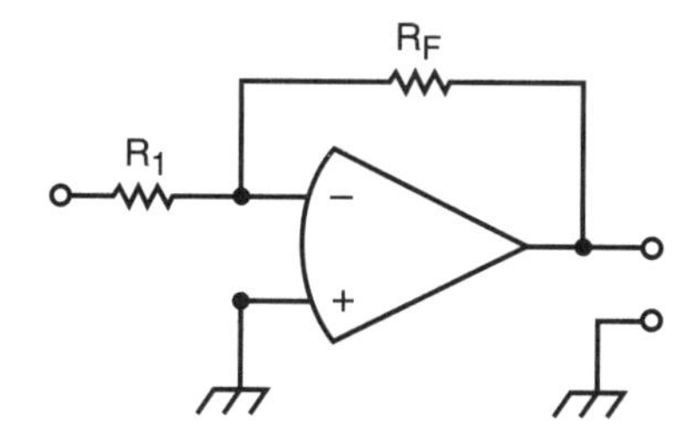


**Figure E6-4**

---

**E6B17 – Question deleted from Element 4 Pool by QPC.**

---

**E6B18 Which of the following circuits is used to recover audio from an FM voice signal?**

A. A doubly balanced mixer
B. A phase-locked loop
C. A differential voltage amplifier
D. A variable frequency oscillator

**ANSWER B:** The phase-locked loop may also operate as an FM detector, similar to the direct frequency modulation circuit in the PLL section of a transmitter. Voltage at the phase detector becomes demodulated audio within the phase-locked loop circuitry on an FM receiver.

**E6B19 What is the capture range of a phase-locked loop circuit?**

A. The frequency range over which the circuit can lock
B. The voltage range over which the circuit can lock
C. The input impedance range over which the circuit can lock
D. The range of time it takes the circuit to lock

**ANSWER A:** When a phase-locked loop circuit drops out of lock, it has exceeded the frequency range over which the circuit can effectively lock. The "capture range" is the frequency range over which a PLL circuit will lock.

**Modern transcievers, like this small hand-held, use PLLs to lock on to the receiving frequency.**

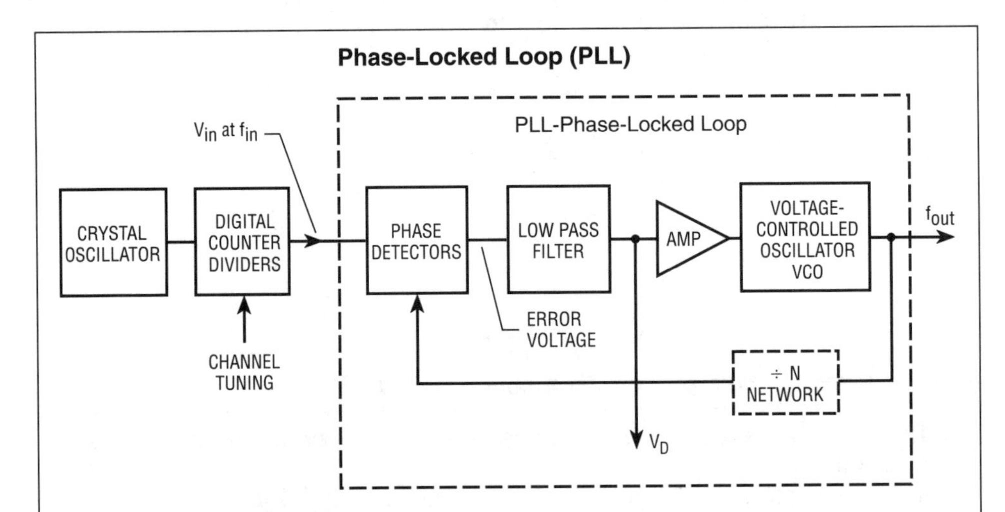


In a phase-locked loop the output frequency of the VCO is locked by phase and frequency to the frequency of the input signal $V_{in}$.

By using a very accurate crystal controlled source and dividing it down (or multiplying it up), the output frequency, $f_{out}$, of the PLL will be controlled very accurately and with rigid stability to the input frequency, $f_{in}$. The phase detector develops an error voltage that is determined by the frequencies and phase of the two inputs. The feedback system wants to reduce the error voltage to zero.

$V_D$ is a voltage proportional to the frequency changes occuring in $V_{in}$; therefore, the PLL can be used as an FM demodulator.

With divide by N network, $f_{out}$ will be $f_{in}$ multiplied by N.

Channel tuning occurs in the Digital Counter Dividers.

**E6B20 How are junction diodes rated?**

A. Maximum forward current and capacitance
B. Maximum reverse current and PIV
C. Maximum reverse current and capacitance
D. Maximum forward current and PIV

**ANSWER D:** When choosing a junction diode, you will need to know how much forward current will be passing through the diode, and the amount of peak inverse voltage (PIV) that the diode must stand in its non-conducting direction.

---

**E6B21 What is one common use for PIN diodes?**

A. As a constant current source
B. As a constant voltage source
C. As an RF switch
D. As a high voltage rectifier

**ANSWER C:** PIN diodes may be used in small handheld transceivers for RF switching. This gets away from mechanical relays for switching.

---

**E6B22 What type of bias is required for an LED to produce luminescence?**

A. Reverse bias
B. Forward bias
C. Zero bias
D. Inductive bias

**ANSWER B:** The LED requires forward bias in order to illuminate. As indicated in symbol 5 of Figure E6-3, the LED is a diode, and it must be forward biased to make it give off light.

---

**E6B23 What is an operational amplifier?**

A. A high-gain, direct-coupled differential amplifier whose characteristics are determined by components external to the amplifier
B. A high-gain, direct-coupled audio amplifier whose characteristics are determined by components external to the amplifier
C. An amplifier used to increase the average output of frequency modulated amateur signals to the legal limit
D. A program subroutine that calculates the gain of an RF amplifier

**ANSWER A:** The operational amplifier (op amp) is a direct-coupled differential amplifier that offers high gain and high input impedance. The "differential" wording refers to the op-amp input design where the output is determined by the difference voltage between the two inputs. Different op-amp circuits and characteristics are determined by the external components connected to the amplifier.

---

**E6B24 What is meant by the term op-amp input-offset voltage?**

A. The output voltage of the op-amp minus its input voltage
B. The difference between the output voltage of the op-amp and the input voltage required in the following stage
C. The potential between the amplifier input terminals of the op-amp in a closed-loop condition
D. The potential between the amplifier input terminals of the op-amp in an open-loop condition

**ANSWER C:** If the op amp has been constructed properly, you should see almost no voltage between the amplifier input terminals when the feedback loop is closed.

---

**E6B25 What is the input impedance of a theoretically ideal op-amp?**

A. 100 ohms
B. 1000 ohms
C. Very low
D. Very high

**ANSWER D:** Remember that the input impedance to the ideal op amp is always H<u>i</u> <u>i</u>N.

---

**E6B26 What is the output impedance of a theoretically ideal op-amp?**

A. Very low
B. Very high
C. 100 ohms
D. 1000 ohms

**ANSWER A:** Remember that the output impedance of the ideal op amp is always Low out.

---

**E6B27 What is a phase-locked loop circuit?**

A. An electronic servo loop consisting of a ratio detector, reactance modulator, and voltage-controlled oscillator
B. An electronic circuit also known as a monostable multivibrator
C. An electronic servo loop consisting of a phase detector, a low-pass filter and voltage-controlled oscillator
D. An electronic circuit consisting of a precision push-pull amplifier with a differential input

**ANSWER C:** Few other circuits have changed the course of Amateur Radio equipment as the phase-locked loop (PLL) circuit. There is only one correct answer with the word phase in it, so look for the word phase when you read phase-locked loop in the question.

---

**E6B28 What functions are performed by a phase-locked loop?**

A. Wide-band AF and RF power amplification
B. Comparison of two digital input signals, digital pulse counter
C. Photovoltaic conversion, optical coupling
D. Frequency synthesis, FM demodulation

**ANSWER D:** In building his first external frequency synthesizer, your author was able to "synthesize" hundreds of channels with just a couple of crystals. How many of you remember being "rock bound" with an old rig?

---

***E6C TTL digital integrated circuits; CMOS digital integrated circuits; gates***

---

**E6C01 What is the recommended power supply voltage for TTL series integrated circuits?**

A. 12 volts
B. 1.5 volts
C. 5 volts
D. 13.6 volts

**ANSWER C:** If you ever worked on TTL (transistor-transistor logic) circuits, you know the importance of those 5-volt regulators that always seem to burn up at the wrong time. TTL circuits run on 5.0 volts.

---

**E6C02 What logic state do the inputs of a TTL device assume if they are left open?**

A. A high-logic state
B. A low-logic state
C. The device becomes randomized and will not provide consistent high or low-logic states
D. Open inputs on a TTL device are ignored

**ANSWER A:** Here are several key words – inputs open and high state. Remember four letters make up the word open, and four letters make up the word high.

---

**E6C03 What level of input voltage is high in a TTL device operating with a 5-volt power supply?**

A. 2.0 to 5.5 volts
B. 1.5 to 3.0 volts
C. 1.0 to 1.5 volts
D. -5.0 to -2.0 volts

**ANSWER A:** If the input voltage is at the high level, depending on the number of circuits connected at the point, by specification the level will be between 2.0 volts and 5.5 volts.

---

**E6C04 What level of input voltage is low in a TTL device operating with a 5-volt power-supply?**

A. -2.0 to -5.5 volts
B. 2.0 to 5.5 volts
C. 0.0 to 0.8 volts
D. -0.8 to 0.4 volts

**ANSWER C:** If the input voltage to a TTL IC is at the low level, it will usually be around +0.2 volts. It can be as high as +0.8 volts, but then the noise margin is down to zero.

---

**E6C05 What is NOT a major advantage of CMOS over other devices?**

A. Small size
B. Low power consumption
C. Low cost
D. Differential output

**ANSWER D:** The CMOS circuit is small, consumes little power, but something not considered a major advantage of the CMOS is differential output.

---

**E6C06 Why do CMOS digital integrated circuits have high immunity to noise on the input signal or power supply?**

A. Larger bypass capacitors are used in CMOS circuit design
B. The input switching threshold is about two times the power supply voltage
C. The input switching threshold is about one-half the power supply voltage
D. Input signals are stronger

**ANSWER C:** The CMOS device is relatively immune to noise because its switching threshold is one-half the power supply voltage. As a result, power supply noise transients or input transients much larger than other logic circuit types will not cause a transition in the input state.

---

**E6C07 In Figure E6-5, what is the schematic symbol for an AND gate?**

A. 1
B. 2
C. 3
D. 4

**ANSWER A:** AND gate looks like the letter D, with two inputs or more, and one output.

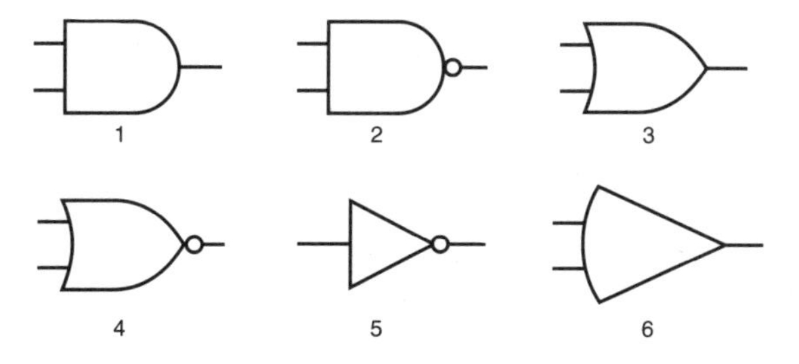


**Figure E6-5**

---

**E6C08 In Figure E6-5, what is the schematic symbol for a NAND gate?**

A. 1
B. 2
C. 3
D. 4

**ANSWER B:** Since the NAND gate has a logic "0" at its output, spot the little tiny "0" on its nose. It looks like the letter D, the same as the AND gate, except for the small circle at its output.

---

**E6C09 In Figure E6-5, what is the schematic symbol for an OR gate?**

A. 2
B. 3
C. 4
D. 6

**ANSWER B:** The word "OR" starts with an "O", and its input is curved like a portion of the letter "O". The OR gate symbol has a curved input and a pointed output.

---

**E6C10 In Figure E6-5, what is the schematic symbol for a NOR gate?**

A. 1
B. 2
C. 3
D. 4

**ANSWER D:** Since the NOR gate will give us a logic "0" at its output if any inputs are logic "1", look for the little "O" on the nose, and a concave input. It looks just like an OR gate, except for the small circle on its nose.

---

**E6C11 In Figure E6-5, what is the schematic symbol for the NOT operation (inverter)?**

A. 2
B. 4
C. 5
D. 6

**ANSWER C:** The NOT gate is triangular in shape and has a little "O" on its nose. It looks altogether different from any of the other logic gates. You will probably have one question based on gate symbols. Make up flashcards, and see if you can easily call out which is which.

---

***E6D Vidicon and cathode-ray tube devices; charge-coupled devices (CCDs); liquid crystal displays (LCDs); toroids: permeability, core material, selecting, winding***

---

**E6D01 How is the electron beam deflected in a vidicon?**

A. By varying the beam voltage
B. By varying the bias voltage on the beam forming grids inside the tube
C. By varying the beam current
D. By varying electromagnetic fields

**ANSWER D:** The electron beam is deflected by varying electromagnetic fields in coils surrounding the tube.

---

**E6D02 What is cathode ray tube (CRT) persistence?**

A. The time it takes for an image to appear after the electron beam is turned on
B. The relative brightness of the display under varying conditions of ambient light
C. The ability of the display to remain in focus under varying conditions
D. The length of time the image remains on the screen after the beam is turned off

**ANSWER D:** The term "persistence" in a cathode-ray tube is the length of time that an image will remain on the screen after the beam gets turned off.

---

**E6D03 If a cathode ray tube (CRT) is designed to operate with an anode voltage of 25,000 volts, what will happen if the anode voltage is increased to 35,000 volts?**

A. The image size will decrease and the tube will produce X-rays
B. The image size will increase and the tube will produce X-rays
C. The image will become larger and brighter
D. There will be no apparent change

**ANSWER A:** If the anode voltage on a CRT is increased from a normal 25,000 to 35,000 volts, the image will slightly shrink, and the tube could produce harmful X-rays.

**E6D04 Exceeding what design rating can cause a cathode ray tube (CRT) to generate X-rays?**

A. The heater voltage
B. The anode voltage
C. The operating temperature
D. The operating frequency

**ANSWER B:** X-rays can be emitted from a cathode-ray tube if the anode voltage is increased beyond the value at which the tube is safely rated.

---

**E6D05 Which of the following is true of a charge-coupled device (CCD)?**

A. Its phase shift changes rapidly with frequency
B. It is a CMOS analog-to-digital converter
C. It samples an analog signal and passes it in stages from the input to the output
D. It is used in a battery charger circuit

**ANSWER C:** A charge-coupled device (CCD) samples an analog signal and passes it in stages from input to output. CCDs are used in digital still and video cameras to capture images.

---

**E6D06 What function does a charge-coupled device (CCD) serve in a modern video camera?**

A. It stores photogenerated charges as signals corresponding to pixels
B. It generates the horizontal pulses needed for electron beam scanning
C. It focuses the light used to produce a pattern of electrical charges corresponding to the image
D. It combines audio and video information to produce a composite RF signal

**ANSWER A:** The charge-coupled device is continually clocking the data out, or the data "dissipates" like a dynamic RAM. The CCD converts photo-generated charges into signals corresponding to pixels.

---

**E6D07 What is a liquid-crystal display (LCD)?**

A. A modern replacement for a quartz crystal oscillator which displays its fundamental frequency
B. A display that uses a crystalline liquid to change the way light is refracted
C. A frequency-determining unit for a transmitter or receiver
D. A display that uses a glowing liquid to remain brightly lit in dim light

**ANSWER B:** A liquid-crystal display, found on most amateur radio handheld units, uses a crystalline liquid to change the way light is refracted from a rear mirror through the liquid.

**Liquid Crystal Displays are popular because of their large size and low power consumption**

---

**E6D08 What material property determines the inductance of a toroidal inductor with a 10-turn winding?**

A. Core load current
B. Core resistance
C. Core reactivity
D. Core permeability

**ANSWER D:** The key words in the question are "material property," and the material that determines the inductance of a toroid is the core permeability of the iron donut.

**Ferrite Toroids**
Courtesy of Palomar Engineers

**E6D09 By careful selection of core material, over what frequency range can toroidal cores produce useful inductors?**

A. From a few kHz to no more than several MHz
B. From 100 Hz to at least 1000 MHz
C. From 100 Hz to no more than 3000 kHz
D. From a few hundred MHz to at least 1000 GHz

**ANSWER B:** If you carefully select the core material, you should be able to go from 100 Hz (nearly DC) to at least 1,000 MHz.

**E6D10 What is one important reason for using powdered-iron toroids rather than ferrite toroids in an inductor?**

A. Powdered-iron toroids generally have greater initial permeabilities
B. Powdered-iron toroids generally have better temperature stability
C. Powdered-iron toroids generally require fewer turns to produce a given inductance value
D. Powdered-iron toroids are easier to use with surface-mount technology

**ANSWER B:** Powdered-iron toroid donuts generally have much better temperature stability than a straight ferrite toroid.

**E6D11 What devices are commonly used as VHF and UHF parasitic suppressors at the input and output terminals of transistorized HF amplifiers?**

A. Electrolytic capacitors
B. Butterworth filters
C. Ferrite beads
D. Steel-core toroids

**ANSWER C:** If you look into the modern VHF/UHF single-band or dual-band amplifier, you will see many leads dressed on a small ferrite bead to minimize parasitics coming down voltage or control lines.

**E6D12 What is a primary advantage of using a toroidal core instead of a solenoidal core in an inductor?**

A. Toroidal cores contain most of the magnetic field within the core material
B. Toroidal cores make it easier to couple the magnetic energy into other components
C. Toroidal cores exhibit greater hysteresis
D. Toroidal cores have lower Q characteristics

**ANSWER A:** The toroidal inductor concentrates the magnetic field within the core material, protecting other nearby components from stray magnetic fields from that inductor.

**E6D13 How many turns will be required to produce a 1-mH inductor using a ferrite toroidal core that has an inductance index ($A_L$) value of 523 millihenrys/1000 turns?**

A. 2 turns
B. 4 turns
C. 43 turns
D. 229 turns

**ANSWER C:** Here is the formula to calculate the number of turns required to produce a specific inductance where you know the inductance index, $A_L$, of the material. $A_L$ is determined and published by the core manufacturer and is an index that is the microhenrys of inductance per 1000 turns or millihenrys of inductance per 1000 turns of single layer wire on the core.

$$N = 1000 \sqrt{\frac{L}{A_L}}$$

where: N is number of turns needed
L is inductance in **millihenrys** of core inductor
$A_L$ (A sub L) is inductance index in **mH per 1000 turns**

$$N = 1000 \sqrt{\frac{1}{523}}$$

$$N = 1000 \sqrt{0.001912} = 1000 \sqrt{19.12 \times 10^{-4}}$$

$$N = 1 \times 10^3 \times 4.37 \times 10^{-2}$$

$$N = 4.37 \times 10^1 = 43.7$$

Here's how to work it out on your calculator. Hit the clear button. Then enter 1 ÷ 523 =. The result is 0.001912. Press $\sqrt{\ }$ and the result is 0.0437264. Press × 1000 = N for the answer 43.7264. Your closest correct answer is 43.

---

**E6D14 How many turns will be required to produce a 5-microhenry inductor using a powdered-iron toroidal core that has an inductance index ($A_L$) value of 40 microhenrys/100 turns?**

A. 35 turns
B. 13 turns
C. 79 turns
D. 141 turns

**ANSWER A:** This time we are talking about calculating the number of turns in a powdered-iron toroidal core. What we just worked with on in the last question as a ferrite toroidal core. The tip-off if you miss the wording "powdered-iron" is that we only have 5 microhenrys after a bunch of turns, rather than millihenrys with a ferrite core. Powdered-iron has better temperature stability, but the overall inductance will be dramatically less. Here's your next formula:

$$N = 100 \sqrt{\frac{L}{A_L}}$$

where: N is number of turns
L is inductance in **microhenrys**
$A_L$ (A sub L) is inductance index in **µH per 100 turns**

$$N = 100 \sqrt{\frac{5}{40}}$$

$$N = 100 \sqrt{0.125} = 100 \sqrt{12.5 \times 10^{-2}}$$

$$N = 3.53 \times 10^{-1} \times 10^2 = 3.53 \times 10^1 = 35.3$$

Now that wasn't so hard, was it? The formulas are almost the same, and luckily on both of these problems if you should reverse the formulas in error, they do not have an incorrect answer just waiting for you to get it wrong.

**Maximum Number of Turns Using Popular Toroidal Cores**

| Wire Size | Core Size T-80 | T-68 | T-50 | T-37 | T-25 |
|---|---|---|---|---|---|
| 12 | 14 | 9 | 6 | 3 | |
| 14 | 18 | 13 | 8 | 5 | 1 |
| 16 | 24 | 17 | 13 | 7 | 2 |
| 18 | 32 | 23 | 18 | 10 | 4 |
| 22 | 53 | 38 | 30 | 19 | 9 |
| 28 | 108 | 80 | 64 | 42 | 23 |
| 32 | 171 | 127 | 103 | 68 | 38 |

**$A_L$ — μH per 100 Turns — for Popular Toroidal Cores**

| CORE SPECS ▶ | MIX 6 | MIX 10 | MIX 12 | MIX 0 | | | |
|---|---|---|---|---|---|---|---|
| Core Size ▼ | *μ = 8.5 2-30 MHz | μ = 6.0 10-100 MHz | μ = 4.0 20-200 MHz | μ = 1.0 50-250 MHz | SIZE (INCHES) OD | ID | Ht |
| T-80 | 45 | 32 | 22 | 8.5 | 0.80 | 0.50 | 0.25 |
| T-68 | 47 | 32 | 21 | 7.5 | 0.68 | 0.37 | 0.19 |
| T-50 | 40 | 31 | 18 | 6.4 | 0.50 | 0.30 | 0.19 |
| T-37 | 30 | 25 | 15 | 4.9 | 0.37 | 0.20 | 0.13 |
| T-25 | 27 | 19 | 12 | 4.5 | 0.25 | 0.12 | 0.10 |

*permeability

**E6D15 What type of CRT deflection is better when high-frequency waves are to be displayed on the screen?**

A. Electromagnetic
B. Tubular
C. Radar
D. Electrostatic

**ANSWER D:** Cathode-ray tube (CRT) electrostatic deflection is better for high-frequency signals. Horizontal and vertical deflection plates, similar to capacitor plates, deflect the electron beam onto the screen according to signal voltages applied to the plates.

---

**E6D16 Which is NOT true of a charge-coupled device (CCD)?**

A. It uses a combination of analog and digital circuitry
B. It can be used to make an audio delay line
C. It can be used as an analog-to-digital converter
D. It samples and stores analog signals

**ANSWER C:** This question asks which is NOT true of a CCD. A charge-coupled device cannot be used as an analog-to-digital converter.

---

**E6D17 What is the principle advantage of liquid-crystal display (LCD) devices?**

A. They consume low power
B. They can display changes instantly
C. They are visible in all light conditions
D. They can be easily interchanged with other display devices

**ANSWER A:** It takes very little current to sustain a liquid crystal display segment.

**E6D18 What is one important reason for using ferrite toroids rather than powdered-iron toroids in an inductor?**

A. Ferrite toroids generally have lower initial permeabilities
B. Ferrite toroids generally have better temperature stability
C. Ferrite toroids generally require fewer turns to produce a given inductance value
D. Ferrite toroids are easier to use with surface mount technology

**ANSWER C:** Another popular feature of using ferrite toroids over powdered-iron toroids is the fact that ferrite toroids require less turns of wire to produce a given inductance value.

***E6E Quartz crystal (frequency determining properties as used in oscillators and filters); monolithic amplifiers (MMICs)***

**E6E01 For single-sideband phone emissions, what would be the bandwidth of a good crystal lattice band-pass filter?**

A. 6 kHz at -6 dB
B. 2.1 kHz at -6 dB
C. 500 Hz at -6 dB
D. 15 kHz at -6 dB

**ANSWER B:** A single sideband emission is normally about 3 kHz wide. Without making the voice sound pinched, a 2.1 kHz filter is what comes with most worldwide sets.

**E6E02 For double-sideband phone emissions, what would be the bandwidth of a good crystal lattice band-pass filter?**

A. 1 kHz at -6 dB
B. 500 Hz at -6 dB
C. 6 kHz at -6 dB
D. 15 kHz at -6 dB

**ANSWER C:** Believe it or not, there are still some portions on the worldwide ham bands where AM, double-sideband, full-carrier phone emission still exists. This is twice as wide as SSB, and would require a 6 kHz band-pass filter with -6 dB skirts to pass the signal.

**E6E03 What is a crystal lattice filter?**

A. A power supply filter made with interlaced quartz crystals
B. An audio filter made with four quartz crystals that resonate at 1-kHz intervals
C. A filter with wide bandwidth and shallow skirts made using quartz crystals
D. A filter with narrow bandwidth and steep skirts made using quartz crystals

**ANSWER D:** The new worldwide set you purchase will probably have some good SSB and CW filters built in. The filters have a nice narrow bandwidth and steep skirts. The filters are made up of quartz crystals. You can buy accessory filters for a tighter response, but many times these will make your received signals sound pinched. Unless you do a lot of CW work, to get started, stay with the filters that come with the set.

**E6E04 What technique is used to construct low-cost, high-performance crystal ladder filters?**

A. Obtain a small quantity of custom-made crystals
B. Choose a crystal with the desired bandwidth and operating frequency to match a desired center frequency
C. Measure crystal bandwidth to ensure at least 20% coupling
D. Measure crystal frequencies and carefully select units with a frequency variation of less than 10% of the desired filter bandwidth

**ANSWER D:** High-performance crystal ladder filters are selected from crystals that have a frequency variation of less than 10 percent of the desired filter bandwidth.

**E6E05 Which of the following factors has the greatest effect in helping determine the bandwidth and response shape of a crystal ladder filter?**

A. The relative frequencies of the individual crystals
B. The DC voltage applied to the quartz crystal
C. The gain of the RF stage preceding the filter
D. The amplitude of the signals passing through the filter

**ANSWER A:** The bandwidth and response of a crystal filter will relate to the frequency of the individual crystals within that filter unit. (Yes, *ladder* filter.)

---

**E6E06 What is the piezoelectric effect?**

A. Physical deformation of a crystal by the application of a voltage
B. Mechanical deformation of a crystal by the application of a magnetic field
C. The generation of electrical energy by the application of light
D. Reversed conduction states when a P-N junction is exposed to light

**ANSWER A:** If you apply a voltage to a quartz crystal, it will vibrate at a specific frequency. This is the reason crystal oscillators are so accurate. They oscillate at the frequency, or harmonics of the frequency, of the crystal, and continue to do so with great accuracy unless the temperature changes outside design limits.

---

**E6E07 What is the characteristic impedance of circuits in which MMICs are designed to work?**

A. 50 ohms
B. 300 ohms
C. 450 ohms
D. 10 ohms

**ANSWER A:** The beauty of the MMIC is that it has a characteristic impedance of 50 ohms, and this is what the microwavers love!

---

**E6E08 What is the typical noise figure of a monolithic microwave integrated circuit (MMIC) amplifier?**

A. Less than 1 dB
B. Approximately 3.5 to 6 dB
C. Approximately 8 to 10 dB
D. More than 20 dB

**ANSWER B:** The monolithic microwave integrated circuit (MMIC) has a typical noise floor from 3.5 to 6 dB, but they offer good gain from 8 dB to as high as 24 dB up to 1,000 MHz, so this typical noise figure is acceptable.

---

**E6E09 What type of amplifier device consists of a small pill sized package with an input lead, an output lead and 2 ground leads?**

A. A junction field-effect transistor (JFET)
B. An operational amplifier integrated circuit (OAIC)
C. An indium arsenide integrated circuit (IAIC)
D. A monolithic microwave integrated circuit (MMIC)

**ANSWER D:** You can spot the MMIC device by its input lead, output lead, and 2 ground leads in a "pill-sized" package.

---

**E6E10 What typical construction technique do amateurs use when building an amplifier for the microwave bands containing a monolithic microwave integrated circuit (MMIC)?**

A. Ground-plane "ugly" construction
B. Microstrip construction
C. Point-to-point construction
D. Wave-soldering construction

**ANSWER B:** Up at microwave frequencies, we use microstrip construction, which is ideal for the MMIC.

---

**Monolithic Microwave Integrated Circuit**

Photo Courtesy of Hewlett Packard Co.

**E6E11 How is the operating bias voltage supplied to a monolithic microwave integrated circuit (MMIC) that uses four leads?**

A. Through a resistor and RF choke connected to the amplifier output lead
B. MMICs require no operating bias
C. Through a capacitor and RF choke connected to the amplifier input lead
D. Directly to the bias-voltage (Vcc IN) lead

**ANSWER A:** The operating bias voltage for the MMIC is developed through a resistor and RF choke connected to the amplifier output lead.

**E6E12 How is the DC power from a voltage source fed to a monolithic microwave integrated circuits (MMIC)?**

A. Through a coupling capacitor
B. Through a PIN diode
C. Through a silicon-controlled rectifier
D. Through a resistor

**ANSWER D:** A resistor in series with a voltage source and MMIC acts as a current limiter and load for the circuit. The resistor is similar to what might be found off of the collector in a transistor.

**E6E13 What supply voltage do monolithic microwave integrated circuits (MMIC) amplifiers typically require?**

A. 1 volt DC
B. 12 volts DC
C. 20 volts DC
D. 120 volts DC

**ANSWER B:** A MMIC has a low noise figure, excellent power gain, and at the linear-regulated 12V voltage source, may draw only 18ma.

**E6E14 What is the most common package for inexpensive monolithic microwave integrated circuit (MMIC) amplifiers?**

A. Beryllium oxide packages
B. Glass packages
C. Plastic packages
D. Ceramic packages

**ANSWER C:** While the military may use ceramic packaging for the MMIC, plastic packaging dramatically decreases the expense of a monolithic microwave integrated circuit.

## Subelement E7 – Practical Circuits [7 Exam Questions — 7 Groups]

### *E7A Digital logic circuits: Flip flops; Astable and monostable multivibrators; Gates (AND, NAND, OR, NOR); Positive and negative logic*

**E7A01 What is a bistable multivibrator circuit?**

A. An "AND" gate
B. An "OR" gate
C. A flip-flop
D. A clock

**ANSWER C:** A bistable multivibrator is a flip-flop. Bistable stands for two stable states.

---

**E7A02 How many output level changes are obtained for every two trigger pulses applied to the input of a "T" flip-flop circuit?**

A. None
B. One
C. Two
D. Four

**ANSWER C:** Two trigger pulses, two output level changes. The T FF toggles from one state to another for each input pulse.

---

**E7A03 The frequency of an AC signal can be divided electronically by what type of digital circuit?**

A. A free-running multivibrator
B. A bistable multivibrator
C. An OR gate
D. An astable multivibrator

**ANSWER B:** A bistable multivibrator can be used to divide the frequency of an AC signal.

---

**E7A04 How many flip-flops are required to divide a signal frequency by 4?**

A. 1
B. 2
C. 4
D. 8

**ANSWER B:** Since a flip-flop has two stable states, you will need two flip-flops to divide a signal frequency by four.

---

**E7A05 What is the characteristic function of an astable multivibrator?**

A. It alternates between two stable states
B. It alternates between a stable state and an unstable state
C. It blocks either a 0 pulse or a 1 pulse and passes the other
D. It alternates between two unstable states

**ANSWER D:** An ASTABLE multivibrator continuously switches back and forth between two unstable states.

---

**E7A06 What is the characteristic function of a monostable multivibrator?**

A. It switches momentarily to the opposite binary state and then returns after a set time to its original state
B. It is a clock that produces a continuous square wave oscillating between 1 and 0
C. It stores one bit of data in either a 0 or 1 state
D. It maintains a constant output voltage, regardless of variations in the input voltage

**ANSWER A:** This type of multivibrator may momentarily be monostable (MO MO). It stays in an original state until triggered to its other state, where it remains for a time usually determined by external components, after which it returns to the original state.

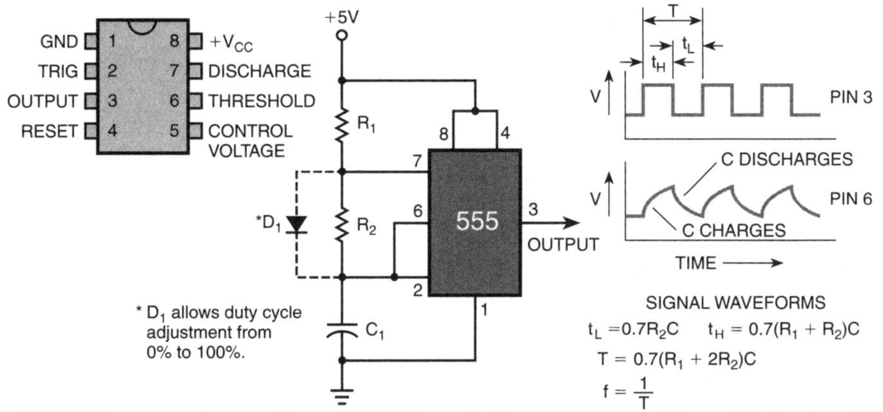


**An IC 555 timer can be used as a bistable, astable, or monostable multivibrator. This schematic shows it wired as an astable multivibrator.**
**The resulting astable voltage waveforms at pins 3 and 6 are shown.**

Source: *Basic Digital Electronics*, Evans, ©1996, Master Publishing, Inc., Lincolnwood, IL

**E7A07 What logical operation does an AND gate perform?**

A. It produces a logic “0” at its output only if all inputs are logic “1”
B. It produces a logic “1” at its output only if all inputs are logic “1”
C. It produces a logic “1” at its output if only one input is a logic “1”
D. It produces a logic “1” at its output if all inputs are logic “0”

**ANSWER B:** The AND gate has two or more inputs with a single output, and produces a logic “1” at its output only if all inputs are logic “1”. Remember AND, all, and the number 1 and the number 1 looking like the letters ALL.

**E7A08 What logical operation does a NAND gate perform?**

A. It produces a logic “0” at its output only when all inputs are logic “0”
B. It produces a logic “1” at its output only when all inputs are logic “1”
C. It produces a logic “0” at its output if some but not all of its inputs are logic “1”
D. It produces a logic “0” at its output only when all inputs are logic “1”

**ANSWER D:** The word NAND reminds me of the word “naw”, meaning no or nothing. The NAND gate produces a logic “0” at its output only when all inputs are logic “1”.

**E7A09 What logical operation does an OR gate perform?**

A. It produces a logic “1” at its output if any input is or all inputs are logic “1”
B. It produces a logic “0” at its output if all inputs are logic “1”
C. It only produces a logic “0” at its output when all inputs are logic “1”
D. It produces a logic “1” at its output if all inputs are logic “0”

**ANSWER A:** The OR gate will produce a logic “1” at its output if any input is, or all inputs are, a logic “1”.

**E7A10 What logical operation does a NOR gate perform?**

A. It produces a logic “0” at its output only if all inputs are logic “0”
B. It produces a logic “1” at its output only if all inputs are logic “1”
C. It produces a logic “0” at its output if any input is or all inputs are logic “1”
D. It produces a logic “1” at its output only when none of its inputs are logic “0”

**ANSWER C:** The NOR gate produces a logic “0” at its output if any or all inputs are logic “1”. With OR gate and NOR gate, spot that word ANY.

**E7A11 What is a truth table?**

A. A table of logic symbols that indicate the high logic states of an op-amp
B. A diagram showing logic states when the digital device's output is true
C. A list of input combinations and their corresponding outputs that characterize the function of a digital device
D. A table of logic symbols that indicates the low logic states of an op-amp

**ANSWER C:** We use a "truth table" in digital circuitry to characterize a digital device's function. If you buy the owner's technical manual on a piece of Amateur Radio gear, you will usually find a page describing the truth table of digital devices used in the equipment.

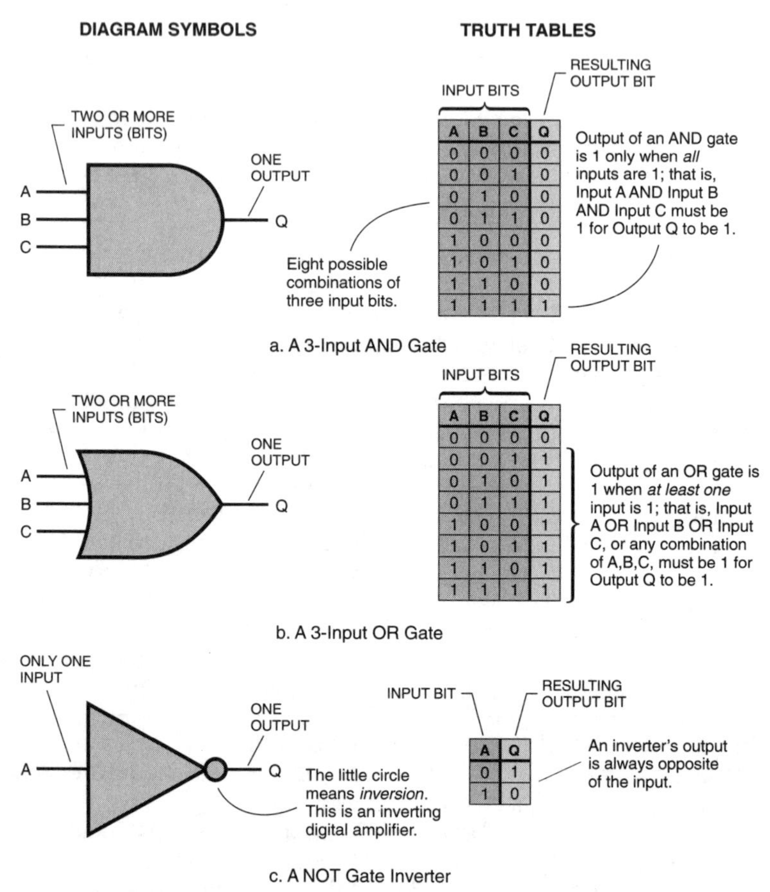


**AND, OR, and NOT Truth Tables**

Source: *Basic Electronics* © 1994, Master Publishing, Inc., Lincolnwood, Illinois

**E7A12 In a positive-logic circuit, what level is used to represent a logic 1?**

A. A low level
B. A positive-transition level
C. A negative-transition level
D. A high level

**ANSWER D:** In a positive-logic circuit, the logic "1" is a high level, or the most positive level. Think of the four letters in high and the four letters in plus (for positive).

**E7A13 In a negative-logic circuit, what level is used to represent a logic 1?**

A. A low level
B. A positive-transition level
C. A negative-transition level
D. A high level

**ANSWER A:** In a negative-logic circuit, the logic "1" is now a low level, or the least negative level.

***E7B Amplifier circuits: Class A, Class AB, Class B, Class C, amplifier operating efficiency (i.e., DC input versus PEP), transmitter final amplifiers; amplifier circuits: tube, bipolar transistor, FET***

**E7B01 For what portion of a signal cycle does a Class AB amplifier operate?**

A. More than 180 degrees but less than 360 degrees
B. Exactly 180 degrees
C. The entire cycle
D. Less than 180 degrees

**ANSWER A:** The Class AB amplifier has some of the properties of both the Class A and the Class B amplifier. It has output for more than 180 degrees of the cycle, but less than 360 degrees of the cycle.

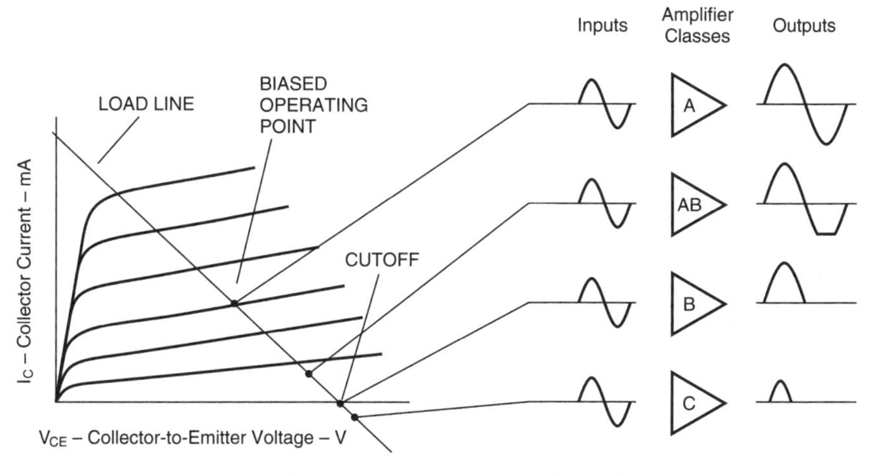


**Various Classes of Transistorized Amplifiers**

**E7B02 Which class of amplifier provides the highest efficiency?**

A. Class A
B. Class B
C. Class C
D. Class AB

**ANSWER C:** The highest efficiency comes out of a Class C amplifier. Since it operates the minimum time out of the cycle, it dissipates less power, but at the same time, produces narrow pulses which can cause loss of linearity.

**E7B03 Where on the load line should a bipolar-transistor, common-emitter Class A power amplifier be operated for best efficiency and stability?**

A. Below the saturation region
B. Above the saturation region
C. At the zero bias point
D. Just below the thermal runaway point

**ANSWER A:** We operate our solid-state power amplifiers just below the saturation point for best linearity and minimum distortion.

**E7B04 How can parasitic oscillations be eliminated from a power amplifier?**

A. By tuning for maximum SWR
B. By tuning for maximum power output
C. By neutralization
D. By tuning the output

**ANSWER C:** We can reduce parasitic oscillations, and clean up our signal, through the steps of neutralization in a power amplifier.

**E7B05 How can even-order harmonics be reduced or prevented in transmitter amplifiers?**

A. By using a push-push amplifier
B. By using a push-pull amplifier
C. By operating Class C
D. By operating Class AB

**ANSWER B:** We reduce even-order harmonics by using a push-pull amplifier.

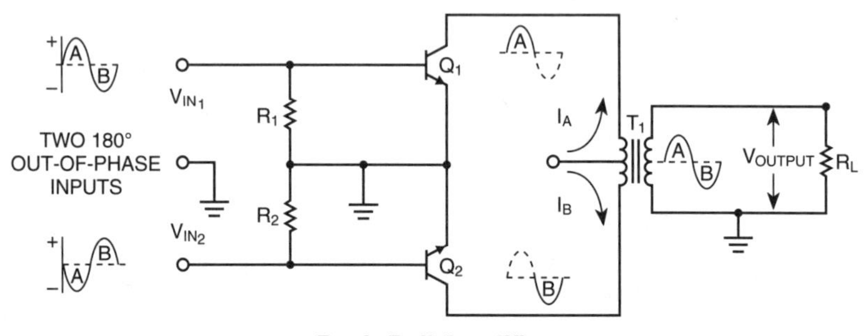


**Push-Pull Amplifier**

**E7B06 What can occur when a nonlinear amplifier is used with a single-sideband phone transmitter?**

A. Reduced amplifier efficiency
B. Increased intelligibility
C. Sideband inversion
D. Distortion

**ANSWER D:** There are some 2-meter and 432 FM (only) power amplifiers out there that sometimes get hooked up into a satellite transmitting station. Weak signal and satellite operation calls for SSB and CW. If you transmit with an FM-only-rated, solid-state power amplifier, it is not linear and will distort your signal. Make sure that any solid-state power amplifier you are going to use on VHF and UHF SSB is rated "linear."

**E7B07 How can a vacuum-tube power amplifier be neutralized?**

A. By increasing the grid drive
B. By feeding back an in-phase component of the output to the input
C. By feeding back an out-of-phase component of the output to the input
D. By feeding back an out-of-phase component of the input to the output

**ANSWER C:** We can keep our power amplified signal clean if the power amplifier is neutralized. For neutralization, follow the instructions in your amplifier manual. Chances are, you will be feeding back an out-of-phase component of the output to the input with neutralizing capacitors that are chosen to obtain the proper amount of feedback.

**E7B08 What is the procedure for tuning a vacuum-tube power amplifier having an output pi-network?**

A. Adjust the loading capacitor to maximum capacitance and then dip the plate current with the tuning capacitor
B. Alternately increase the plate current with the tuning capacitor and dip the plate current with the loading capacitor

C. Adjust the tuning capacitor to maximum capacitance and then dip the plate current with the loading capacitor
D. Alternately increase the plate current with the loading capacitor and dip the plate current with the tuning capacitor

**ANSWER D:** If you're going to be running a big power amplifier on the HF bands, chances are it will still use a vacuum tube for kilowatts of power output. Solid-state power amplifiers require no tuning, but most solid-state power amps only put out about 800 watts max. With a tube power amplifier, alternately tune and increase the plate current with a loading capacitor, and then dip the plate current with a tuning capacitor. Do this for short periods of time, and do it on a frequency that no one else is transmitting on. I first tune up into a dummy load to get me close in settings, and then switch over to the antenna, select a frequency which no one is using, announce, "I'm going to quickly tune up," and then QUICKLY TUNE UP.

---

**E7B09 In Figure E7-1, what is the purpose of R1 and R2?**
A. Load resistors
B. Fixed bias
C. Self bias
D. Feedback

**ANSWER B:** This reminds you a little bit of Thevenin's Theorem doesn't it? R1 and R2 form a voltage divider to set the voltage and the current at the base of the transistor.

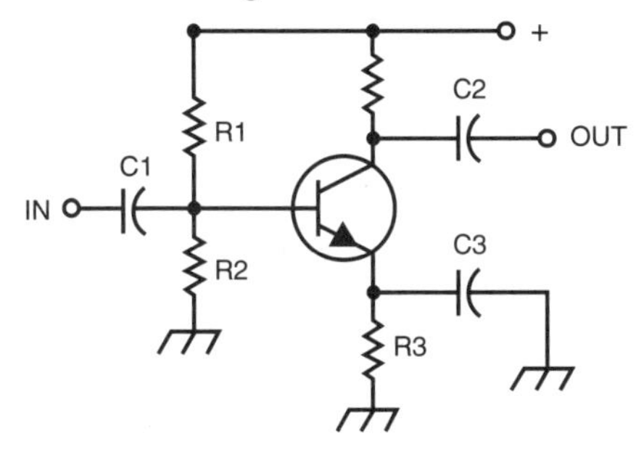


Figure E7-1

---

**E7B10 In Figure E7-1, what is the purpose of C3?**
A. AC feedback
B. Input coupling
C. Power supply decoupling
D. Emitter bypass

**ANSWER D:** One end of C3 is connected to the emitter of the transistor, the other end is grounded. It is across (in parallel with) R3. It bypasses AC signals around R3 to ground, so it is classified as an *emitter bypass* (key words) capacitor.

---

**E7B11 In Figure E7-1, what is the purpose of R3?**
A. Fixed bias
B. Emitter bypass
C. Output load resistor
D. Self bias

**ANSWER D:** Current through R3 produces a voltage drop that affects the base-emitter bias voltage. Since the transistor's collector and base current (the emitter current) determine the voltage at the emitter, it is called *self bias* (key word).

---

**E7B12 What type of circuit is shown in Figure E7-1?**
A. Switching voltage regulator
B. Linear voltage regulator
C. Common emitter amplifier
D. Emitter follower amplifier

**ANSWER C:** In a common emitter amplifier, the input signal is applied between base and emitter. In this circuit, the emitter is at AC ground because of C3. The output is taken from the collector.

**E7B13 In Figure E7-1, what is the purpose of C1?**

A. Decoupling
B. Output coupling
C. Self bias
D. Input coupling

**ANSWER D:** C1 is in series with the input, so it is an *input coupling* (key words) capacitor. It will pass AC (the input signal), but block DC from getting to the voltage divider or the base of the transistor.

**E7B14 In Figure E7-2, what is the purpose of R?**

A. Emitter load
B. Fixed bias
C. Collector load
D. Voltage regulation

**ANSWER A:** One end of the resistor R is connected to the emitter, the other end to ground. It is classified as an *emitter load* (key words) resistor in this circuit because the output is taken from the emitter.

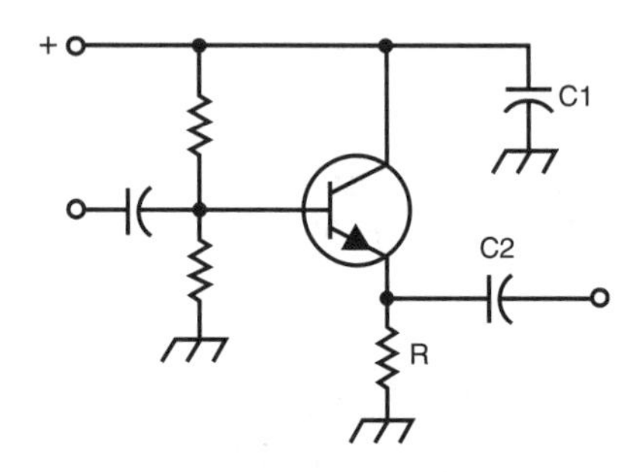


**Figure E7-2**

**E7B15 In Figure E7-2, what is the purpose of C2?**

A. Output coupling
B. Emitter bypass
C. Input coupling
D. Hum filtering

**ANSWER A:** Notice the little circle after C2? This denotes output of the circuit, and C2 is the *output coupling* (key words) capacitor.

**E7B16 What is the purpose of D1 in the circuit shown in Figure E7-3?**

A. Line voltage stabilization
B. Voltage reference
C. Peak clipping
D. Hum filtering

**ANSWER B:** D1 is a voltage regulator Zener diode, which is used to provide a voltage reference for Q1. If the load current increases (R2 gets smaller), the output voltage would tend to decrease. When this voltage change is compared to the reference voltage across D1 by Q1, Q1 adjusts its voltage drop from collector to emitter to raise the output voltage to the designed value with the new load current.

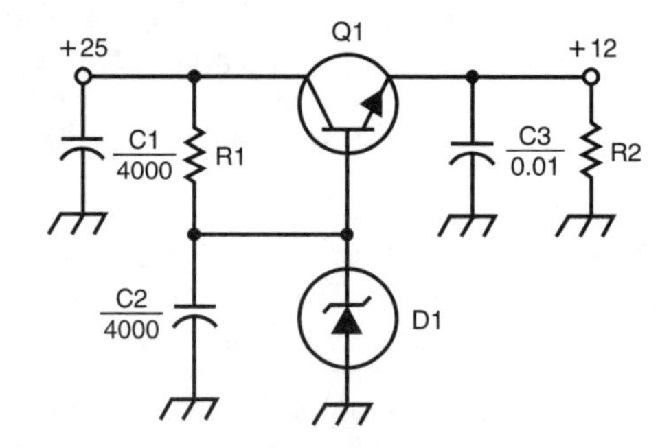


**Figure E7-3**

**E7B17 What is the purpose of Q1 in the circuit shown in Figure E7-3?**

A. It increases the output ripple
B. It provides a constant load for the voltage source
C. It increases the current-handling capability
D. It provides D1 with current

**ANSWER C:** This transistor can handle greater current than a Zener diode. They sometimes call these transistors "series-pass transistors."

---

**E7B18 What is the purpose of C2 in the circuit shown in Figure E7-3?**

A. It bypasses hum around D1
B. It is a brute force filter for the output
C. To self resonate at the hum frequency
D. To provide fixed DC bias for Q1

**ANSWER A:** C2 provides additional filtering for the ripple frequency. It *bypasses* (key word) voltage changes around the Zener diode, D1, to keep the reference voltage stable.

---

**E7B19 What type of circuit is shown in Figure E7-3?**

A. Switching voltage regulator
B. Grounded emitter amplifier
C. Linear voltage regulator
D. Emitter follower

**ANSWER C:** Components D1 and Q1 give the answer away – a linear voltage regulator. This should give you a good, solid, regulated voltage output, even though the load current may be changing.

---

**E7B20 What is the purpose of C1 in the circuit shown in Figure E7-3?**

A. It resonates at the ripple frequency
B. It provides fixed bias for Q1
C. It decouples the output
D. It filters the supply voltage

**ANSWER D:** C1 is an electrolytic filter capacitor which removes ripple frequency (60-cycle or 120-cycle) from the input voltage.

---

**E7B21 What is the purpose of C3 in the circuit shown in Figure E7-3?**

A. It prevents self-oscillation
B. It provides brute force filtering of the output
C. It provides fixed bias for Q1
D. It clips the peaks of the ripple

**ANSWER A:** You might see a lot of C3-type circuits in your power supply. This is an RF bypass capacitor, keeping any stray RF energy from getting into the circuit, which could cause self-oscillation.

---

**E7B22 What is the purpose of R1 in the circuit shown in Figure E7-3?**

A. It provides a constant load to the voltage source
B. It couples hum to D1
C. It supplies current to D1
D. It bypasses hum around D1

**ANSWER C:** R1 is connected to the power supply input voltage and supplies the bias current to operate D1 and Q1 at a stable operating point for the voltage regulator output voltage and current desired.

**E7B23 What is the purpose of R2 in the circuit shown in Figure E7-3?**

A. It provides fixed bias for Q1
B. It provides fixed bias for D1
C. It decouples hum from D1
D. It provides a constant minimum load for Q1

**ANSWER D:** Finally, R2 connected to the 12-volt output helps maintain voltage regulation by keeping a constant minimum load for the transistor Q1, and acts as a bleeder resistor to discharge capacitors after the power supply has been turned off.

***E7C Impedance-matching networks: Pi, L, Pi-L; filter circuits: constant K, M-derived, band-stop, notch, crystal lattice, pi-section, T-section, L-section, Butterworth, Chebyshev, elliptical; filter applications (audio, IF, digital signal processing {DSP})***

**E7C01 How are the capacitors and inductors of a low-pass filter pi-network arranged between the network's input and output?**

A. Two inductors are in series between the input and output and a capacitor is connected between the two inductors and ground
B. Two capacitors are in series between the input and output and an inductor is connected between the two capacitors and ground
C. An inductor is in parallel with the input, another inductor is in parallel with the output, and a capacitor is in series between the two
D. A capacitor is in parallel with the input, another capacitor is in parallel with the output, and an inductor is in series between the two

**ANSWER D:** The pi-network will feature a capacitor in parallel with the input, and a capacitor in parallel with the output, with an inductor in series between the two. It looks like the Greek letter $\pi$.

**E7C02 What is an L-network?**

A. A network consisting entirely of four inductors
B. A network consisting of an inductor and a capacitor
C. A network used to generate a leading phase angle
D. A network used to generate a lagging phase angle

**ANSWER B:** An L-network has both an inductor and a capacitor.

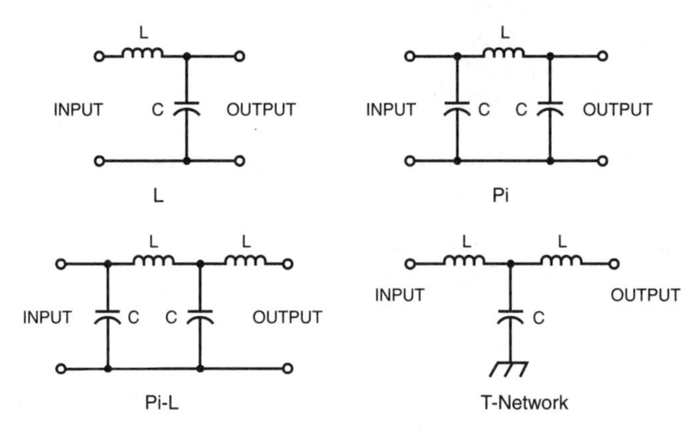


**Matching Networks**

**E7C03 A T-network with series capacitors and a parallel (shunt) inductor has which of the following properties?**

A. It transforms impedances and is a low-pass filter
B. It transforms reactances and is a low-pass filter
C. It transforms impedances and is a high-pass filter
D. It transforms reactances and is a narrow bandwidth notch filter

**ANSWER C:** The T-network transforms impedances and is also used as a high-pass filter.

**E7C04 What advantage does a pi-L-network have over a pi-network for impedance matching between the final amplifier of a vacuum-tube type transmitter and a multiband antenna?**

A. Greater harmonic suppression
B. Higher efficiency
C. Lower losses
D. Greater transformation range

**ANSWER A:** Worldwide ham sets into a multi-band antenna system can generate harmonics. The pi-L-network always gives us greater harmonic suppression inside the transceiver.

**E7C05 How does a network transform one impedance to another?**

A. It introduces negative resistance to cancel the resistive part of an impedance
B. It introduces transconductance to cancel the reactive part of an impedance
C. It cancels the reactive part of an impedance and changes the resistive part
D. Network resistances substitute for load resistances

**ANSWER C:** In older worldwide sets, you could vary the matching network to cancel the reactive (X) part of an impedance, and change the value of the resistive part (R) of an impedance. Newer transistorized sets have a fixed output, and there is no manual operator control of the fixed matching devices. However, manufacturers have come to the rescue in antenna matching by providing new worldwide ham sets with built-in automatic antenna tuners.

**E7C06 Which filter type is described as having ripple in the passband and a sharp cutoff?**

A. A Butterworth filter
B. An active LC filter
C. A passive op-amp filter
D. A Chebyshev filter

**ANSWER D:** If you look at the word "Chebyshev," it almost looks like it has a ripple, doesn't it?

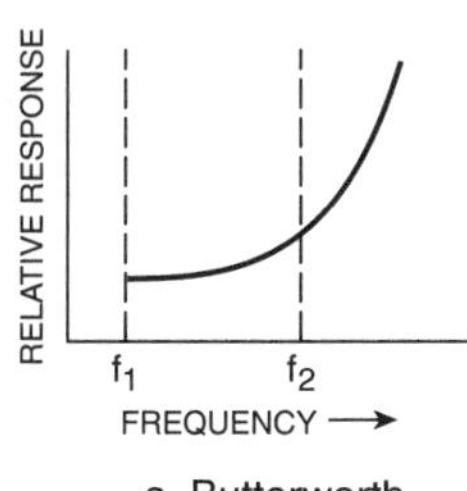

a. Butterworth

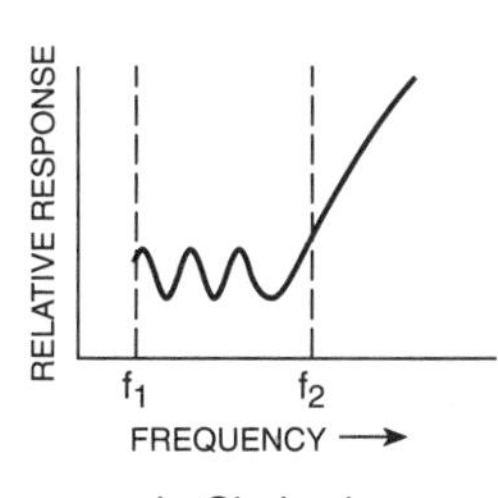

b. Chebyshev

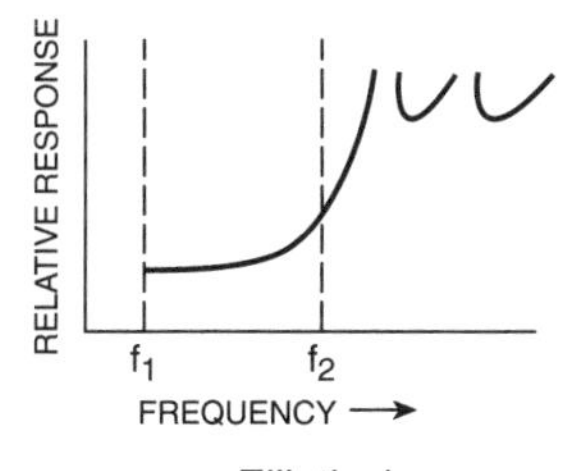

c. Elliptical

**Comparison of Low-Pass Filter Circuits**

**E7C07 What are the distinguishing features of an elliptical filter?**

A. Gradual passband rolloff with minimal stop-band ripple
B. Extremely flat response over its passband, with gradually rounded stop-band corners

C. Extremely sharp cutoff, with one or more infinitely deep notches in the stop band
D. Gradual passband rolloff with extreme stop-band ripple

**ANSWER C:** The elliptical filter has an immediate and extremely sharp cutoff, with one or more deep notches in the stop band.

---

**E7C08 What kind of audio filter would you use to attenuate an interfering carrier signal while receiving an SSB transmission?**

A. A band-pass filter
B. A notch filter
C. A pi-network filter
D. An all-pass filter

**ANSWER B:** To attenuate and block a specific whistle (or "heterodyne") on an SSB receiver, switch on the notch filter, and ever so carefully sweep through the signal until it disappears. Tune slowly! A notch filter can require mighty critical adjustment!

---

**E7C09 What characteristic do typical SSB receiver IF filters lack that is important to digital communications?**

A. Steep amplitude-response skirts
B. Passband ripple
C. High input impedance
D. Linear phase response

**ANSWER D:** Using your voice communications transceiver to receive the digital modes may require more than a terminal node controller. The SSB receiver probably has a 2.8 kHz SSB filter that lacks good linear phase response.

---

**E7C10 What kind of digital signal processing audio filter might be used to remove unwanted noise from a received SSB signal?**

A. An adaptive filter
B. A crystal-lattice filter
C. A Hilbert-transform filter
D. A phase-inverting filter

**ANSWER A:** The "adaptive filter" found in digital signal processing (DSP) circuits removes unwanted noise from incoming SSB signals.

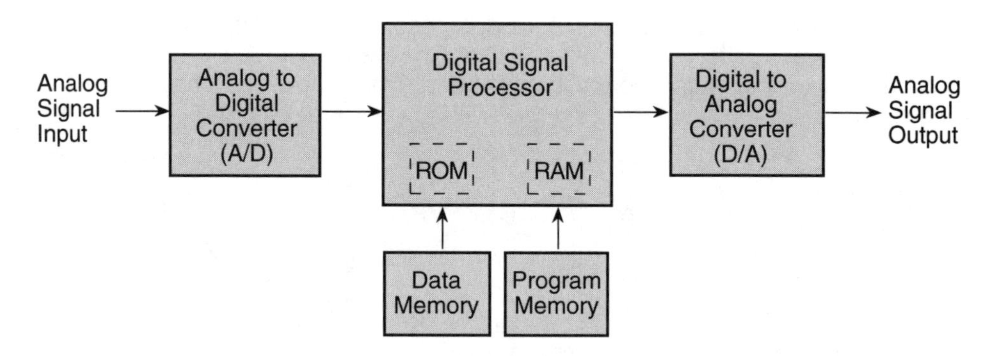


**Block diagram of a basic digital signal processing (DSP) system**

---

**E7C11 What kind of digital signal processing filter might be used in generating an SSB signal?**

A. An adaptive filter
B. A notch filter
C. A Hilbert-transform filter
D. An elliptical filter

**ANSWER C:** In a DSP unit for transmit, here we might use the Hilbert-transform filter – think of "transform" for SSB transmit.

**Digital Signal Processor**

**E7C12 Which type of filter would be the best to use in a 2-meter repeater duplexer?**

A. A crystal filter
B. A cavity filter
C. A DSP filter
D. An L-C filter

**ANSWER B:** For a 2-meter repeater duplexer system, we need big filters. We call them cavity filters because they are hollowed out to form extremely narrow-band tuned resonant circuits.

**4-Cavity Duplexer for 2-Meter Band**
Courtesy of WACOM Products, Inc.

**E7C13 What is a pi-network?**

A. A network consisting entirely of four inductors or four capacitors
B. A Power Incidence network
C. An antenna matching network that is isolated from ground
D. A network consisting of one inductor and two capacitors or two inductors and one capacitor

**ANSWER D:** The pi-network consists of three components, looking like the Greek symbol p. The legs could be either capacitors or coils, and the horizontal line at the top could be either a coil or a capacitor. (Look again at the illustration at E7C02.)

**E7C14 What is a pi-L-network?**
A. A Phase Inverter Load network
B. A network consisting of two inductors and two capacitors
C. A network with only three discrete parts
D. A matching network in which all components are isolated from ground

**ANSWER B:** A slightly more complex circuit is the pi-L network, consisting of two inductors and two capacitors.

**E7C15 Which type of network provides the greatest harmonic suppression?**
A. L-network
B. Pi-network
C. Pi-L-network
D. Inverse Pi network

**ANSWER C:** The network with the most components will provide the greatest harmonic suppression. The pi-L-network is found in most worldwide ham sets.

***E7D Oscillators: types, applications, stability; voltage-regulator circuits: discrete, integrated and switched mode***

**E7D01 What are three major oscillator circuits often used in Amateur Radio equipment?**
A. Taft, Pierce and negative feedback
B. Colpitts, Hartley and Taft
C. Taft, Hartley and Pierce
D. Colpitts, Hartley and Pierce

**ANSWER D:** Colpitts oscillators have a capacitor just like C in the name. Hartley is tapped, and a Pierce oscillator uses a crystal.

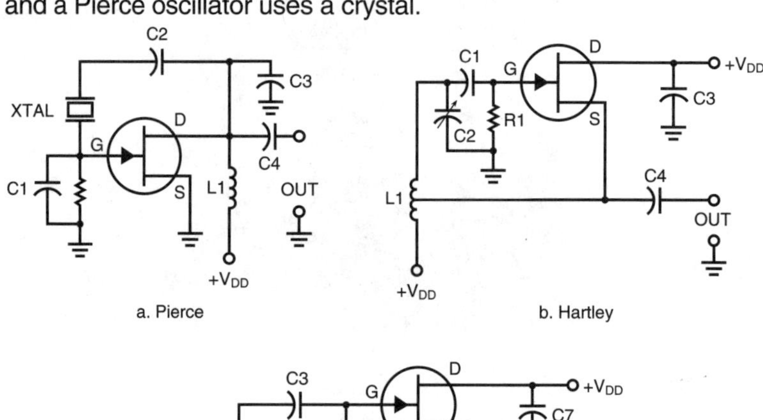


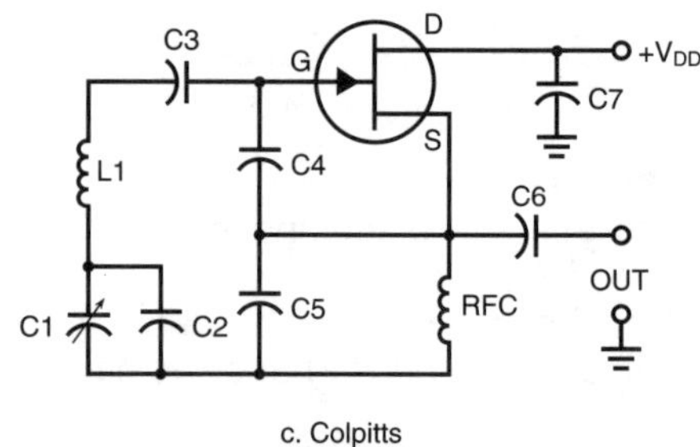


Oscillators

**E7D02 What condition must exist for a circuit to oscillate?**

A. It must have a gain of less than 1
B. It must be neutralized
C. It must have positive feedback sufficient to overcome losses
D. It must have negative feedback sufficient to cancel the input

**ANSWER C:** In order to keep an oscillator oscillating, it must have positive feedback sufficient to overcome its natural losses. To keep your kids out there swinging on their swing set, you have to give them a push every so often. That's exactly what positive feedback does, but on every cycle.

---

**E7D03 How is the positive feedback coupled to the input in a Hartley oscillator?**

A. Through a tapped coil
B. Through a capacitive divider
C. Through link coupling
D. Through a neutralizing capacitor

**ANSWER A:** Hartley is always tapped. The tapped coil provides inductive coupling for positive feedback.

---

**E7D04 How is the positive feedback coupled to the input in a Colpitts oscillator?**

A. Through a tapped coil
B. Through link coupling
C. Through a capacitive divider
D. Through a neutralizing capacitor

**ANSWER C:** On the Colpitts oscillator, we use a capacitive divider to provide feedback.

---

**E7D05 How is the positive feedback coupled to the input in a Pierce oscillator?**

A. Through a tapped coil
B. Through link coupling
C. Through a neutralizing capacitor
D. Through a quartz crystal

**ANSWER D:** Every Pierce oscillator has a quartz crystal, and we use the quartz crystal to obtain positive feedback.

---

**E7D06 Which type of oscillator circuits are commonly used in a VFO?**

A. Pierce and Zener
B. Colpitts and Hartley
C. Armstrong and deForest
D. Negative feedback and Balanced feedback

**ANSWER B:** Since the Colpitts oscillator uses a big capacitor in a variable frequency oscillator (VFO), it's the most common oscillator circuit for older VFO radios. Newer radios, even though they say they have a VFO, are really digitally controlled via an optical reader. They just look like they have a big capacitor behind that big tuning dial!

---

**E7D07 Why is a very stable reference oscillator normally used as part of a phase-locked loop (PLL) frequency synthesizer?**

A. Any amplitude variations in the reference oscillator signal will prevent the loop from locking to the desired signal
B. Any phase variations in the reference oscillator signal will produce phase noise in the synthesizer output
C. Any phase variations in the reference oscillator signal will produce harmonic distortion in the modulating signal
D. Any amplitude variations in the reference oscillator signal will prevent the loop from changing frequency

**ANSWER B:** The reference oscillator must have little or no phase variations because phase variations will produce phase noise in the synthesizer output. Excessive phase noise can degrade transmitter and receiver performance.

**E7D08 What is one characteristic of a linear electronic voltage regulator?**

A. It has a ramp voltage as its output
B. The pass transistor switches from the "off" state to the "on" state
C. The control device is switched on or off, with the duty cycle proportional to the line or load conditions
D. The conduction of a control element is varied in direct proportion to the load current to maintain a constant output voltage

**ANSWER D:** A linear electronic voltage regulator varies the conduction of a circuit in direct proportion to variations in the *line voltage* (key words) to or the *load current* (key words) from the device. You will find in your base station power supplies, which utilize a big heavy transformer, a sophisticated voltage regulation circuit.

**Voltage Regulator IC Mounted in Circuit Board**

**E7D09 What is one characteristic of a switching electronic voltage regulator?**

A. The conduction of a control element is varied in direct proportion to the line voltage or load current
B. It provides more than one output voltage
C. The control device is switched on or off, with the duty cycle automatically adjusted to maintain a constant average output voltage
D. It gives a ramp voltage at its output

**ANSWER C:** The switching voltage regulator actually switches the control device completely *on or off* (key words).

**E7D10 What device is typically used as a stable reference voltage in a linear voltage regulator?**

A. A Zener diode
B. A tunnel diode
C. An SCR
D. A varactor diode

**ANSWER A:** For voltage regulation and a stable reference voltage, a Zener diode is found in most ham sets.

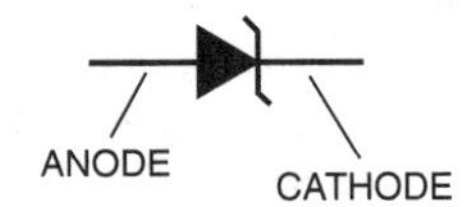


Here is the schematic symbol of a Zener diode. Since a diode only passes energy in one direction, look for that one-way arrow, plus a "Z" indicating it is a Zener diode. Doesn't that vertical line look like a tiny "Z".

**Zener Diode**

**E7D11 What type of linear regulator is used in applications requiring efficient use of the primary power source?**

A. A constant current source
B. A series regulator
C. A shunt regulator
D. A shunt current source

**ANSWER B:** A series regulator normally runs cool. It's more efficient than a shunt regulator. Everything passes through the series regulator.

---

**E7D12 What type of linear voltage regulator is used in applications requiring a constant load on the unregulated voltage source?**

A. A constant current source
B. A series regulator
C. A shunt current source
D. A shunt regulator

**ANSWER D:** In circuits where the load on the unregulated input source must be kept constant, we use the slightly less-efficient shunt regulator. These can sometimes get quite warm, and are usually mounted to the chassis of the equipment. The chassis acts as a heat sink.

---

**E7D13 Which of the following Zener diode voltages will result in the best temperature stability for a voltage reference?**

A. 2.4 volts
B. 3.0 volts
C. 5.6 volts
D. 12.0 volts

**ANSWER C:** 5.6 volts is considered the ideal voltage for a zero change in the temperature coefficient of a Zener diode.

---

**E7D14 What are the important characteristics of a three-terminal regulator?**

A. Maximum and minimum input voltage, minimum output current and voltage
B. Maximum and minimum input voltage, maximum and minimum output current and maximum output voltage
C. Maximum and minimum input voltage, minimum output current and maximum output voltage
D. Maximum and minimum input voltage, minimum output voltage and Maximum input and output current

**ANSWER B:** Two important characteristics of a 3-terminal regulator are maximum and minimum input voltage, which are given in all answers. The other three are minimum output current and minimum and maximum output voltage. Only one answer indicates *maximum output current and voltage* (key words), and these are important considerations when choosing a 3-terminal regulator for a particular circuit.

---

**E7D15 What type of voltage regulator limits the voltage drop across its junction when a specified current passes through it in the reverse-breakdown direction?**

A. A Zener diode
B. A three-terminal regulator
C. A bipolar regulator
D. A pass-transistor regulator

**ANSWER A:** We use Zener diodes as voltage regulators.

---

***E7E Modulators: reactance, phase, balanced; detectors; mixer stages; frequency synthesizers***

---

**E7E01 How is an F3E FM-phone emission produced?**

A. With a balanced modulator on the audio amplifier
B. With a reactance modulator on the oscillator
C. With a reactance modulator on the final amplifier
D. With a balanced modulator on the oscillator

**ANSWER B:** The *reactance modulator* (key words) causes the oscillator to vary frequency in accordance with the modulation. It's the *oscillator* (key word) that is modulated.

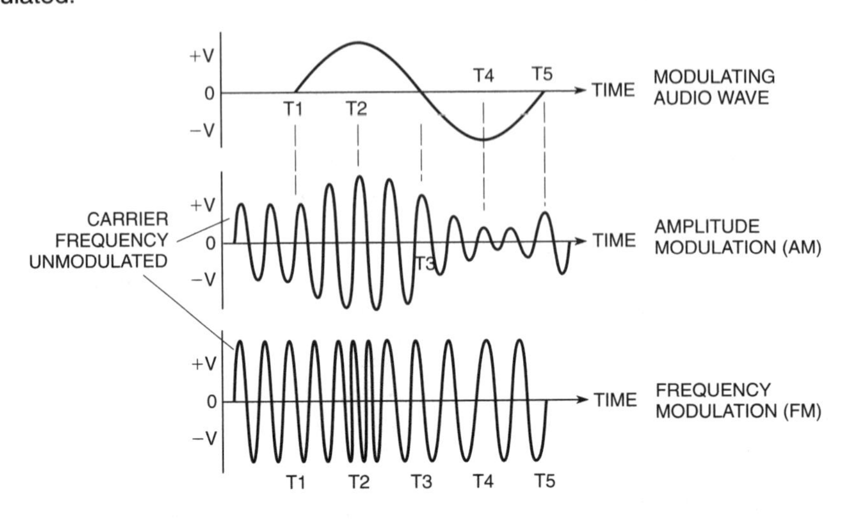


**Modulation**

**E7E02 How does a reactance modulator work?**

A. It acts as a variable resistance or capacitance to produce FM signals
B. It acts as a variable resistance or capacitance to produce AM signals
C. It acts as a variable inductance or capacitance to produce FM signals
D. It acts as a variable inductance or capacitance to produce AM signals

**ANSWER C:** The *reactance* (key word) modulator is found in an FM transceiver, and varies the *inductance or capacitance* (key words) to produce the FM signal. Watch out for answer A!

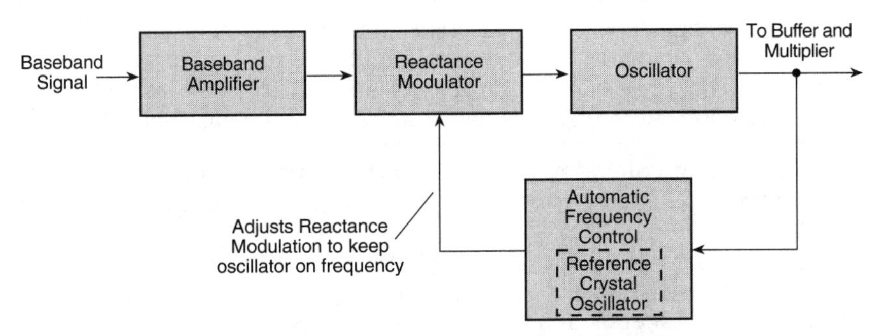


**Block diagram for lgenerating FM modulation**

**E7E03 How does a phase modulator work?**

A. It varies the tuning of a microphone preamplifier to produce PM signals
B. It varies the tuning of an amplifier tank circuit to produce AM signals
C. It varies the tuning of an amplifier tank circuit to produce PM signals
D. It varies the tuning of a microphone preamplifier to produce AM signals

**ANSWER C:** Phase modulation varies the tuning of an amplifier tank circuit to produce FM signals. It is NOT in the microphone preamp circuit!

**E7E04 How can a single-sideband phone signal be generated?**

A. By using a balanced modulator followed by a filter
B. By using a reactance modulator followed by a mixer
C. By using a loop modulator followed by a mixer
D. By driving a product detector with a DSB signal

**ANSWER A:** In a single-sideband transceiver, a lower or upper sideband filter removes the unwanted sideband.

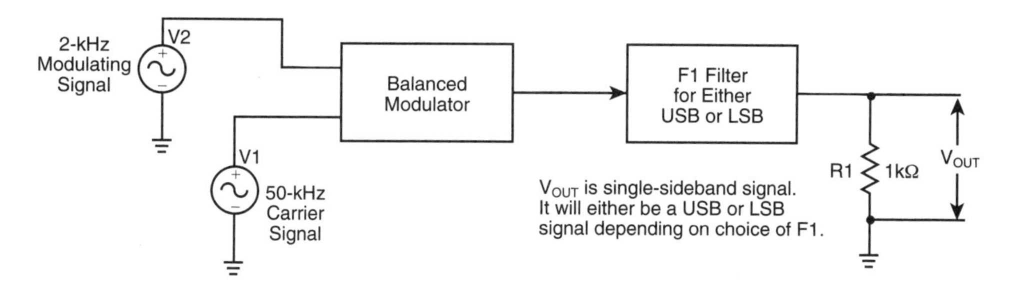


**Generating an SSB signal**

**E7E05 What audio shaping network is added at a transmitter to proportionally attenuate the lower audio frequencies, giving an even spread to the energy in the audio band?**

A. A de-emphasis network
B. A heterodyne suppressor
C. An audio prescaler
D. A pre-emphasis network

**ANSWER D:** Pre-emphasis networks in modern FM transmitters help improve the signal-to-noise ratio.

**E7E06 What audio shaping network is added at a receiver to restore proportionally attenuated lower audio frequencies?**

A. A de-emphasis network
B. A heterodyne suppressor
C. An audio prescaler
D. A pre-emphasis network

**ANSWER A:** In order to restore proportionally-attenuated lower audio frequencies, a de-emphasis network is found in the communications receiver.

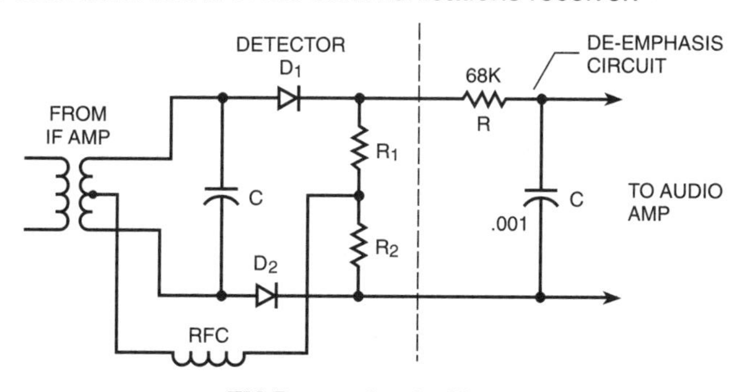


**FM De-emphasis Circuit**

**E7E07 What is the mixing process?**

A. The elimination of noise in a wideband receiver by phase comparison
B. The elimination of noise in a wideband receiver by phase differentiation

C. The recovery of the intelligence from a modulated RF signal
D. The combination of two signals to produce sum and difference frequencies

**ANSWER D:** Inside your ham radio transceiver are several stages of mixers that combine two signals to produce sum and difference frequencies.

---

**E7E08 What are the principal frequencies that appear at the output of a mixer circuit?**

A. Two and four times the original frequency
B. The sum, difference and square root of the input frequencies
C. The original frequencies and the sum and difference frequencies
D. 1.414 and 0.707 times the input frequency

**ANSWER C:** Out of the mixer comes your original two frequencies, and the sum and difference frequencies. The more elaborate the transceiver, the more mixing stages found in the sets. This helps filter out unwanted signals or phantom signals that could cause interference.

---

**E7E09 What occurs in a receiver when an excessive amount of signal energy reaches the mixer circuit?**

A. Spurious mixer products are generated
B. Mixer blanking occurs
C. Automatic limiting occurs
D. A beat frequency is generated

**ANSWER A:** Some VHF power amplifiers may incorporate a hefty pre-amplifier circuit for use on receiving which boosts incoming signal levels. The pre-amp turned on could generate spurious mixer products as a result of the high signal level. The spurious products are not really signals that you are trying to receive. If you use a handheld with a power amp, turn the pre-amp off.

---

**E7E10 What type of frequency synthesizer circuit uses a stable voltage-controlled oscillator, programmable divider, phase detector, loop filter and a reference frequency source?**

A. A direct digital synthesizer
B. A hybrid synthesizer
C. A phase-locked loop synthesizer
D. A diode-switching matrix synthesizer

**ANSWER C:** Almost all new VHF and UHF transceivers use phase-locked loop synthesizers. But the latest switch for high frequency manufacturers is to direct digital synthesis for lower phase noise.

---

**E7E11 What type of frequency synthesizer circuit uses a phase accumulator, lookup table, digital to analog converter and a low-pass antialias filter?**

A. A direct digital synthesizer
B. A hybrid synthesizer
C. A phase-locked loop synthesizer
D. A diode-switching matrix synthesizer

**ANSWER A:** The direct digital synthesizer (DDS) takes an extremely low noise floor and gets it even lower yet for hams with monster antennas and minimum outside noise to appreciate.

**E7E12 What are the main blocks of a direct digital frequency synthesizer?**

A. A variable-frequency crystal oscillator, phase accumulator, digital to analog converter and a loop filter
B. A stable voltage-controlled oscillator, programmable divider, phase detector, loop filter and a digital to analog converter
C. A variable-frequency oscillator, programmable divider, phase detector and a low-pass antialias filter
D. A phase accumulator, lookup table, digital to analog converter and a low-pass antialias filter

**ANSWER D:** Here is what a block diagram of direct-digital-frequency synthesizer (DDS) looks like:

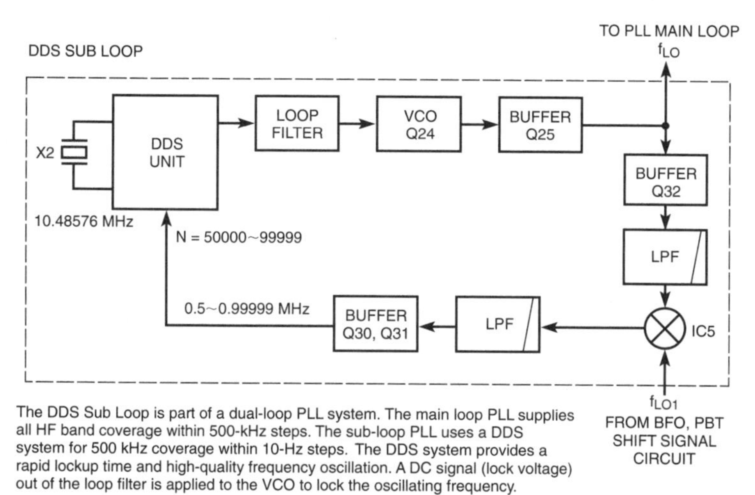


**DDS Sub Loop of a Transmitter Dual-Loop PLL**

Courtesy of ICOM America, Inc.

---

**E7E13 What information is contained in the lookup table of a direct digital frequency synthesizer?**

A. The phase relationship between a reference oscillator and the output waveform
B. The amplitude values that represent a sine-wave output
C. The phase relationship between a voltage-controlled oscillator and the output waveform
D. The synthesizer frequency limits and frequency values stored in the radio memories

**ANSWER B:** The look-up table in a DDS system has an amplitude value that represents a sine-wave output.

---

**E7E14 What are the major spectral impurity components of direct digital synthesizers?**

A. Broadband noise
B. Digital conversion noise
C. Spurs at discrete frequencies
D. Nyquist limit noise

**ANSWER C:** Unwanted components of a DDS output would be spurs at discrete frequencies – unwanted emissions. These are easier to filter than broadband noise.

**E7E15 What are the major spectral impurity components of phase-locked loop synthesizers?**

A. Broadband noise
B. Digital conversion noise
C. Spurs at discrete frequencies
D. Nyquist limit noise

**ANSWER A:** Broadband noise is the nemesis of a PLL synthesizer.

**E7E16 What is the process of detection?**

A. The masking of the intelligence on a received carrier
B. The recovery of the intelligence from a modulated RF signal
C. The modulation of a carrier
D. The mixing of noise with a received signal

**ANSWER B:** Your radio's detector *recovers intelligence* (key words) from a modulated RF signal.

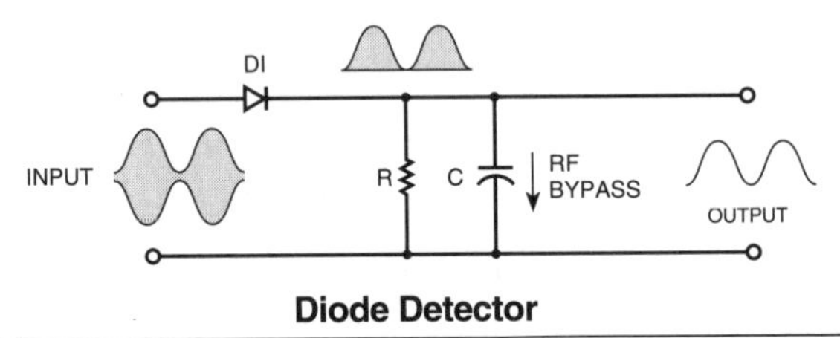


**Diode Detector**

**E7E17 What is the principle of detection in a diode detector?**

A. Rectification and filtering of RF
B. Breakdown of the Zener voltage
C. Mixing with noise in the transition region of the diode
D. The change of reactance in the diode with respect to frequency

**ANSWER A:** Since a diode only conducts for half of the AC signal, it may be used as a rectifier. By filtering out the radio frequency energy after rectification, detection is accomplished – the modulating signal is what remains.

**E7E18 What does a product detector do?**

A. It provides local oscillations for input to a mixer
B. It amplifies and narrows bandpass frequencies
C. It mixes an incoming signal with a locally generated carrier
D. It detects cross-modulation products

**ANSWER C:** A product detector is found in SSB receivers. It mixes the incoming signal with a beat frequency oscillator signal. The beat frequency oscillator signal is a locally-generated carrier that is mixed with the incoming signal.

**E7E19 How are FM-phone signals detected?**

A. With a balanced modulator
B. With a frequency discriminator
C. With a product detector
D. With a phase splitter

**ANSWER B:** FM signals must be detected differently than AM signals because frequency changes must be detected rather than amplitude changes. We use a reactance modulator to transmit FM. We use a *frequency discriminator* (key words) to detect an FM phone signal.

**E7E20 What is a frequency discriminator?**

A. An FM generator
B. A circuit for filtering two closely adjacent signals
C. An automatic band-switching circuit
D. A circuit for detecting FM signals

**ANSWER D:** Anytime you see the words frequency discriminator you know that you are dealing with an FM transceiver.

---

**E7E21 How can an FM-phone signal be produced?**

A. By modulating the supply voltage to a Class-B amplifier
B. By modulating the supply voltage to a Class-C amplifier
C. By using a reactance modulator on an oscillator
D. By using a balanced modulator on an oscillator

**ANSWER C:** The reactance modulator is found in an FM transceiver to develop an FM phone signal.

---

## *E7F Digital frequency divider circuits; frequency marker generators; frequency counters*

---

**E7F01 What is the purpose of a prescaler circuit?**

A. It converts the output of a JK flip-flop to that of an RS flip-flop
B. It multiplies an HF signal so a low-frequency counter can display the operating frequency
C. It prevents oscillation in a low-frequency counter circuit
D. It divides an HF signal so a low-frequency counter can display the operating frequency

**ANSWER D:** You will find a prescaler in frequency counters to divide down HF and VHF signals so they can be counted and displayed on a low-frequency counter.

---

**E7F02 How many states does a decade counter digital IC have?**

A. 2
B. 10
C. 20
D. 100

**ANSWER B:** Studying this material may seem like it is taking a decade, right? A decade is 10. Maybe a decade of days, or a decade of weeks, but, hopefully, not a decade of months.

---

**E7F03 What is the function of a decade counter digital IC?**

A. It produces one output pulse for every ten input pulses
B. It decodes a decimal number for display on a seven-segment LED display
C. It produces ten output pulses for every input pulse
D. It adds two decimal numbers

**ANSWER A:** A decade counter digital IC gives one output pulse for every 10 input pulses.

---

**E7F04 What additional circuitry is required in a 100-kHz crystal-controlled marker generator to provide markers at 50 and 25 kHz?**

A. An emitter-follower
B. Two frequency multipliers
C. Two flip-flops
D. A voltage divider

**ANSWER C:** We would use two flip-flops to take a 100-kHz signal and provide markers at 50 kHz and 25 kHz.

**E7F05 If a 1-MHz oscillator is used with a divide-by-ten circuit to make a marker generator, what will the output be?**

A. A 1-MHz sinusoidal signal with harmonics every 100 kHz
B. A 100-kHz signal with harmonics every 100 kHz
C. A 1-MHz square wave with harmonics every 1 MHz
D. A 100-kHz signal modulated by a 10-kHz signal

**ANSWER B:** One MHz is 1,000 kHz. If the oscillator is a "divide-by-ten" circuit, there will be a 100 kHz signal with harmonics every 100 kHz.

---

**E7F06 What is a crystal-controlled marker generator?**

A. A low-stability oscillator that sweeps through a band of frequencies
B. An oscillator often used in aircraft to determine the craft's location relative to the inner and outer markers at airports
C. A high-stability oscillator whose output frequency and amplitude can be varied over a wide range
D. A high-stability oscillator that generates a series of reference signals at known frequency intervals

**ANSWER D:** The crystal-controlled marker generator is a free-running oscillator that generates a series of reference signals at known frequency intervals.

---

**E7F07 What type of circuit does NOT make a good marker generator?**

A. A sinusoidal crystal oscillator
B. A crystal oscillator followed by a class C amplifier
C. A TTL device wired as a crystal oscillator
D. A crystal oscillator and a frequency divider

**ANSWER A:** It is much easier to obtain harmonics from square waves than it would be from a sinusoidal oscillator circuit. While a sinusoidal crystal oscillator would give you one terrific fundamental frequency, it would not give you the harmonics necessary to make it a good marker generator.

---

**E7F08 What is the purpose of a marker generator?**

A. To add audio markers to an oscilloscope
B. To provide a frequency reference for a phase locked loop
C. To provide a means of calibrating a receiver's frequency settings
D. To add time signals to a transmitted signal

**ANSWER C:** The purpose of a marker generator is to provide a means of calibrating receivers, as well as showing a calibration mark on monitor scopes and spectrum analyzers. For this question and answer, just look for "calibrating a receiver."

---

**E7F09 What does the accuracy of a frequency counter depend on?**

A. The internal crystal reference
B. A voltage-regulated power supply with an unvarying output
C. Accuracy of the AC input frequency to the power supply
D. Proper balancing of the power-supply diodes

**ANSWER A:** All frequency counters will depend on an internal crystal reference. This crystal oscillator is calibrated to the National Standards Institute WWV and WWVH signals using tiny capacitors.

---

**E7F10 How does a frequency counter determine the frequency of a signal?**

A. It counts the total number of pulses in a circuit
B. It monitors a WWV reference signal for comparison with the measured signal

C. It counts the number of input pulses in a specific period of time
D. It converts the phase of the measured signal to a voltage which is proportional to the frequency

**ANSWER C:** There are several manufacturers of excellent amateur radio frequency counters – accurate enough to also be used in the commercial radio service. A frequency counter counts the number of input pulses in a specific period of time. You can usually set the sample rate by pushing a button on the front of the counter.

**Frequency Counter**
Courtesy of OptoElectronics

**E7F11 What is the purpose of a frequency counter?**
A. To indicate the frequency of the strongest input signal which is within the counter's frequency range
B. To generate a series of reference signals at known frequency intervals
C. To display all frequency components of a transmitted signal
D. To compare the difference between the input and a voltage-controlled oscillator and produce an error voltage

**ANSWER A:** The frequency counter is a near-field receiver, capable of indicating the frequency of a near-field "strongest input" signal which is within the counter's frequency range. Most counters will easily cover from low frequency all the way through 2 GHz ultra-high frequency.

***E7G Active audio filters: characteristics; basic circuit design; preselector applications***

**E7G01 What determines the gain and frequency characteristics of an op-amp RC active filter?**
A. The values of capacitances and resistances built into the op-amp
B. The values of capacitances and resistances external to the op-amp
C. The input voltage and frequency of the op-amp's DC power supply
D. The output voltage and smoothness of the op-amp's DC power supply

**ANSWER B:** One of the nice features of an op-amp is that it will operate as an RC filter – resistors (R) and capacitors (C) are in a circuit outside the op-amp. They are not built into the op-amp – they are external.

**E7G02 What causes ringing in a filter?**

A. The slew rate of the filter
B. The bandwidth of the filter
C. The filter shape, as measured in the frequency domain
D. The gain of the filter

**ANSWER C:** The shape of an extremely tight filter, measured in the frequency domain, may lead to ringing to the point where it could cover up extremely weak signals. We use extremely tight filters in a moon-bounce receiver on VHF and UHF. The change from mechanical filters to digital signal processing has helped minimize ringing.

---

**E7G03 What are the advantages of using an op-amp instead of LC elements in an audio filter?**

A. Op-amps are more rugged and can withstand more abuse than can LC elements
B. Op-amps are fixed at one frequency
C. Op-amps are available in more varieties than are LC elements
D. Op-amps exhibit gain rather than insertion loss

**ANSWER D:** Here we are, back to that wonderful op-amp which exhibits high gain with negligible insertion loss because its input impedance is also high.

---

**E7G04 What type of capacitors should be used in a high-stability op-amp RC active filter circuit?**

A. Electrolytic
B. Disc ceramic
C. Polystyrene
D. Paper dielectric

**ANSWER C:** The tiny op-amp filter should use polystyrene capacitors because they are very stable and will not vary when the circuit begins to heat up.

---

**E7G05 How can unwanted ringing and audio instability be prevented in a multisection op-amp RC audio filter circuit?**

A. Restrict both gain and Q
B. Restrict gain, but increase Q
C. Restrict Q, but increase gain
D. Increase both gain and Q

**ANSWER A:** In order to keep an op-amp from going into oscillation, the gain and the Q are restricted and set by the feedback circuit.

---

**E7G06 What parameter must be selected when selecting the resistor and capacitor values for an RC active filter using an op-amp?**

A. Filter bandwidth
B. Desired current gain
C. Temperature coefficient
D. Output-offset overshoot

**ANSWER A:** Resistor and capacitor values for an RC active filter, using an operational amplifier, will have greatest effect on the filter bandwidth.

---

**E7G07 The design of a preselector involves a trade-off between bandwidth and what other factor?**

A. The amount of ringing
B. Insertion loss
C. The number of parts
D. The choice of capacitors or inductors

**ANSWER B:** A receiver preselector may assist in reducing out-of-band intermodulation and interference. However, you must look at the slight sacrifice of receiver sensitivity due to the insertion loss of the preselector.

**E7G08 When designing an op-amp RC active filter for a given frequency range and Q, what steps are typically followed when selecting the external components?**

A. Standard capacitor values are chosen first, the resistances are calculated, then resistors of the nearest standard value are used
B. Standard resistor values are chosen first, the capacitances are calculated, then capacitors of the nearest standard value are used
C. Standard resistor and capacitor values are used, the circuit is tested, then additional resistors are added to make any adjustments
D. Standard resistor and capacitor values are used, the circuit is tested, then additional capacitors are added to make any adjustments

**ANSWER A:** When designing an RC active filter for a specific frequency range and selectivity, always choose a readily-available capacitor value on which to base your op-amp filter calculations. Then calculate resistance and work around a standard resistance value, if possible, that won't require you to develop complex series/parallel resistor circuits. Obtaining specific values of resistors by using standard values in parallel and series combinations is much easier to do than finding special values of capacitors.

---

**E7G09 When designing an op-amp RC active filter for a given frequency range and Q, why are the external capacitance values usually chosen first, then the external resistance values calculated?**

A. An op-amp will perform as an active filter using only standard external capacitance values
B. The calculations are easier to make with known capacitance values rather than with known resistance values
C. Capacitors with unusual capacitance values are not widely available, so standard values are used to begin the calculations
D. The equations for the calculations can only be used with known capacitance values

**ANSWER C:** If you ended up with a well-designed RC filter, but with capacitor values that are not available, you would have to redesign everything. Avoid this surprise at the end by developing the design of your op-amp RC filter at the beginning with capacitors of commonly available values.

---

**E7G10 What are the principal uses of an op-amp RC active filter in amateur circuitry?**

A. High-pass filters used to block RFI at the input to receivers
B. Low-pass filters used between transmitters and transmission lines
C. Filters used for smoothing power-supply output
D. Audio filters used for receivers

**ANSWER D:** In amateur radio transceivers, op-amp circuits are used in the audio section of receivers.

---

**E7G11 Where should an op-amp RC active audio filter be placed in an amateur receiver?**

A. In the IF strip, immediately before the detector
B. In the audio circuitry immediately before the speaker or phone jack
C. Between the balanced modulator and frequency multiplier
D. In the low-level audio stages

**ANSWER D:** We use an op amp in the low-level audio stages in an amateur receiver.

---

## Subelement E8 – Signals and Emissions [4 Exam Questions — 4 Groups]

***E8A AC waveforms: sine wave, square wave, sawtooth wave; AC measurements: peak, peak-to-peak and root-mean-square (RMS) value, peak-envelope-power (PEP) relative to average***

**E8A01 Starting at a positive peak, how many times does a sine wave cross the zero axis in one complete cycle?**

A. 180 times
B. 4 times
C. 2 times
D. 360 times

**ANSWER C:** In a complete cycle, the sine wave begins at the axis, goes up to a maximum, then down through the axis once to a maximum in the opposite direction, and then back to the zero axis for a total of two crossings, completing one cycle.

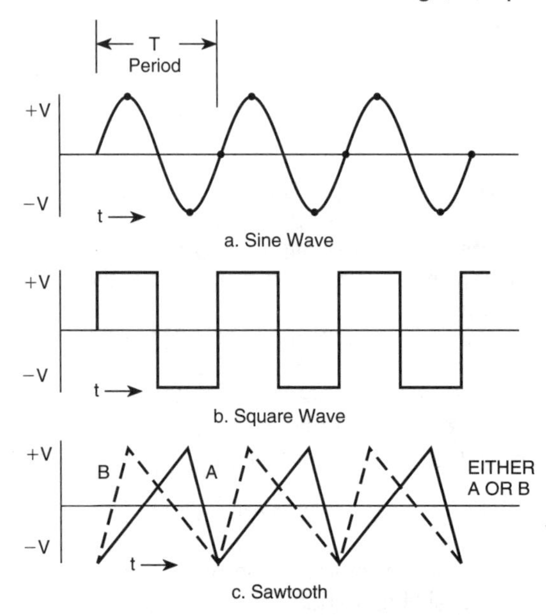


**Sine, Square and Sawtooth Waveforms**

**E8A02 What is a wave called that abruptly changes back and forth between two voltage levels and remains an equal time at each level?**

A. A sine wave
B. A cosine wave
C. A square wave
D. A sawtooth wave

**ANSWER C:** A square waveform has steep waveform edges, in which the waveform changes rapidly from one voltage level to another. The times that the waveform is at each level are equal. The key words are abruptly changes back and forth; when the waveform does that, then it is a square wave. See E8A01.

**E8A03 What sine waves added to a fundamental frequency make up a square wave?**

A. A sine wave 0.707 times the fundamental frequency
B. All odd and even harmonics
C. All even harmonics
D. All odd harmonics

**ANSWER D:** The square wave is a tough one to rid of harmonics, because the fundamental frequency and all odd harmonics may be found in a square wave. But only the odd harmonics – 1, 3, 5, 7, etc. Square people are odd, aren't they?

---

**E8A04 What type of wave is made up of a sine wave of a fundamental frequency and all its odd harmonics?**

A. A square wave
B. A sine wave
C. A cosine wave
D. A tangent wave

**ANSWER A:** If you're odd, you're probably square.

---

**E8A05 What is a sawtooth wave?**

A. A wave that alternates between two values and spends an equal time at each level
B. A wave with a straight line rise time faster than the fall time (or vice versa)
C. A wave that produces a phase angle tangent to the unit circle
D. A wave whose amplitude at any given instant can be represented by a point on a wheel rotating at a uniform speed

**ANSWER B:** A sawtooth wave has a straight line rise, and an immediate fall, or vice versa. It looks like a saw blade if you look at it on an oscilloscope. See E8A01.

---

**E8A06 What type of wave has a rise time significantly faster than the fall time (or vice versa)?**

A. A cosine wave
B. A square wave
C. A sawtooth wave
D. A sine wave

**ANSWER C:** The sawtooth wave can be backwards, too – an immediate rise time significantly faster than the fall time. See E8A01.

---

**E8A07 What type of wave is made up of sine waves of a fundamental frequency and all harmonics?**

A. A sawtooth wave
B. A square wave
C. A sine wave
D. A cosine wave

**ANSWER A:** If it has all the *harmonics* (key word), the answer is a sawtooth wave.

---

**E8A08 What is the peak voltage at a common household electrical outlet?**

A. 240 volts
B. 170 volts
C. 120 volts
D. 340 volts

**ANSWER B:** When you measure your house line voltage with a volt meter, you are measuring RMS. If you were to look at it on an oscilloscope, and look at the electrical peaks, it would be 1.414 times higher than 120 volts, or a total of 170 volts peak.

---

**E8A09 What is the peak-to-peak voltage at a common household electrical outlet?**

A. 240 volts
B. 120 volts
C. 340 volts
D. 170 volts

**ANSWER C:** If you really want to impress your friends about your house line voltage, tell them that you have 340 volts coming out of the socket. If they bet you don't, tell them you are stating the voltage as peak-to-peak, which is twice as much as peak voltage.

---

**E8A10 What is the RMS voltage at a common household electrical power outlet?**

A. 120-V AC
B. 340-V AC
C. 85-V AC
D. 170-V AC

**ANSWER A:** Household AC line voltage is rated at 120-V AC RMS.

---

**E8A11 What is the RMS value of a 340-volt peak-to-peak pure sine wave?**

A. 120-V AC
B. 170-V AC
C. 240-V AC
D. 300-V AC

**ANSWER A:** 340 volts peak-to-peak divided in half is 170 volts peak. Multiply the peak voltage by 0.707, and you end up with 120 volts AC, RMS.

---

**E8A12 What is the equivalent to the root-mean-square value of an AC voltage?**

A. The AC voltage found by taking the square of the average value of the peak AC voltage
B. The DC voltage causing the same heating in a given resistor as the peak AC voltage
C. The DC voltage causing the same heating in a given resistor as the RMS AC voltage of the same value
D. The AC voltage found by taking the square root of the average AC value

**ANSWER C:** You sometimes see this as RMS, also called effective value of an AC voltage. Since AC voltage is constantly, continuously changing in amplitude, we must look at its root mean square value before we can come up with the same heating effect in a given resistor as a constant DC voltage.

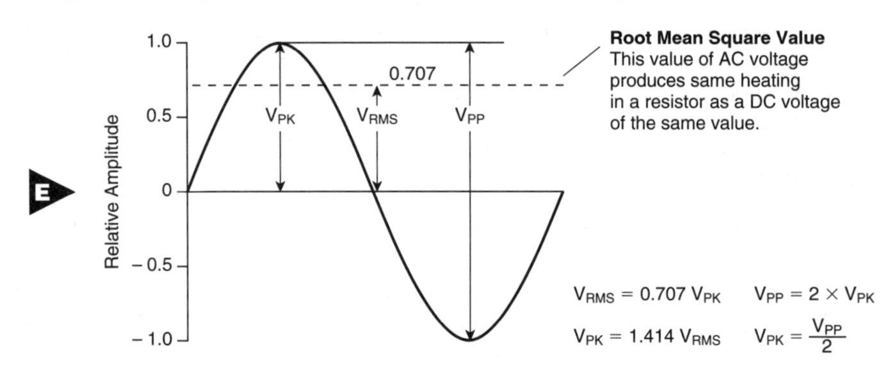


**RMS ($V_{RMS}$), Peak ($V_{PK}$), and Peak-to-Peak ($V_{PP}$) Voltage**

---

**E8A13 What would be the most accurate way of measuring the RMS voltage of a complex waveform?**

A. By using a grid dip meter
B. By measuring the voltage with a D'Arsonval meter
C. By using an absorption wavemeter
D. By measuring the heating effect in a known resistor

**ANSWER D:** In a very complex wave form, one way to determine its effective value is to measure the heating effect in a known resistor, compared to a DC known voltage.

---

**E8A14 For many types of voices, what is the approximate ratio of PEP to average power during a modulation peak in a single-sideband phone signal?**

A. 2.5 to 1
B. 25 to 1
C. 1 to 1
D. 100 to 1

**ANSWER A:** When you talk about power output from your SSB worldwide set, we normally refer to power as peak envelope power. Depending on your modulation, the ratio of PEP to average power is about 2.5 to 1. You can remember 2.5 because that's very close to 2.5 kHz of bandwidth for a properly-modulated SSB signal.

**E8A15 In a single-sideband phone signal, what determines the PEP-to-average power ratio?**

A. The frequency of the modulating signal
B. The speech characteristics
C. The degree of carrier suppression
D. The amplifier power

**ANSWER B:** It's not all that easy to determine the ratio of peak envelope power to average power in your worldwide SSB set because your speech characteristics may vary from those of another operator when using the same equipment.

**On SSB transmit, peak envelope power (PEP) is determined by your speech characteristics.**

---

**E8A16 What is the approximate DC input power to a Class B RF power amplifier stage in an FM-phone transmitter when the PEP output power is 1500 watts?**

A. 900 watts
B. 1765 watts
C. 2500 watts
D. 4500 watts

**ANSWER C:** You can determine input power by dividing PEP output power by the efficiency of the amplifier. Because efficiency is normally rated in a percentage, divide percentage efficiency by 100 to get the decimal value. In this problem, divide 1500 watts PEP output power by 0.6 because the efficiency of a Class B amplifier is approximately 60 percent. They don't tell you that – you must remember that a Class B amp is a bit more efficient than a class AB amplifier that has an efficiency of 50%. 1500 divided by 0.6 gives a power input of approximately 2500 watts, and for the common Class B amplifier, 2500 watts in will normally provide about 1500 watts out. Here is the formula to calculate Peak Envelope Power:

**Peak Envelope Power**

$PEP = P_{DC} \times \text{Efficiency}$

PEP = Peak envelope power in **watts**

$P_{DC}$ = Input DC power in **watts**

| Amplifier Class | Efficiency |
|---|---|
| C | 80% |
| B | 60% |
| AB | 50% |
| A | <50% |

**E8A17 What is the approximate DC input power to a Class AB RF power amplifier stage in an unmodulated carrier transmitter when the PEP output power is 500 watts?**

A. 250 watts
B. 600 watts
C. 1000 watts
D. 1500 watts

**ANSWER C:** The Class AB amplifier is a real work horse, and is one of the most common amps found in worldwide ham radio shacks. It's 50 percent efficient. If you're putting out 500 watts, your input power is approximately 1000 watts. Remember, Class C amplifier efficiency is about 80 percent; Class B is approximately 60 percent; and Class AB is approximately 50 percent. These values will not be given on your test, so be sure to have them memorized before the big examination.

**E8A18 What is the period of a wave?**

A. The time required to complete one cycle
B. The number of degrees in one cycle
C. The number of zero crossings in one cycle
D. The amplitude of the wave

**ANSWER A:** The period of a single wave is the time required to complete one cycle of that wave. See illustration at E8A01.

***E8B FCC emission designators versus emission types; modulation symbols and transmission characteristics; modulation methods; modulation index; deviation ratio; pulse modulation: width; position***

**E8B01 What is emission A3C?**

A. Facsimile
B. RTTY
C. ATV
D. Slow Scan TV

**ANSWER A:** Three character signals and emissions designators may still show up on paperwork going to and from the FCC, or paperwork to local frequency coordinators. When you see the number 3 in the middle of an emission designator, it usually means voice or other analog information. The A means amplitude modulation, and the C is what gives away this answer – C for faCsimile. See the ITU Emission Designators on page 200.

**E8B02 What type of emission is produced when an AM transmitter is modulated by a facsimile signal?**

A. A3F
B. A3C
C. F3F
D. F3C

**ANSWER B:** Remember, C for faCsimile. Put this together with A for Amplitude modulated and you have the correct answer. While facsimile is not real big on the ham bands, most ham sets will accept weather facsimile programs when tied into a computer, and read out FAXed weather charts quite well.

**E8B03 What does a facsimile transmission produce?**

A. Tone-modulated telegraphy
B. A pattern of printed characters designed to form a picture
C. Printed pictures by electrical means
D. Moving pictures by electrical means

**ANSWER C:** Just like your telephone FAX, ham radio facsimile will give you *printed pictures by electrical means* (key words).

**Weather fax is just one way that mariners use ham radio**

**E8B04 What is emission F3F?**

A. Modulated CW
B. Facsimile
C. RTTY
D. Television

**ANSWER D:** Television. F for television, and F for the frequency modulated part of the TV signal.

**E8B05 What type of emission is produced when an SSB transmitter is modulated by a slow-scan television signal?**

A. J3A
B. F3F
C. A3F
D. J3F

**ANSWER D:** Single sideband always has that Jiggly sound, doesn't it? Emission designators for SSB will always start out with a J3. And since we're talking about television, it will be J3F.

**E8B06 If the first symbol of an ITU emission designator is J, representing a single-sideband, suppressed-carrier signal, what information about the emission is described?**

A. The nature of any signal multiplexing
B. The type of modulation of the main carrier
C. The maximum permissible bandwidth
D. The maximum signal level, in decibels

**ANSWER B:** The first symbol of an ITU emission designator is the type of modulation of the main carrier.

**E8B07 If the second symbol of an ITU emission designator is 1, representing a single channel containing quantized, or digital information, what information about the emission is described?**

A. The maximum transmission rate, in bauds
B. The maximum permissible deviation
C. The nature of signals modulating the main carrier
D. The type of information to be transmitted

**ANSWER C:** Here the second symbol is the number 1 – this is the NATURE of the signals modulating the main carrier.

## ITU Emission Designators

**First Symbol – Modulation system**

A – Double sideband AM
C – Vestigial sideband AM
D – Amplitude/angle modulated
F – Frequency modulation
G – Phase modulation
H – Single sideband/full carrier
J – Single sideband/suppressed carrier
K – AM pulse
L – Pulse modulated in width/duration
M – Pulse modulated in position/phase
N – Unmodulated carrier
P – Unmodulated pulses
Q – Angle modulated during pulse
R – Single sideband/reduced or variable level carrier
V – Combination of pulse emissions
W – Other types of pulses

**Second Symbol – Nature of signal modulating carrier**

0 – No modulation
1 – Digital data without modulated subcarrier
2 – Digital data on modulated subcarrier
3 – Analog modulated
7 – Two or more channels of digital data
8 – Two or more channels of analog data
9 – Combination of analog and digital information
X – Other

**Third Symbol – Information to be conveyed**

A – Manually received telegraphy
B – Automatically received telegraphy
C – Facsimile (FAX)
D – Digital information
E – Voice telephony
F – Video/television
N – No information
W – Combination of these
X – Other

The signal may be "assembled" from the different elements of the emission designators.
For example, an amplitude-modulated voice single sideband signal is J3E: J = single sideband; 3 = analog, single channel; E = telephony.

1. **CW** - International Morse code telegraphy emissions having designators with A, C, H, J or R as the first symbol, 1 as the second symbol, A or B as the third symbol.
2. **DATA** - Telemetry, telecommand and computer communications emissions having designators with A, C, D, F, G, H, J or R as the first symbol: 1 as the second symbol; D as the third symbol; and also emission J2D. Only a digital code of a type specifically authorized in the §Part 97.3 rules may be transmitted.
3. **IMAGE** - Facsimile and television emissions having designators with A, C, D, F, G, H, J or R as the first symbol; 1, 2 or 3 as the second symbol; C or F as the third symbol; and also emissions having B as the first symbol; 7, 8 or 9 as the second symbol; W as the third symbol.
4. **MCW (Modulated carrier wave)** - Tone-modulated international Morse code telegraphy emissions having designators with A, C, D, F, G, H or R as the first symbol; 2 as the second symbol; A or B as the third symbol.
5. **PHONE** - Speech and other sound emissions having designators with A, C, D, F, G, H, J or R as the first symbol; 1, 2 or 3 as the second symbol; E as the third symbol. Also speech emissions having B as the first symbol; 7, 8 or 9 as the second symbol; E as the third symbol. MCW for the purpose of performing the station identification procedure, or for providing telegraphy practice interspersed with speech, or incidental tones for the purpose of selective calling, or alerting or to control the level of a demodulated signal may also be considered phone.
6. **PULSE** - Emissions having designators with K, L, M, P, Q, V or W as the first symbol; , 1, 2, 3, 7, 8, 9 or X as the second symbol; A, B, C, D, E, F, N, W or X as the third symbol.
7. **RTTY (Radioteletype)** - Narrow-band, direct-printing telegraphy emissions having designators with A, C, D, F, G, H, J or R as the first symbol; 1 as the second symbol; B as the third symbol; and also emission J2B. Only a digital code of a type specifically authorized in the §Part 97.3 rules may be transmitted.
8. **SS (Spread Spectrum)** - Emissions using bandwidth-expansion modulation emissions having designators with A, C, D, F, G, H, J or R as the first symbol; X as the second symbol; X as the third symbol. Only a SS emission of a type specifically authorized in §Part 97.3 rules may be transmitted.
9. **TEST** - Emissions containing no information having the designators with N as the third symbol. Test does not include pulse emissions with no information or modulation unless pulse emissions are also authorized in the frequency band.

**E8B08 If the third symbol of an ITU emission designator is D, representing data transmission, telemetry or telecommand, what information about the emission is described?**

A. The maximum transmission rate, in bauds
B. The maximum permissible deviation
C. The nature of signals modulating the main carrier
D. The type of information to be transmitted

**ANSWER D:** The third symbol in an ITU emission designator is the type of INFORMATION to be transmitted.

---

**E8B09 – Question deleted from Element 4 Pool by QPC.**

---

**E8B10 How does the modulation index of a phase-modulated emission vary with RF carrier frequency (the modulated frequency)?**

A. It increases as the RF carrier frequency increases
B. It decreases as the RF carrier frequency increases
C. It varies with the square root of the RF carrier frequency
D. It does not depend on the RF carrier frequency

**ANSWER D:** This is one of those answers that has a "no" in it that makes it correct. When you calculate modulation index, you do *not* (key word) look at the actual RF carrier frequency as part of your calculations.

---

**E8B11 In an FM-phone signal having a maximum frequency deviation of 3000 Hz either side of the carrier frequency, what is the modulation index when the modulating frequency is 1000 Hz?**

A. 3 C. 3000
B. 0.3 D. 1000

**ANSWER A:** 3000 divided by 1000 gives you a modulation index of 3.

$$\text{Modulation Index } (\chi) = \frac{\text{Peak Deviation (D)}}{\text{Modulation Frequency } (m)} \text{ or } \chi = \frac{D}{m}$$

---

**E8B12 What is the modulation index of an FM-phone transmitter producing a maximum carrier deviation of 6 kHz when modulated with a 2-kHz modulating frequency?**

A. 6000 C. 2000
B. 3 D. 1/3

**ANSWER B:** 6 divided by 2 gives us a modulation index of 3.

---

**E8B13 What is the deviation ratio of an FM-phone signal having a maximum frequency swing of plus or minus 5 kHz and accepting a maximum modulation rate of 3 kHz?**

A. 60 C. 0.6
B. 0.167 D. 1.67

**ANSWER D:** Divide the maximum carrier frequency swing by the maximum modulation frequency. 3 kHz modulation frequency into 5 kHz frequency swing gives you a deviation ratio of 1.67, answer D. Deviation ratio is the ratio of the maximum carrier swing to your highest audio modulating frequency when you speak into the microphone.

$$\text{Deviation Ratio} = \frac{\text{Maximum Carrier Frequency Deviation in kHz}}{\text{Maximum Modulation Frequency in kHz}}$$

**E8B14 In a pulse-width modulation system, why is the transmitter's peak power much greater than its average power?**

A. The signal duty cycle is less than 100%
B. The signal reaches peak amplitude only when voice modulated
C. The signal reaches peak amplitude only when voltage spikes are generated within the modulator
D. The signal reaches peak amplitude only when the pulses are also amplitude modulated

**ANSWER A:** Since pulse modulation is not a continuous burst of power, the duty cycle is always less than 100 percent.

**E8B15 What is one way that voice is transmitted in a pulse-width modulation system?**

A. A standard pulse is varied in amplitude by an amount depending on the voice waveform at that instant
B. The position of a standard pulse is varied by an amount depending on the voice waveform at that instant
C. A standard pulse is varied in duration by an amount depending on the voice waveform at that instant
D. The number of standard pulses per second varies depending on the voice waveform at that instant

**ANSWER C:** Remember pulse-width and duration. A standard pulse is varied in duration by an amount depending on the modulating waveform of the voice signal.

**E8B16 – Question deleted from Element 4 Pool by QPC.**

**E8B17 Which of the following describe the three most-used symbols of an ITU emission designator?**

A. Type of modulation, transmitted bandwidth and modulation code designator
B. Bandwidth of the modulating signal, nature of the modulating signal and transmission rate of signals
C. Type of modulation, nature of the modulating signal and type of information to be transmitted
D. Power of signal being transmitted, nature of multiplexing and transmission speed

**ANSWER C:** The three most-used symbols of an ITU emission, like J3E, would be *type* of modulation (J), *nature* of the modulating signal (3), and *type* of information to be transmitted (E). Remember TNT.

**E8B18 If the first symbol of an ITU emission designator is G, representing a phase-modulated signal, what information about the emission is described?**

A. The nature of any signal multiplexing
B. The maximum permissible deviation
C. The nature of signals modulating the main carrier
D. The type of modulation of the main carrier

**ANSWER D:** The first letter always represents the type of modulation of the main carrier.

---

**E8B19 In a pulse-position modulation system, what parameter does the modulating signal vary?**

A. The number of pulses per second
B. Both the frequency and amplitude of the pulses
C. The duration of the pulses
D. The time at which each pulse occurs

**ANSWER D:** This question is about a pulse-position modulation system. The position is the time at which each pulse occurs. The modulating signal varies the *time* when pulses occur.

---

**E8B20 In a pulse-width modulation system, what parameter does the modulating signal vary?**

A. Pulse frequency
B. Pulse duration
C. Pulse amplitude
D. Pulse intensity

**ANSWER B:** It is the *duration* of the pulse that the modulating signal varies.

---

**E8B21 How are the pulses of a pulse-modulated signal usually transmitted?**

A. A pulse of relatively short duration is sent; a relatively long period of time separates each pulse
B. A pulse of relatively long duration is sent; a relatively short period of time separates each pulse
C. A group of short pulses are sent in a relatively short period of time; a relatively long period of time separates each group
D. A group of short pulses are sent in a relatively long period of time; a relatively short period of time separates each group

**ANSWER A:** During pulse modulation, the duration of each pulse is relatively short, and there is a long period of time between each of these separate pulses. The long separation time is between each pulse.

---

**E8B22 In an FM-phone signal, what is the term for the ratio between the deviation of the frequency modulated signal and the modulating frequency?**

A. FM compressibility
B. Quieting index
C. Percentage of modulation
D. Modulation index

**ANSWER D:** Be careful with this one. Since they don't say the highest modulating frequency, but say only "modulating frequency," they are looking for the modulation index. Notice that there is not any answer that says "deviation ratio" to get you in trouble.

---

**E8B23 What is meant by deviation ratio?**

A. The ratio of the audio modulating frequency to the center carrier frequency
B. The ratio of the maximum carrier frequency deviation to the highest audio modulating frequency
C. The ratio of the carrier center frequency to the audio modulating frequency
D. The ratio of the highest audio modulating frequency to the average audio modulating frequency

**ANSWER B:** It's important to set the deviation ratio properly on your FM transceiver to prevent splatter, and a signal bandwidth that is too wide. Deviation ratio is the ratio of the maximum carrier swing to your highest audio modulating frequency when you speak into the microphone. (See E8B13.)

**E8B24 What is the deviation ratio of an FM-phone signal having a maximum frequency swing of plus or minus 7.5 kHz and accepting a maximum modulation rate of 3.5 kHz?**

A. 2.14
B. 0.214
C. 0.47
D. 47

**ANSWER A:** To calculate deviation ratio of an FM signal, simply divide the maximum carrier frequency swing by the maximum modulation frequency. The small number goes into the larger number. 3.5 kHz into 7.5 kHz ends up with a deviation ratio of 2.14. Simple, huh?

***E8C Digital signals: including CW; digital signal information rate vs. bandwidth; spread-spectrum communications***

**E8C01 What digital code consists of elements having unequal length?**

A. ASCII
B. AX.25
C. Baudot
D. Morse code

**ANSWER D:** Morse Code is made up of unequal length elements. Dits are short, and dahs are long.

**E8C02 What are some of the differences between the Baudot digital code and ASCII?**

A. Baudot uses four data bits per character, ASCII uses seven; Baudot uses one character as a shift code, ASCII has no shift code
B. Baudot uses five data bits per character, ASCII uses seven; Baudot uses two characters as shift codes, ASCII has no shift code
C. Baudot uses six data bits per character, ASCII uses seven; Baudot has no shift code, ASCII uses two characters as shift codes
D. Baudot uses seven data bits per character, ASCII uses eight; Baudot has no shift code, ASCII uses two characters as shift codes

**ANSWER B:** Baudot = 5 bits per character. ASCII = 7 bits per character. Baudot uses two characters as a shift code, whereas ASCII has no shift code. Since ASCII has seven bits, it has enough code combinations to represent both upper case and lower case character sets without a shift code.

**E8C03 What is one advantage of using the ASCII code for data communications?**

A. It includes built-in error-correction features
B. It contains fewer information bits per character than any other code
C. It is possible to transmit both upper and lower case text
D. It uses one character as a shift code to send numeric and special characters

**ANSWER C:** ASCII code allows upper and lower case text to be transmitted; Baudot does not.

**E8C04 What digital communications system is well suited for meteor-scatter communications at times other than during meteor showers?**

A. ACSSB
B. Computerized high speed CW (HSCW)
C. AMTOR
D. Spread spectrum

**ANSWER B:** During big meteor showers, like the Leonids in 2001, many types of modulation were bouncing off the ionized trails. But this question asks about meteor-scatter OTHER THAN during meteor showers, and here is where high-speed, computerized CW works best.

---

**E8C05 What type of error control system does Mode A AMTOR use?**

A. Each character is sent twice
B. The receiving station checks the calculated frame check sequence (FCS) against the transmitted FCS
C. The receiving station checks the calculated frame parity against the transmitted parity
D. The receiving station automatically requests repeats when needed

**ANSWER D:** In mode A AMTOR, if the signal is not received correctly, the receiving station automatically requests repeats. AMTOR is that chirping sound you hear on high frequency.

---

**E8C06 What type of error control system does Mode B AMTOR use?**

A. Each character is sent twice
B. The receiving station checks the calculated frame check sequence (FCS) against the transmitted FCS
C. The receiving station checks the calculated frame parity against the computer-sequencing clock
D. The receiving station automatically requests repeats when needed

**ANSWER A:** In mode B AMTOR, all characters are sent twice.

---

**E8C07 What is the necessary bandwidth of a 13-WPM international Morse code emission A1A transmission?**

A. Approximately 13 Hz
B. Approximately 26 Hz
C. Approximately 52 Hz
D. Approximately 104 Hz

**ANSWER C:** For determining the bandwidth of a Morse code signal, bandwidth equals:

$$BW_{CW} = \text{baud rate} \times \text{wpm} \times \text{fading factor}$$

For CW, the baud rate is 0.8 and the fading factor is 5, therefore, multiply 0.8 times 13 wpm times 5. The answer works out to be 52 Hz.

---

**E8C08 What is the necessary bandwidth for a 170-hertz shift, 300-baud ASCII emission J2D transmission?**

A. 0 Hz
B. 0.3 kHz
C. 0.5 kHz
D. 1.0 kHz

**ANSWER C:** Digital bandwidth is:

$$BW = \text{baud rate} + (1.2 \times \text{f shift})$$
$$BW = 300 + (1.2 \times 170) = 504\ \text{Hz} = 0.504\ \text{kHz}.$$

**E8C09 What is the necessary bandwidth of a 1000-Hz shift, 1200-baud ASCII emission F1D transmission?**

A. 1000 Hz
B. 1200 Hz
C. 440 Hz
D. 2400 Hz

**ANSWER D:** Use the formula in question E8C08. Take your baud rate at 1200 and add it to (1.2 × 1000). This gives you 1200 plus 1200, or 2400 Hz which matches answer D.

---

**E8C10 What is the necessary bandwidth of a 4800-Hz frequency shift, 9600-baud ASCII emission F1D transmission?**

A. 15.36 kHz
B. 9.6 kHz
C. 4.8 kHz
D. 5.76 kHz

**ANSWER A:** Digital bandwidth again – use the formula at question E8C08. In this problem, the baud rate is 9600 and the frequency shift is 4800. Bandwidth = 9600 + (1.2 × 4800) = 15360. The answer is 15,360 Hz, and when the decimal point is moved three places to the left, the bandwidth works out to be 15.36 kHz, matching answer A.

---

**E8C11 What term describes a wide-bandwidth communications system in which the RF carrier varies according to some predetermined sequence?**

A. Amplitude compandored single sideband
B. AMTOR
C. Time-domain frequency modulation
D. Spread-spectrum communication

**ANSWER D:** Spread spectrum communications are becoming more popular. You probably don't realize it, but your new 900-MHz cordless phone uses spread-spectrum communications. The frequency hops in a pre-arranged sequence, eliminating interference and eavesdropping.

---

**E8C12 What spread-spectrum communications technique alters the center frequency of a conventional carrier many times per second in accordance with a pseudo-random list of channels?**

A. Frequency hopping
B. Direct sequence
C. Time-domain frequency modulation
D. Frequency compandored spread-spectrum

**ANSWER A:** Spread spectrum communications alters the center frequency of an RF carrier in a pseudo-random manner causing the frequency to "hop" around from channel to channel. This is termed "frequency hopping."

---

**E8C13 What spread-spectrum communications technique uses a very fast binary bit stream to shift the phase of an RF carrier?**

A. Frequency hopping
B. Direct sequence
C. Binary phase-shift keying
D. Phase compandored spread-spectrum

**ANSWER B:** The term "direct sequence" is used to describe a system where a fast binary stream shifts the phase of an RF carrier.

**E8C14 What controls the spreading sequence of an amateur spread-spectrum transmission?**

A. A frequency-agile linear amplifier
B. A crystal-controlled filter linked to a high-speed crystal switching mechanism
C. A binary linear-feedback shift register
D. A binary code which varies if propagation changes

**ANSWER C:** Spread spectrum is relatively new for amateur radio. However, you probably are using it right now on that new 900-MHz cordless phone. The spreading sequence in spread spectrum comes from a binary linear-feedback shift register.

---

**E8C15 What makes spread-spectrum communications resistant to interference?**

A. Interfering signals are removed by a frequency-agile crystal filter
B. Spread-spectrum transmitters use much higher power than conventional carrier-frequency transmitters
C. Spread-spectrum transmitters can hunt for the best carrier frequency to use within a given RF spectrum
D. Only signals using the correct spreading sequence are received

**ANSWER D:** Just like that new cordless 900-MHz spread-spectrum telephone, spread-spectrum communications are resistant to interference because only those signals with the correct spreading sequence are received.

**900MHz cordless phones use spread-spectrum technology.**

---

**E8C16 What reduces interference from spread-spectrum transmitters to conventional communications in the same band?**

A. A spread-spectrum transmitter avoids channels within the band which are in use by conventional transmitters
B. Spread-spectrum signals appear only as low-level noise in conventional receivers
C. Spread-spectrum signals change too rapidly to be detected by conventional receivers
D. Special crystal filters are needed in conventional receivers to detect spread-spectrum signals

**ANSWER B:** Some of the answers, if studied individually, may be correct in this particular problem. But the question asks why so little interference with conventional radios within the same band, so Answer B dealing with low-level noise is the best answer.

***E8D Peak amplitude (positive and negative); peak-to-peak values: measurements; Electromagnetic radiation; wave polarization; signal-to-noise (S/N) ratio***

**E8D01 What is the term for the amplitude of the maximum positive excursion of a signal as viewed on an oscilloscope?**

A. Peak-to-peak voltage
B. Inverse peak negative voltage
C. RMS voltage
D. Peak positive voltage

**ANSWER D:** The maximum positive excursion for the amplitude of a signal on an oscilloscope is called the peak positive voltage.

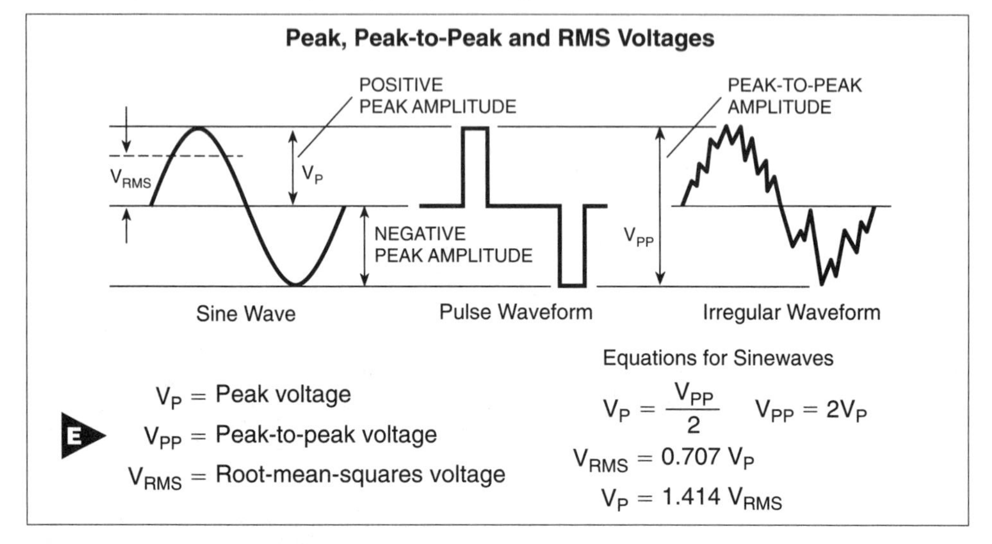


**E8D02 What is the easiest voltage amplitude dimension to measure by viewing a pure sine wave signal on an oscilloscope?**

A. Peak-to-peak voltage
B. RMS voltage
C. Average voltage
D. DC voltage

**ANSWER A:** Looking at a pure sine wave signal on a scope, it is easy to identify the peak-to-peak voltage measured from the maximum positive excursion of the signal to the maximum negative excursion of the signal.

**E8D03 What is the relationship between the peak-to-peak voltage and the peak voltage amplitude in a symmetrical waveform?**

A. 1:1
B. 2:1
C. 3:1
D. 4:1

**ANSWER B:** The relationship between peak-to-peak voltage and peak voltage is 2:1. It is important to realize that the waveform must be symmetrical – and a sine wave certainly is. $V_{PP} = 2V_P$.

**E8D04 What input-amplitude parameter is valuable in evaluating the signal-handling capability of a Class A amplifier?**

A. Peak voltage
B. RMS voltage
C. An average reading power output meter
D. Resting voltage

**ANSWER A:** In a class A amplifier, peak voltage provides a double-check on the capability and linearity of that amplifier.

---

**E8D05 What is the PEP output of a transmitter that has a maximum peak of 30 volts to a 50-ohm load as observed on an oscilloscope?**

A. 4.5 watts
B. 9 watts
C. 16 watts
D. 18 watts

**ANSWER B:** The peak envelope power, PEP, is equal to the average power output. The first step in solving this problem is to take the peak voltage and reduce it to average by multiplying $30 \times 0.707 = 21.21$ volts.

Average volts × average current = average power
Average power, $P = E^2/R$
$P = (21.21)^2 \div 50 = 8.99$ watts

---

**E8D06 If an RMS reading AC voltmeter reads 65 volts on a sinusoidal waveform, what is the peak-to-peak voltage?**

A. 46 volts
B. 92 volts
C. 130 volts
D. 184 volts

**ANSWER D:** $V_p = 1.414 \times V_{RMS}$, and $V_{pp} = 2V_p$. We will first change 65 volts RMS to peak voltage by multiplying $65 \times 1.414 = 91.91$. Since they want peak-to-peak voltage, multiply this by 2, and you end up with 184 volts.

---

**E8D07 What is the advantage of using a peak-reading voltmeter to monitor the output of a single-sideband transmitter?**

A. It would be easy to calculate the PEP output of the transmitter
B. It would be easy to calculate the RMS output power of the transmitter
C. It would be easy to calculate the SWR on the transmission line
D. It would be easy to observe the output amplitude variations

**ANSWER A:** Using a peak-reading voltmeter to monitor power output of an SSB transceiver is an easy way to calculate the actual peak envelope power output without having to go through a calculation for RMS volts to peak volts.

---

**E8D08 What is an electromagnetic wave?**

A. Alternating currents in the core of an electromagnet
B. A wave consisting of two electric fields at right angles to each other
C. A wave consisting of an electric field and a magnetic field at right angles to each other
D. A wave consisting of two magnetic fields at right angles to each other

**ANSWER C:** When you think of an electromagnetic radio wave, think of it as a combination of two fields at right angles traveling in waves through the atmosphere at the same time. One field is vertically polarized and the other is horizontally polarized. If the antenna is vertically polarized, the electric field will be vertical, and the magnetic field will be horizontal.

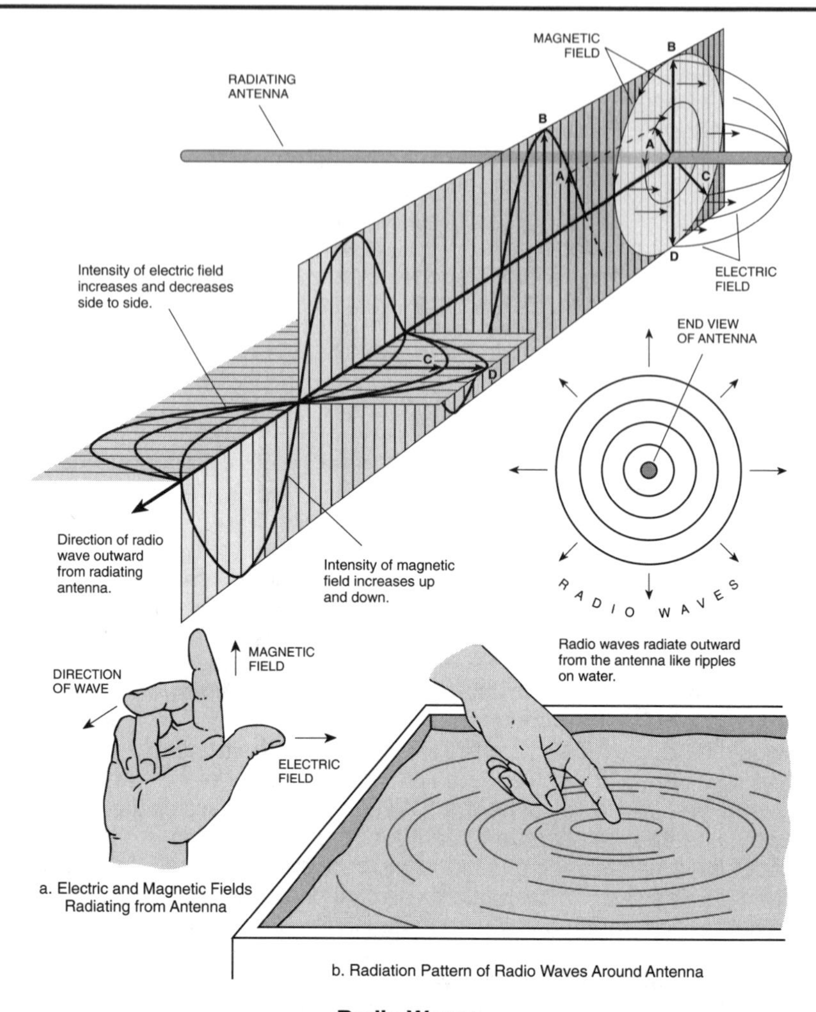


a. Electric and Magnetic Fields Radiating from Antenna

b. Radiation Pattern of Radio Waves Around Antenna

**Radio Waves**

Source: *Basic Electronics* © 1994, Master Publishing, Inc., Lincolnwood, Illinois

---

**E8D09 Which of the following best describes electromagnetic waves traveling in free space?**

A. Electric and magnetic fields become aligned as they travel
B. The energy propagates through a medium with a high refractive index
C. The waves are reflected by the ionosphere and return to their source
D. Changing electric and magnetic fields propagate the energy across a vacuum

**ANSWER D:** When we talk about "free space," we are talking about energy traveling across a *vacuum* (key words).

---

**E8D10 What is meant by circularly polarized electromagnetic waves?**

A. Waves with an electric field bent into a circular shape
B. Waves with a rotating electric field
C. Waves that circle the Earth
D. Waves produced by a loop antenna

**ANSWER B:** Some satellite enthusiasts will use circular polarization to improve their transmission and reception through an orbiting satellite. With circular polarization, the *electric field rotates* (key words). It could be right-hand circular, or left-hand circular, depending on the antenna feed system.

---

**E8D11 What is the polarization of an electromagnetic wave if its magnetic field is parallel to the surface of the Earth?**

A. Circular
B. Horizontal
C. Elliptical
D. Vertical

**ANSWER D:** Watch out for this one – here they are asking about the magnetic field which is parallel to the surface of the earth. If the magnetic field is parallel, the electric field is vertical, so polarization is vertical.

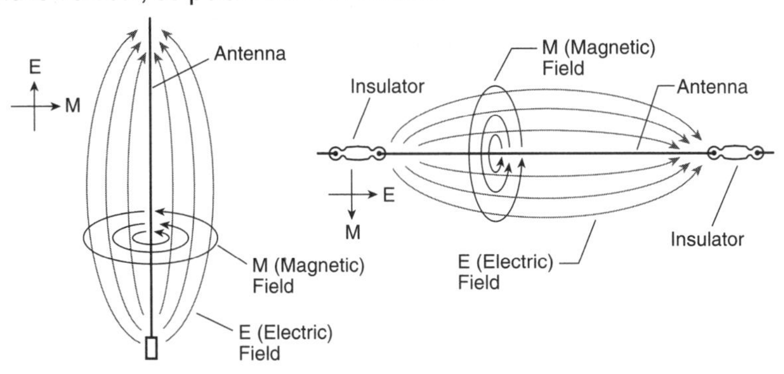


**Horizontal and Vertical Polarization**

---

**E8D12 What is the polarization of an electromagnetic wave if its magnetic field is perpendicular to the surface of the Earth?**

A. Horizontal
B. Circular
C. Elliptical
D. Vertical

**ANSWER A:** If the magnetic field is perpendicular (vertical) to the surface of the earth, the electric lines of force would be horizontal, and this would give us horizontal polarization.

---

**E8D13 What is the primary source of noise that can be heard in an HF-band receiver with an antenna connected?**

A. Detector noise
B. Induction motor noise
C. Receiver front-end noise
D. Atmospheric noise

**ANSWER D:** When you hook up an HF transceiver to your antenna, most of the noise received is atmospheric noise.

---

**E8D14 At approximately what speed do electromagnetic waves travel in free space?**

A. 300 million meters per second
B. 468 million meters per second
C. 186,300 feet per second
D. 300 million miles per second

**ANSWER A:** Electromagnetic radio waves and light waves travel at 300 million meters per second in free space.

---

**E8D15 To ensure you do not exceed the maximum allowable power, what kind of meter would you use to monitor the output signal of a properly adjusted single-sideband transmitter?**

A. An SWR meter reading in the forward direction
B. A modulation meter
C. An average reading wattmeter
D. A peak-reading wattmeter

**ANSWER D:** The peak-reading watt meter momentarily holds the highest amount of output signal in order for you to double check that you are not exceeding the maximum allowable power. Many newer, worldwide ham sets have a peak-reading power output meter with an LCD display built in.

---

**E8D16 What is the average power dissipated by a 50-ohm resistive load during one complete RF cycle having a peak voltage of 35 volts?**

A. 12.2 watts
B. 9.9 watts
C. 24.5 watts
D. 16 watts

**ANSWER A:** The first step in solving this problem is to take the peak voltage and reduce it to average voltage by multiplying 35 volts × 0.707 = 24.745 volts.

Average volts × average current = average power

Average power, $P = E^2/R$

$P = (24.745)^2 \div 50 = 12.2463$ watts

---

**E8D17 If an RMS reading voltmeter reads 34 volts on a sinusoidal waveform, what is the peak voltage?**

A. 123 volts
B. 96 volts
C. 55 volts
D. 48 volts

**ANSWER D:** In this question, 34 volts RMS is multiplied by 1.414 = 48.076 volts. Remember, to go from average to peak, multiple by 1.414. Going from peak to average, you would multiply by 0.707, as we did in the earlier problem.

## Subelement E9 – Antennas [5 Exam Questions — 5 Groups]

***E9A Isotropic radiators: definition; used as a standard for comparison; radiation pattern; basic antenna parameters: radiation resistance and reactance (including wire dipole, folded dipole), gain, beamwidth, efficiency***

**E9A01 Which of the following describes an isotropic radiator?**

A. A grounded radiator used to measure earth conductivity
B. A horizontal radiator used to compare Yagi antennas
C. A theoretical radiator used to compare other antennas
D. A spacecraft radiator used to direct signals toward the earth

**ANSWER C:** An isotropic radiator is simply theoretical; and in every April 1st issue of many magazines, advertisers try to sell "isotropic radiators" to the unsuspecting ham! We use it as a theoretical reference to calculate the gain or loss of real antenna systems.

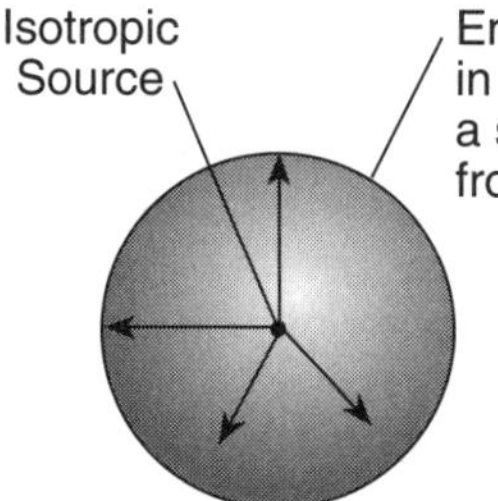


**Isotropic Radiator Pattern**

**E9A02 When is it useful to refer to an isotropic radiator?**

A. When comparing the gains of directional antennas
B. When testing a transmission line for standing-wave ratio
C. When directing a transmission toward the tropical latitudes
D. When using a dummy load to tune a transmitter

**ANSWER A:** When the gain of directional antennas and collinear arrays are compared, they are usually compared to an isotropic radiator.

**E9A03 How much gain does a 1/2-wavelength dipole have over an isotropic radiator?**

A. About 1.5 dB
B. About 2.1 dB
C. About 3.0 dB
D. About 6.0 dB

**ANSWER B:** The halfwave dipole will have approximately 2.1 dB gain over the theoretical isotropic radiator.

**E9A04 Which of the following antennas has no gain in any direction?**

A. Quarter-wave vertical
B. Yagi
C. Half-wave dipole
D. Isotropic radiator

**ANSWER D**: It is the theoretical isotropic radiator that, on paper, has equal signal dispersion in all directions, and no gain in any direction.

**E9A05 Which of the following describes the radiation pattern of an isotropic radiator?**

A. A teardrop in the vertical plane
B. A circle in the horizontal plane
C. A sphere with the antenna in the center
D. Crossed polarized with a spiral shape

**ANSWER C:** The isotropic radiation pattern extends out as the surface of a sphere with the make-believe antenna in the center. (See illustration at E9A01.)

---

**E9A06 Why would one need to know the feed point impedance of an antenna?**

A. To match impedances for maximum power transfer
B. To measure the near-field radiation density from a transmitting antenna
C. To calculate the front-to-side ratio of the antenna
D. To calculate the front-to-back ratio of the antenna

**ANSWER A:** When the antenna (the load) has the same characteristic impedance as the feedline, maximum power will be transferred. A mismatch may result in high SWR, transmit power feedback, and potential transmitter heat-up. Most coaxial cable presents a 50-ohm impedance.

---

**E9A07 What factors determine the radiation resistance of an antenna?**

A. Transmission-line length and antenna height
B. Antenna location with respect to nearby objects and the conductors' length/diameter ratio
C. It is a physical constant and is the same for all antennas
D. Sunspot activity and time of day

**ANSWER B:** Radiation resistance of a mobile antenna may be severely compromised if the antenna is mounted too low. The location of the antenna with respect to nearby objects, as well as the length-to-diameter ratio of the conductors, is very important.

---

**E9A08 What is the term for the ratio of the radiation resistance of an antenna to the total resistance of the system?**

A. Effective radiated power
B. Radiation conversion loss
C. Antenna efficiency
D. Beamwidth

**ANSWER C:** The ratio of the radiation resistance of an antenna to the total resistance of the system times 100 is the percent antenna efficiency.

---

**E9A09 What is included in the total resistance of an antenna system?**

A. Radiation resistance plus space impedance
B. Radiation resistance plus transmission resistance
C. Transmission-line resistance plus radiation resistance
D. Radiation resistance plus ohmic resistance

**ANSWER D:** To calculate the total resistance of an antenna system, it is radiation resistance plus ohmic resistances within the coils and conductors.

---

**E9A10 What is a folded dipole antenna?**

A. A dipole one-quarter wavelength long
B. A type of ground-plane antenna
C. A dipole whose ends are connected by a one-half wavelength piece of wire
D. A hypothetical antenna used in theoretical discussions to replace the radiation resistance

**ANSWER C:** The folded dipole is a fun one to build out of television twin-lead cable. As shown in the diagram, the twin-lead folded dipole is constructed by using a half-wavelength piece of twin lead. Scrape the insulation from the wire ends and twist the wires together on each end. In the center of one of the leads in the twin-lead, cut the lead and scrape some of the insulation off to expose bare wires on the end of each quarter-wave piece. Feed these wires with standard 300-ohm twin-lead, or connect a balun transformer to the bare wires to permit 50-ohm or 75-ohm coax cable to feed the folded dipole.

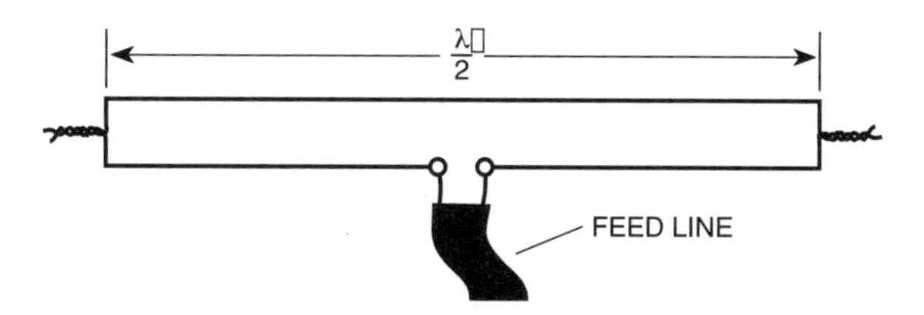


**Folded Dipole Antenna Made from 300-Ohm Twin-Lead**

---

**E9A11 What is meant by antenna gain?**

A. The numerical ratio relating the radiated signal strength of an antenna to that of another antenna
B. The numerical ratio of the signal in the forward direction to the signal in the back direction
C. The numerical ratio of the amount of power radiated by an antenna compared to the transmitter output power
D. The final amplifier gain minus the transmission-line losses (including any phasing lines present)

**ANSWER A:** We normally rate antenna gain figures in dB. This is the numerical ratio relating the radiated signal strength of your antenna to that of another antenna down the street. Generally, the ham with the higher gain antenna will always get better signal reports.

---

**E9A12 What is meant by antenna bandwidth?**

A. Antenna length divided by the number of elements
B. The frequency range over which an antenna can be expected to perform well
C. The angle between the half-power radiation points
D. The angle formed between two imaginary lines drawn through the ends of the elements

**ANSWER B:** If you operate on the lower bands, such as 40 meters, antenna bandwidth is an important consideration. Bandwidth is the frequency range over which you can expect your antenna to perform well. Small antennas on 40 meters don't have a very wide bandwidth. Monster antennas for 40 meters have good bandwidth.

---

**E9A13 How can the approximate beamwidth of a beam antenna be determined?**

A. Note the two points where the signal strength of the antenna is down 3 dB from the maximum signal point and compute the angular difference
B. Measure the ratio of the signal strengths of the radiated power lobes from the front and rear of the antenna

C. Draw two imaginary lines through the ends of the elements and measure the angle between the lines
D. Measure the ratio of the signal strengths of the radiated power lobes from the front and side of the antenna

**ANSWER A:** On directional antennas, beamwidth is an important consideration. Beamwidth is always described by the angle between the two points where the signal strength is down 3 dB from the maximum signal point. The "tighter" your beam pattern, the smaller the beamwidth.

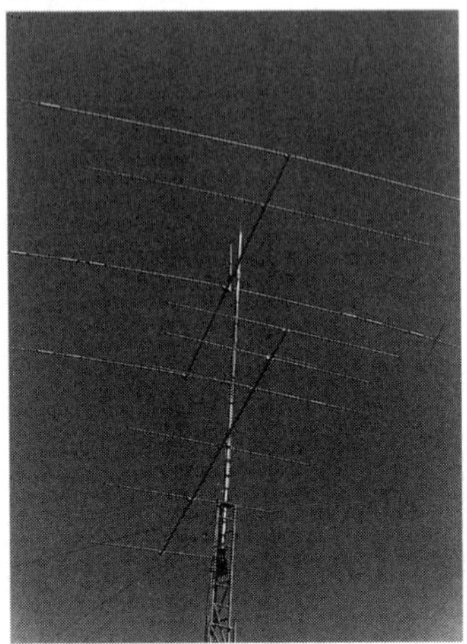

**Beamwidth and bandwidth are both important considerations in antenna selection.**

**E9A14 How is antenna efficiency calculated?**

A. (radiation resistance / transmission resistance) × 100%
B. (radiation resistance / total resistance) × 100%
C. (total resistance / radiation resistance) × 100%
D. (effective radiated power / transmitter output) × 100%

**ANSWER B:** The efficiency of your antenna is an important consideration when you plan to upgrade your antenna system. Efficiency equals the radiation resistance divided by the total resistance, multiplied by 100 percent. You want to keep your ohmic resistance as low as possible to make your antenna more efficient. Generally, bigger antenna elements have greater radiation resistance, lower ohmic resistance, higher efficiency, and increased bandwidth. Bigger is always better!

**E9A15 How can the efficiency of an HF grounded vertical antenna be made comparable to that of a half-wave dipole antenna?**

A. By installing a good ground radial system
B. By isolating the coax shield from ground
C. By shortening the vertical
D. By lengthening the vertical

**ANSWER A:** Most vertical antennas are one-quarter wavelength long, and use a mirror image counterpoise to compare to that of a halfwave antenna. A good ground radial system will give your signal more punch, lower the noise floor, lower your take-off angle, and give you good DX.

**E9A16 What theoretical reference antenna provides a comparison for antenna measurements?**

A. Quarter-wave vertical
B. Yagi
C. Bobtail curtain
D. Isotropic radiator

**ANSWER D:** The isotropic radiator is a theoretical antenna that works equally well in all directions. Its pattern is spherical around the antenna, as shown in the illustration at E9A01, and no matter where you are the field strength from the antenna will be the same. In practice, it is impossible to build an isotropic radiator, but it is used as a theoretical standard of comparison for antenna systems.

---

**E9A17 How much gain does an antenna have over a 1/2-wavelength dipole when it has 6 dB gain over an isotropic radiator?**

A. About 3.9 dB
B. About 6.0 dB
C. About 8.1 dB
D. About 10.0 dB

**ANSWER A:** The halfwave dipole has about 2.1 dB gain over an isotropic radiator. If an antenna under measurement has 6 dB gain over an isotropic radiator,
6 dB − 2.1 dB = about 3.9 dB gain over the dipole.

---

**E9A18 How much gain does an antenna have over a 1/2-wavelength dipole when it has 12 dB gain over an isotropic radiator?**

A. About 6.1 dB
B. About 9.9 dB
C. About 12.0 dB
D. About 14.1 dB

**ANSWER B:** Here is a bigger antenna that has 12 dB gain over an isotropic radiator, so 12 dB gain − 2.1 dB = about 9.9 dB over the dipole.

---

**E9A19 Which of the following describes the directivity of an isotropic radiator?**

A. Directivity in the E plane
B. Directivity in the H plane
C. Directivity in the Z plane
D. No directivity at all

**ANSWER D:** The isotropic radiator is an imaginary antenna that radiates equally in all directions with no directivity in any one direction.

---

**E9A20 What is meant by the radiation resistance of an antenna?**

A. The combined losses of the antenna elements and feed line
B. The specific impedance of the antenna
C. The equivalent resistance that would dissipate the same amount of power as that radiated from an antenna
D. The resistance in the atmosphere that an antenna must overcome to be able to radiate a signal

**ANSWER C:** Radiation resistance is an important term because it compares the power radiated from an antenna to the equivalent resistance that would dissipate the same amount of power as heat. The higher the radiation resistance of an antenna, the better its performance. Don't confuse radiation resistance with simple DC resistance – they are exactly opposite. Radiation resistance is the equivalent resistance that would dissipate the same amount of power as that radiated from the antenna.

***E9B Free-space antenna patterns: E and H plane patterns (i.e., azimuth and elevation in free-space); gain as a function of pattern; antenna design (computer modeling of antennas)***

**E9B01 What determines the free-space polarization of an antenna?**

A. The orientation of its magnetic field (H Field)
B. The orientation of its free-space characteristic impedance
C. The orientation of its electric field (E Field)
D. Its elevation pattern

**ANSWER C:** Keep in mind that it is the electric lines of force that determine the polarization of an antenna. This is called the orientation of the electric field, E Field.

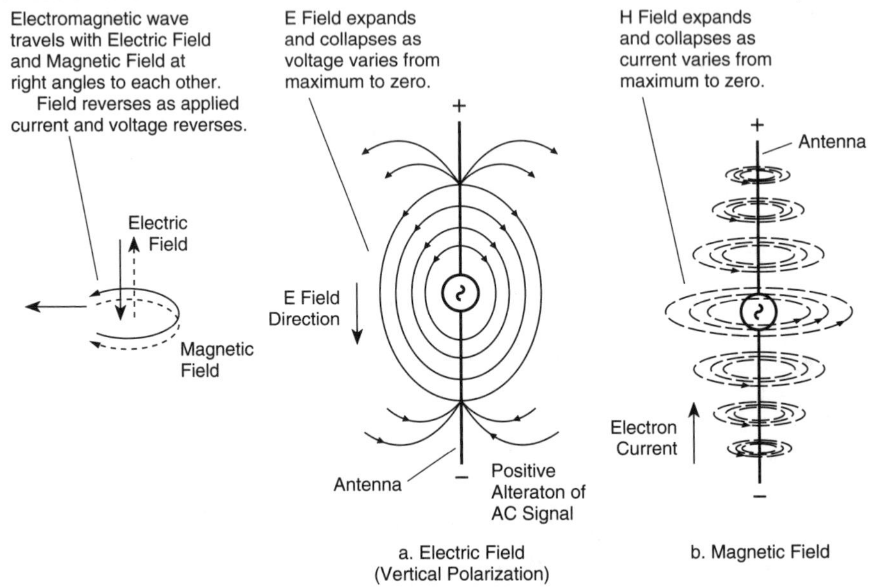


**Propagating Electromagnetic Waves**



**E9B02 In the free-space H-Field radiation pattern shown in Figure E9-1, what is the 3-dB beamwidth?**

A. 75 degrees
B. 50 degrees
C. 25 degrees
D. 30 degrees

**ANSWER B:** Looking carefully at the diagram in Figure E9-1, notice that the main lobe of the pattern intersects the 3dB points at approximately plus and minus 25 degrees. Hence, the 3-dB beamwidth is approximately 50 degrees.

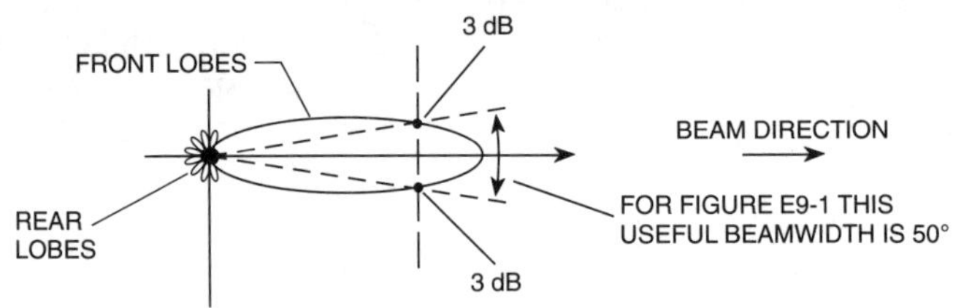


**Directional Radiation Pattern of a Yagi Beam**

**E9B03 In the free-space H-Field pattern shown in Figure E9-1, what is the front-to-back ratio?**

A. 36 dB
B. 18 dB
C. 24 dB
D. 14 dB

**ANSWER B:** If you look carefully at Figure E9-1, the rear lobe is about -18 dB down from the front lobe.

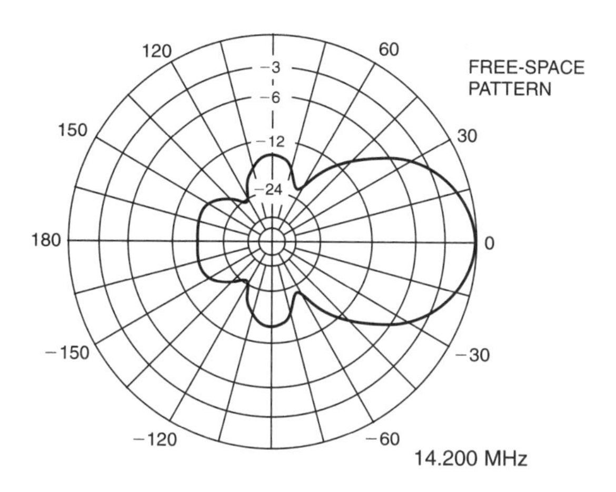


**Figure E9-1**

---

**E9B04 In the free-space H-field pattern shown in Figure E9-1, what is the front-to-side ratio?**

A. 12 dB
B. 14 dB
C. 18 dB
D. 24 dB

**ANSWER B:** Study Figure E9-1 carefully, and go with 14 dB as the front-to-side ratio. Looking at the figure, this is not an exact measurement, so just stick 14 dB in your brain before the test. Figure E9-1 will look exactly like it does here in the book when your VE team gives it to you as part of your test papers at your upcoming Extra Class exam.

---

**E9B05 What information is needed to accurately evaluate the gain of an antenna?**

A. Radiation resistance
B. E-Field and H-Field patterns
C. Loss resistance
D. All of these choices

**ANSWER D:** Information necessary to evaluate the gain of an antenna would include radiation resistance, E- and H-field patterns, loss resistance, and one additional very important consideration, boom length!

---

**E9B06 Which is NOT an important reason to evaluate a gain antenna across the whole frequency band for which it was designed?**

A. The gain may fall off rapidly over the whole frequency band
B. The feed-point impedance may change radically with frequency
C. The rearward pattern lobes may vary excessively with frequency
D. The dielectric constant may vary significantly

**ANSWER D:** The varying dielectric constant of an antenna is NOT an important reason to evaluate the gain of an antenna over an entire band.

**E9B07 What usually occurs if a Yagi antenna is designed solely for maximum forward gain?**

A. The front-to-back ratio increases
B. The feed-point impedance becomes very low
C. The frequency response is widened over the whole frequency band
D. The SWR is reduced

**ANSWER B:** Maximum forward gain will sometimes drop the feedpoint impedance to an extremely low level requiring matching transformers between the feedpoint and feed line.

---

**E9B08 If the boom of a Yagi antenna is lengthened and the elements are properly retuned, what usually occurs?**

A. The gain increases
B. The SWR decreases
C. The front-to-back ratio increases
D. The gain bandwidth decreases rapidly

**ANSWER A:** Longer boom, more gain!

---

**E9B09 What type of computer program is commonly used for modeling antennas?**

A. Graphical analysis
B. Method of Moments
C. Mutual impedance analysis
D. Calculus differentiation with respect to physical properties

**ANSWER B:** The American Radio Relay League has a software program called "Method of Moments," and this is a great way for modeling a proposed antenna on the computer screen.

---

**E9B10 What is the principle of a Method of Moments analysis?**

A. A wire is modeled as a series of segments, each having a distinct value of current
B. A wire is modeled as a single sine-wave current generator
C. A wire is modeled as a series of points, each having a distinct location in space
D. A wire is modeled as a series of segments, each having a distinct value of voltage across it

**ANSWER A:** A wire is modeled as a series of segments, each segment having a distinct value of antenna current.

---

***E9C Phased vertical antennas; radiation patterns; beverage antennas; rhombic antennas: resonant; terminated; radiation pattern; antenna patterns: elevation above real ground, ground effects as related to polarization, take-off angles as a function of height above ground***

---

**E9C01 What is the radiation pattern of two 1/4-wavelength vertical antennas spaced 1/2-wavelength apart and fed 180 degrees out of phase?**

A. Unidirectional cardioid
B. Omnidirectional
C. Figure-8 broadside to the antennas
D. Figure-8 end-fire in line with the antennas

**ANSWER D:** Two quarter-wave vertical antennas spaced one-half wavelength apart and fed 180 degrees out of phase results in a figure-8 radiation pattern with best radiation in line with the antennas.

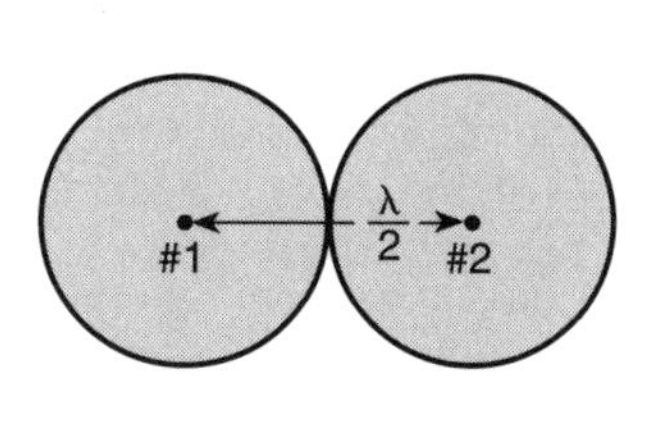


a. Two λ/4 Verticals Fed 180° Out of Phase (Figure 8 End-Fire In-Line)

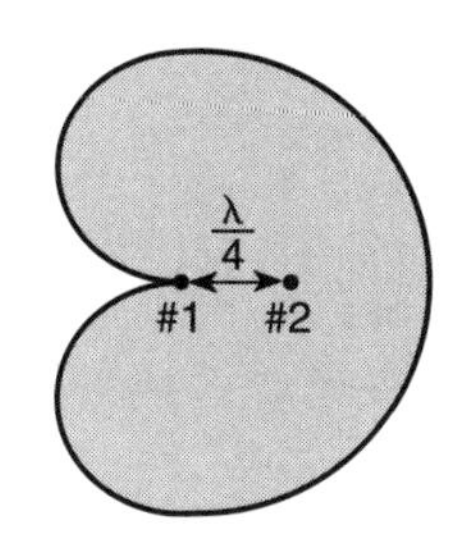


b. Two λ/4 Verticals Fed 90° Out of Phase (Cardiod)

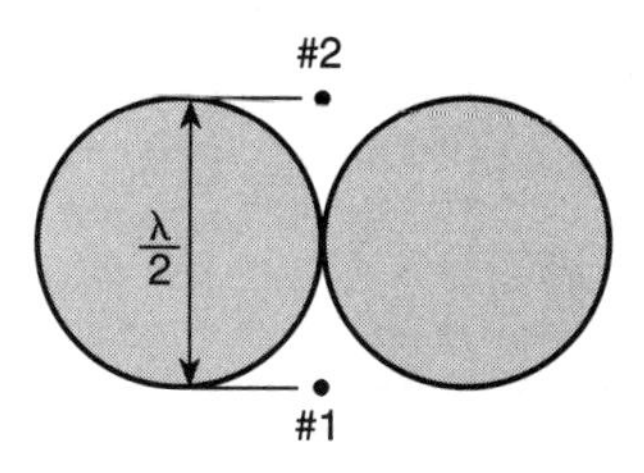


c. Two λ/4 Verticals Fed In Phase (Figure 8 Broadside)

**Antenna Radiation Patterns**

---

**E9C02 What is the radiation pattern of two 1/4-wavelength vertical antennas spaced 1/4-wavelength apart and fed 90 degrees out of phase?**

A. Unidirectional cardioid
B. Figure-8 end-fire
C. Figure-8 broadside
D. Omnidirectional

**ANSWER A:** Remember when we talked about RDF antennas for direction finding? Two quarter-wave verticals spaced one-quarter wavelength apart and fed 90 degrees out of phase results in a cardioid pattern. This is called "unidirectional."

---

**E9C03 What is the radiation pattern of two 1/4-wavelength vertical antennas spaced 1/2-wavelength apart and fed in phase?**

A. Omnidirectional
B. Cardioid unidirectional
C. Figure-8 broadside to the antennas
D. Figure-8 end-fire in line with the antennas

**ANSWER C:** Two quarter-waves spaced one-half wavelength apart and fed in phase will also result in a figure-8 pattern, but maximum gain will be broadside to the antennas.

---

**E9C04 What is the radiation pattern of two 1/4-wavelength vertical antennas spaced 1/4-wavelength apart and fed 180 degrees out of phase?**

A. Omnidirectional
B. Cardioid unidirectional
C. Figure-8 broadside to the antennas
D. Figure-8 end-fire in line with the antennas

**ANSWER D:** Now there are two quarter-wave verticals spaced only one quarter-wavelength apart, and fed 180 degrees out of phase. The result will be an end-fire figure-8 pattern in line with the antennas, just as if they were one-half wavelength apart. The pattern will be less distinct.

**E9C05 What is the radiation pattern for two 1/4-wavelength vertical antennas spaced 1/8-wavelength apart and fed 180 degrees out of phase?**

A. Omnidirectional
B. Cardioid unidirectional
C. Figure-8 broadside to the antennas
D. Figure-8 end-fire in line with the antennas

**ANSWER D:** The same two quarter-wavelength vertical antennas are spaced only one-eighth wavelength apart, and still fed 180 degrees out of phase. The result is still an end-fire figure-8 pattern in line with the antennas, but the pattern will be relatively undefined. It will begin to approach that of a simple quarter-wavelength vertical antenna.

---

**E9C06 What is the radiation pattern for two 1/4-wavelength vertical antennas spaced 1/4-wavelength apart and fed in phase?**

A. Substantially unidirectional
B. Elliptical
C. Cardioid unidirectional
D. Figure-8 end-fire in line with the antennas

**ANSWER B:** Two quarter-wavelength verticals now spaced one quarter-wavelength apart, and fed in phase, will begin to develop an elliptical pattern.

---

**E9C07 Which of the following is the best description of a resonant rhombic antenna?**

A. Unidirectional; four-sided, each side a half-wavelength long; terminated in a resistance equal to its characteristic impedance
B. Bidirectional; four-sided, each side approximately one wavelength long; open at the end opposite the transmission line connection
C. Four-sided; an LC network at each vertex except for the transmission connection; tuned to resonate at the operating frequency
D. Four-sided, each side of a different physical length; traps at each vertex for changing resonance according to band usage

**ANSWER B:** The resonant rhombic antenna is a bidirectional antenna where each side is equal to one wavelength. It is open at one end, with the transmission line connected to the other end. You need a lot of wire for a resonant rhombic on 80 meters!

---

**E9C08 What are the advantages of a terminated rhombic antenna?**

A. Wide frequency range, high gain and high front-to-back ratio
B. High front-to-back ratio, compact size and high gain
C. Unidirectional radiation pattern, high gain and compact size
D. Bidirectional radiation pattern, high gain and wide frequency range

**ANSWER A:** A terminated rhombic antenna with the right termination resistor will offer wide frequency range, high gain, and high front-to-back ratio. This is what you want.

---

**E9C09 What are the disadvantages of a terminated rhombic antenna for the HF bands?**

A. A large area for proper installation and a narrow bandwidth
B. A large area for proper installation and a low front-to-back ratio
C. A large area and four sturdy supports for proper installation
D. A large amount of aluminum tubing and a low front-to-back ratio

**ANSWER C:** The big disadvantage for the terminated rhombic antenna for high frequency is the very large area it will occupy, and the necessity for a minimum of 4 sturdy supports for proper installation. You better have plenty of real estate!

## Rhombic Antenna

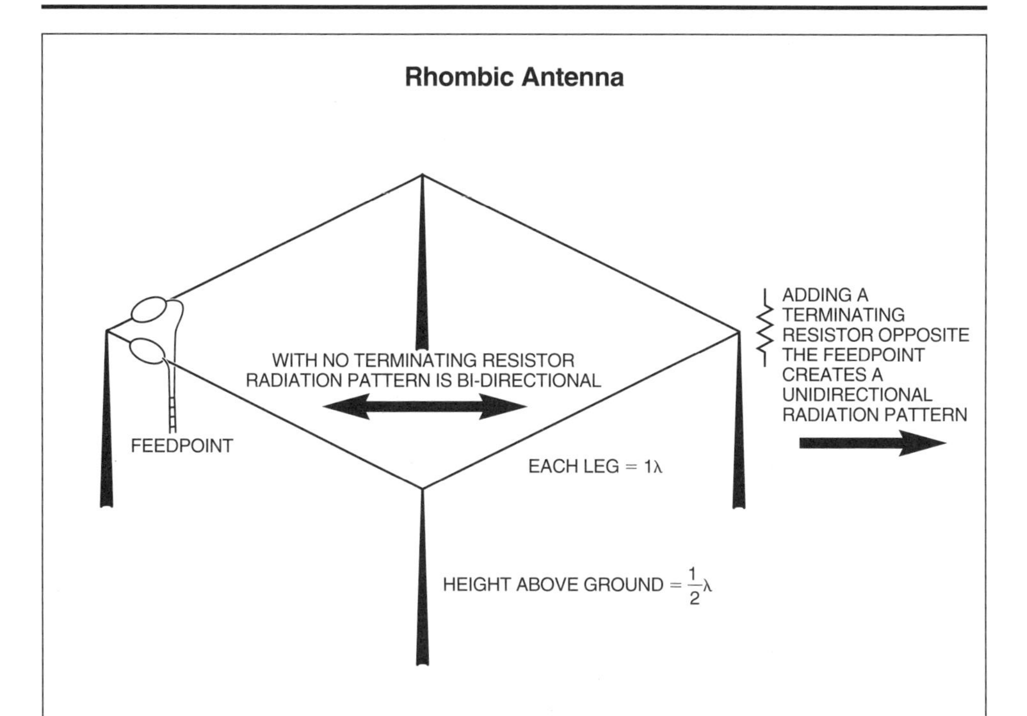


The resonant rhombic antenna is bidirectional, radiating off the unterminated end and radiating in the direction of the transmission line. There is no terminating resistor. Feedpoint impedance varies considerably depending on the input frequency. Each leg is a minimum of one wavelength long, and the rhombic requires multiple poles to keep the radiating wire relatively straight.
The non-resonant rhombic antenna will use a non-inductive resistor to terminate the opposite end from the feedpoint. Feedpoint impedance is typically 600-900 ohms. The terminating non-inductive resistor should be around 800 ohms. The non-resonant rhombic radiates in a substantially uni-directional path toward the terminated resistor. But like its brother, the non-resonant rhombic requires plenty of real estate plus multiple poles to keep each leg absolutely straight.

**E9C10 What is the effect of a terminating resistor on a rhombic antenna?**

A. It reflects the standing waves on the antenna elements back to the transmitter
B. It changes the radiation pattern from essentially bidirectional to essentially unidirectional
C. It changes the radiation pattern from horizontal to vertical polarization
D. It decreases the ground loss

**ANSWER B:** A terminating resistor at the opposite end of the rhombic from the feedline changes the radiation pattern from essentially bidirectional to unidirectional.

**E9C11 What type of antenna pattern over real ground is shown in Figure E9-2?**

A. Elevation pattern
B. Azimuth pattern
C. E-Plane pattern
D. Polarization pattern

**ANSWER A:** Figure E9-2 shows the takeoff elevation patterns of an antenna over a real ground source.

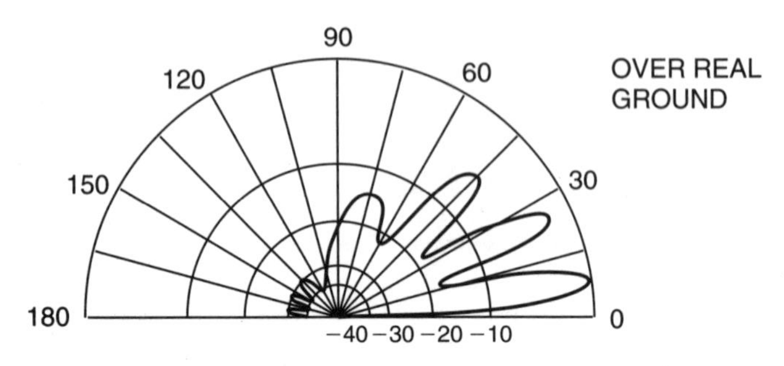


**Figure E9-2**

**E9C12 In the H field antenna radiation pattern shown in Figure E9-2, what is the elevation angle of the peak response?**

A. 45 degrees
B. 75 degrees
C. 7.5 degrees
D. 25 degrees

**ANSWER C:** The main zero dB lobe is about 7.5 degrees up, with two other significant lobes at higher elevations.

**E9C13 In the H field antenna radiation pattern shown in Figure E9-2, what is the front-to-back ratio?**

A. 15 dB
B. 28 dB
C. 3 dB
D. 24 dB

**ANSWER B:** The antenna front-to-back ratio in Figure E9-2 looks to be about 28 to 30 dB. This is the ratio of forward "front" radiation to the slight amount of unwanted radiation in the reverse direction from the back of the antenna. This ratio is the same for transmit as well as receive.

**E9C14 In the H field antenna radiation pattern shown in Figure E9-2, how many elevation lobes appear in the forward direction?**

A. 4
B. 3
C. 1
D. 7

**ANSWER A:** There are three big lobes, PLUS one smaller lobe in the forward direction.

**E9C15 How is the far-field elevation pattern of a vertically polarized antenna affected by being mounted over seawater versus rocky ground?**

A. The low-angle radiation decreases
B. The high-angle radiation increases
C. Both the high- and low-angle radiation decrease
D. The low-angle radiation increases

**ANSWER D:** Here we go – mount it over salt water and achieve a low-angle radiation increase. This gives you a longer range of skywave signals.

**E9C16 If only a modest on-ground radial system can be used with an eighth-wavelength-high, inductively loaded vertical antenna, what would be the best compromise to minimize near-field losses?**

A. 4 radial wires, 1 wavelength long
B. 8 radial wires, a half-wavelength long
C. A wire-mesh screen at the antenna base, an eighth-wavelength square
D. 4 radial wires, 2 wavelengths long

**ANSWER C:** If you plan to build your own on-ground vertical antenna system, you can minimize NEAR field losses by putting a wire mesh screen at the antenna base, in an 1/8-wavelength square. The wire mesh screen will match the electrical wavelength of the antenna.

---

**E9C17 What is one characteristic of a Beverage antenna?**

A. For best performance it must not exceed 1/4 wavelength in length at the desired frequency
B. For best performance it must be mounted more than 1 wavelength above ground at the desired frequency
C. For best performance it should be configured as a four-sided loop
D. For best performance it should be longer than one wavelength

**ANSWER D:** The Beverage antenna is a tremendous signal-grabber and radiator, but for best performance, it really needs to be L O N G! The longer the better for a Beverage antenna.

---

**E9C18 How would the electric field be oriented for a Yagi with three elements mounted parallel to the ground?**

A. Vertically
B. Horizontally
C. Right-hand elliptically
D. Left-hand elliptically

**ANSWER B:** When we talk about Yagi antenna polarization, we always refer to the electric field. The Yagi with elements mounted parallel to the ground is horizontally polarized.

---

**E9C19 What strongly affects the shape of the far-field, low-angle elevation pattern of a vertically polarized antenna?**

A. The conductivity and dielectric constant of the soil
B. The radiation resistance of the antenna
C. The SWR on the transmission line
D. The transmitter output power

**ANSWER A:** Vertically-polarized antennas work best when mounted above an enormous ground plane, such as sea water. The conductivity and dielectric constant of the soil will strongly affect the shape of the far field low-angle radiation pattern.

---

**E9C20 Why are elevated-radial counterpoises popular with vertically polarized antennas?**

A. They reduce the far-field ground losses
B. They reduce the near-field ground losses, compared to on-ground radial systems using more radials
C. They reduce the radiation angle
D. None of these choices is correct

**ANSWER B:** Many amateur operators will use an elevated-radial counterpoise system to reduce the near-field ground losses, compared to on-ground radial systems using more radials.

---

**E9C21 What is a terminated rhombic antenna?**

A. An antenna resonant at approximately double the frequency of the intended band of operation
B. An open-ended bidirectional antenna
C. A unidirectional antenna terminated in a resistance equal to its characteristic impedance
D. A horizontal triangular antenna consisting of two adjacent sides and the long diagonal of a resonant rhombic antenna

**ANSWER C:** Back to the rhombic antenna – if it is terminated, it is a uni-directional antenna terminated in a resistance equal to its characteristic impedance.

***E9D Space and satellite communications antennas: gain; beamwidth; tracking; losses in real antennas and matching: resistivity losses, losses in resonating elements (loading coils, matching networks, etc. {i.e., mobile, trap}); SWR bandwidth; efficiency***

**E9D01 What factors determine the receiving antenna gain required at an amateur satellite station in earth operation?**

A. Height, transmitter power and antennas of satellite
B. Length of transmission line and impedance match between receiver and transmission line
C. Preamplifier location on transmission line and presence or absence of RF amplifier stages
D. Height of earth antenna and satellite orbit

**ANSWER A:** If you plan to receive satellites, you will need to determine height, satellite transmitting power, and the antennas of the satellite.

**E9D02 What factors determine the EIRP required by an amateur satellite station in earth operation?**

A. Satellite antennas and height, satellite receiver sensitivity
B. Path loss, earth antenna gain, signal-to-noise ratio
C. Satellite transmitter power and orientation of ground receiving antenna
D. Elevation of satellite above horizon, signal-to-noise ratio, satellite transmitter power

**ANSWER A:** The effective isotropic radiated power (EIRP) required by an amateur station in earth operation is calculated to the type of satellite antennas, satellite height, and the satellite receiver sensitivity.

**E9D03 What is the approximate beamwidth of a symmetrical pattern antenna with a gain of 20 dB as compared to an isotropic radiator?**

A. 10 degrees
B. 20 degrees
C. 45 degrees
D. 60 degrees

**ANSWER B:** Beamwidth (in degrees) is equal to a fixed value of 203 divided by the square root of 10, raised to a power represented by the antenna gain, in dB, divided by 10:

$$\text{Beamwidth} = \frac{203}{\sqrt{10^X}} \text{ in } \textbf{degrees} \qquad X = \frac{\text{Antenna gain in } \textbf{dB}}{10}$$

In this question, the antenna gain is 20 dB, and X = 20 ÷ 10 = 2

$$\text{Beamwidth} = \frac{203}{\sqrt{10^2}} = \frac{203}{10} = 20.3 \text{ degrees}$$

The beamwidth for this antenna is 20.3° rounded to 20 degrees matching answer B.

**E9D04 How does the gain of a parabolic dish antenna change when the operating frequency is doubled?**

A. Gain does not change
B. Gain is multiplied by 0.707
C. Gain increases 6 dB
D. Gain increases 3 dB

**ANSWER C:** From 1270 MHz on up, you may wish to use a parabolic dish antenna for microwave work. If you double the frequency, the gain of the dish increases by 6 dB, a 4 times increase.

**Parabolic Antenna**

**E9D05 How is circular polarization produced using linearly polarized antennas?**

A. Stack two Yagis, fed 90 degrees out of phase, to form an array with the respective elements in parallel planes
B. Stack two Yagis, fed in phase, to form an array with the respective elements in parallel planes
C. Arrange two Yagis perpendicular to each other, with the driven elements in the same plane, fed 90 degrees out of phase
D. Arrange two Yagis perpendicular to each other, with the driven elements in the same plane, fed in phase

**ANSWER C:** To get circular polarization out of a pair of Yagis, arrange the Yagis perpendicular to each other, with the driven elements in the same plane, and fed 90 degrees out of phase.

**E9D06 How does the beamwidth of an antenna vary as the gain is increased?**

A. It increases geometrically
B. It increases arithmetically
C. It is essentially unaffected
D. It decreases

**ANSWER D:** The parabolic dish concentrates that beamwidth as gain increases. The higher the gain of a dish, the narrower the beam width. Beamwidth decreases as gain increases.

**E9D07 Why does a satellite communications antenna system for earth operation need to have rotators for both azimuth and elevation control?**

A. In order to track the satellite as it orbits the earth
B. Because the antennas are large and heavy
C. In order to point the antenna above the horizon to avoid terrestrial interference
D. To rotate antenna polarization along the azimuth and elevate the system towards the satellite

**ANSWER A:** You need both an azimuth as well as an elevation rotor control to track the satellites up in orbit. We hope you also will join AMSAT to support amateur satellites – visit www.amsat.org to learn more.

---

**E9D08 For a shortened vertical antenna, where should a loading coil be placed to minimize losses and produce the most effective performance?**

A. Near the center of the vertical radiator
B. As low as possible on the vertical radiator
C. As close to the transmitter as possible
D. At a voltage node

**ANSWER A:** To minimize coil losses and to increase performance of a loaded vertical antenna, the loading coil should be near the center of the vertical radiator, not at the base.

---

**E9D09 Why should an HF mobile antenna loading coil have a high ratio of reactance to resistance?**

A. To swamp out harmonics
B. To maximize losses
C. To minimize losses
D. To minimize the Q

**ANSWER C:** When you see those giant open-air coils halfway up the mast on a mobile whip antenna, you are looking at high Q coils which minimize losses. The bigger the coil, the lower the losses.

**Mobile Antenna Loading Coils**

**E9D10 What is a disadvantage of using a trap antenna?**

A. It will radiate harmonics
B. It can only be used for single-band operation
C. It is too sharply directional at lower frequencies
D. It must be neutralized

**ANSWER A:** Trap antennas have one big disadvantage – on older equipment, they can radiate harmonics. Since ham bands may be harmonically related, it's easy to see that the second harmonic of 7 MHz can radiate from a trap antenna quite nicely because the antenna also has perfect resonance on 14 MHz. So watch out for harmonics on a trap antenna.

---

**E9D11 How must the driven element in a 3-element Yagi be tuned to use a hairpin matching system?**

A. The driven element reactance is capacitive
B. The driven element reactance is inductive
C. The driven element resonance is lower than the operating frequency
D. The driven element radiation resistance is higher than the characteristic impedance of the transmission line

**ANSWER A:** For the hairpin matching system to work well, the driven element reactance must be capacitive, $X_C$.

---

**E9D12 What is the equivalent lumped-constant network for a hairpin matching system on a 3-element Yagi?**

A. Pi network
B. Pi-L network
C. L network
D. Parallel-resonant tank

**ANSWER C:** We would use an L-network to match a 3-element Yagi for a hairpin matching system.

---

**E9D13 What happens to the bandwidth of an antenna as it is shortened through the use of loading coils?**

A. It is increased
B. It is decreased
C. No change occurs
D. It becomes flat

**ANSWER B:** When you shorten a vertical antenna with a loading coil, bandwidth decreases, and losses increase.

---

**E9D14 What is an advantage of using top loading in a shortened HF vertical antenna?**

A. Lower Q
B. Greater structural strength
C. Higher losses
D. Improved radiation efficiency

**ANSWER D:** We use top loading and helical loading on a shortened HF vertical antenna to improve radiation efficiency. With this in mind, it's easy to understand why a little 96-inch CB whip antenna with an automatic antenna tuner inside the trunk of your vehicle will not perform as well as a resonant center-loaded, helical-loaded, or top-loaded vertical antenna.

---

**E9D15 What is the approximate input terminal impedance at the center of a folded dipole antenna?**

A. 300 ohms
B. 72 ohms
C. 50 ohms
D. 450 ohms

**ANSWER A:** The impedance at the center of a folded dipole may be a whopping 300 ohms. You'll need an impedance-matching transformer to feed it with 50-ohm coax; however, ordinary TV twin-lead would match very well.

**E9D16 Why is a loading coil often used with an HF mobile antenna?**

A. To improve reception
B. To lower the losses
C. To lower the Q
D. To tune out the capacitive reactance

**ANSWER D:** Since $X_L$ must equal $X_C$ in order for a whip antenna to achieve resonance, we need to add a loading coil on an antenna for lower frequency operation to tune out capacitive reactance.

**E9D17 What is an advantage of using a trap antenna?**

A. It has high directivity in the higher-frequency bands
B. It has high gain
C. It minimizes harmonic radiation
D. It may be used for multi-band operation

**ANSWER D:** The trap antenna may be used for multiband operation, from a single feed line. There are trap dipoles, trap verticals, and trap beam antennas for multiband operation.

**E9D18 What happens at the base feed-point of a fixed length HF mobile antenna as the frequency of operation is lowered?**

A. The resistance decreases and the capacitive reactance decreases
B. The resistance decreases and the capacitive reactance increases
C. The resistance increases and the capacitive reactance decreases
D. The resistance increases and the capacitive reactance increases

**ANSWER B:** As we operate lower on the bands, the base feedpoint resistance decreases (lower decreases), and the capacitive reactance ($X_C$) increases. You can visualize the correct answer by thinking the lower you operate in frequency, the more coil (inductive reactance, $X_L$) you will need to offset and equal the added capacitive reactance, which increases.

**E9D19 What is the beamwidth of a symmetrical pattern antenna with a gain of 30 dB as compared to an isotropic radiator?**

A. 3.2 degrees
B. 6.4 degrees
C. 37 degrees
D. 60 degrees

**ANSWER B:** Let's use the formula shown at E9D03. In this question the antenna gain is 30 dB and X = 30 ÷ 10 = 3; therefore,

$$\text{Beamwidth} = \frac{203}{\sqrt{10}^{3}} = \frac{203}{10\sqrt{10}} = \frac{203}{31.63} = 6.42°$$

The beamwidth for this antenna is 6.42°, matching answer B

**E9D20 What is the beamwidth of a symmetrical pattern antenna with a gain of 15 dB as compared to an isotropic radiator?**

A. 72 degrees
B. 52 degrees
C. 36 degrees
D. 3.6 degrees

**ANSWER C:** Same formula again. The gain is 15 dB and X = 15 ÷ 10 = 1.5. The beamwidth is:

$$\text{Beamwidth} = \frac{203}{\sqrt{10}^{\,1.5}} = \frac{203}{3.16^{1.5}} = \frac{203}{5.62} = 36.1°$$

You may not know how to find $3.16^{1.5}$. What you can do is estimate. $3.16^2$ is 10; therefore, the value you are looking for is between 3.16 and 10. Let's call it Y; therefore, Y = 3.16 × Z. Z is somewhere between 0 and 3.16. Since the power is 1.5, you could choose half of 3.16 or 1.58. However, the power function is logarithmic so 0.5 is weighted past the linear one half point. Let's choose Z = 1.7, therefore, Y = 3.16 × 1.7 = 5.37. With 5.37, the beamwidth = 203 ÷ 5.37 = 37.8°. In this question, you would be close enough to choose the correct answer C of 36.1°.

**E9D21 What is the beamwidth of a symmetrical pattern antenna with a gain of 12 dB as compared to an isotropic radiator?**

A. 34 degrees
B. 45 degrees
C. 58 degrees
D. 51 degrees

**ANSWER D:** Same formula, but now the gain is 12 dB and X = 12 ÷ 10 = 1.2.

$$\text{Beamwidth} = \frac{203}{\sqrt{10}^{\,1.2}} = \frac{203}{3.16^{1.2}} = \frac{203}{3.98} = 51°$$

If you estimate $3.16^{1.2}$ = 3.16 × 1.2 = 3.79, the beamwidth would be 53.6°. This would get you close enough to choose answer D, 51 degrees.

Let's review what's been done in the last three questions. A big 20 dB dish will provide a tight pattern of 20 degrees – 20/20. A giant 30 dB dish will narrow the beam down to about 6.5 degrees. But a much smaller 15 dB dish will widen the beam to a rather broad pattern of 36 degrees, and a 12 dB dish even a wider pattern of 51 degrees. You can almost visualize these patterns.

***E9E Matching antennas to feed lines; characteristics of open and shorted feed lines: 1/8 wavelength; 1/4 wavelength; 1/2 wavelength; feed lines: coax versus open-wire; velocity factor; electrical length; transformation characteristics of line terminated in impedance not equal to characteristic impedance; use of antenna analyzers***

**E9E01 What system matches a high-impedance transmission line to a lower impedance antenna by connecting the line to the driven element in two places, spaced a fraction of a wavelength each side of element center?**

A. The gamma matching system
B. The delta matching system
C. The omega matching system
D. The stub matching system

**ANSWER B:** The delta matching system is a popular way to change the impedance of the transmission line to the impedance of the antenna input.

**E9E02 What system matches an unbalanced feed line to an antenna by feeding the driven element both at the center of the element and at a fraction of a wavelength to one side of center?**

A. The gamma matching system
B. The delta matching system
C. The omega matching system
D. The stub matching system

**ANSWER A:** A gamma match is used with coax cable that forms an unbalanced feed system going to a tube that acts as a capacitor in series with the connection point.

**E9E03 What impedance matching system uses a short perpendicular section of transmission line connected to the feed line near the antenna?**

A. The gamma matching system
B. The delta matching system
C. The omega matching system
D. The stub matching system

**ANSWER D:** When a short section of transmission line is connected to the antenna feed line near the antenna, it is called stub matching. Perhaps you used to cut a stub to cancel out interference in your TV receiver.

**Impedance Matching Stubs**

**Quarter wavelength ($\frac{\lambda}{4}$) stubs (or multiples of odd numbers of $\frac{\lambda}{4}$ s)**

**Less than $\frac{1}{4}\lambda$ stub**

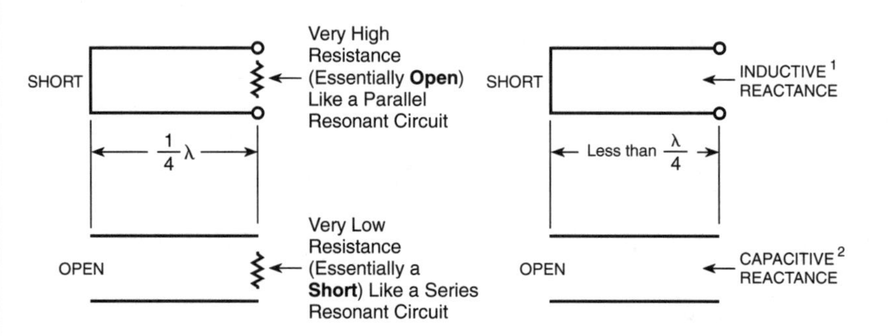


**Half-wavelength ($\frac{\lambda}{2}$) stubs (or multiples of $\frac{\lambda}{2}$ s)**

**Greater than $\frac{\lambda}{4}$, less than $\frac{\lambda}{2}$**

Think of it as two $\frac{\lambda}{4}$ stubs in series.

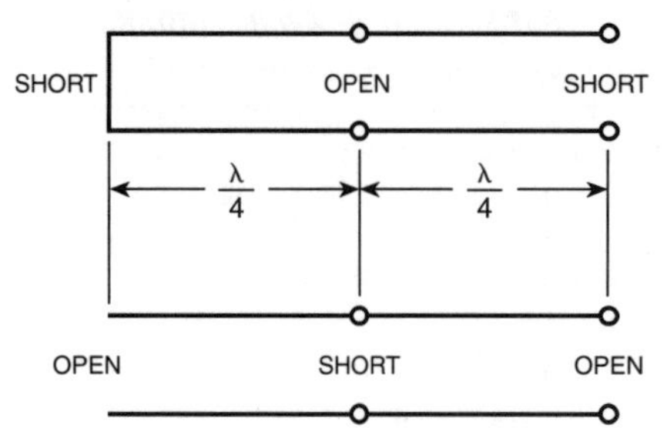


Think of it as a $\frac{\lambda}{4}$ stub plus one less than $\frac{\lambda}{4}$.

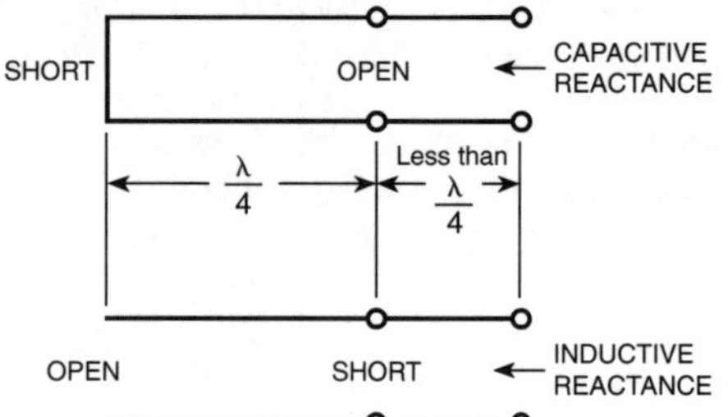


Notes

1. Since a short at end makes the diagram look like a long wire, this should remind the reader that impedance is an inductive reactance.
2. Since an open at end makes the diagram look like two parallel plates, this should remind the reader that impedance is a capacitive reactance.

**E9E04 What should be the approximate capacitance of the resonating capacitor in a gamma matching circuit on a Yagi beam antenna for the 20-meter band?**

A. 14 pF
B. 140 pF
C. 1400 pF
D. 0.14 pF

**ANSWER B:** The rule-of-thumb is to use about 7 picofarads per meter of wavelength, so 7 times 20 is 140 picofarads.

---

**E9E05 What should be the approximate capacitance of the resonating capacitor in a gamma matching circuit on a Yagi beam antenna for the 10-meter band?**

A. 0.2 pF
B. 0.7 pF
C. 700 pF
D. 70 pF

**ANSWER D:** Again, use about 7 picofarads per meter of wavelength for gama matching on a Yagi beam. For 10 meters, $7 \times 10 = 70$ pF.

---

**E9E06 What is the velocity factor of a transmission line?**

A. The ratio of the characteristic impedance of the line to the terminating impedance
B. The index of shielding for coaxial cable
C. The velocity of the wave on the transmission line multiplied by the velocity of light in a vacuum
D. The velocity of the wave on the transmission line divided by the velocity of light in a vacuum

**ANSWER D:** Radio waves travel in free space at 300 million meters per second. But in coaxial cable transmission lines, and other types of feed lines, the radio waves move slower. Velocity factor is the velocity of the radio wave on the transmission line divided by the velocity in free space.

---

**E9E07 What determines the velocity factor in a transmission line?**

A. The termination impedance
B. The line length
C. Dielectrics in the line
D. The center conductor resistivity

**ANSWER C:** There are some great low-loss coaxial cable types out there in radio land with different dielectrics separating the center conductor and the outside shield. It's the dielectric that makes a big difference in the velocity factor.

---

**E9E08 Why is the physical length of a coaxial cable transmission line shorter than its electrical length?**

A. Skin effect is less pronounced in the coaxial cable
B. The characteristic impedance is higher in a parallel feed line
C. The surge impedance is higher in a parallel feed line
D. RF energy moves slower along the coaxial cable

**ANSWER D:** Radio frequency energy moves *slightly slower* (key words) inside coaxial cable than it would in free space.

**E9E09 What is the typical velocity factor for a coaxial cable with polyethylene dielectric?**

A. 2.70
B. 0.66
C. 0.30
D. 0.10

**ANSWER B:** Most coax cable has a velocity factory of 0.66. This is not stated on the outside jacket, and may not even be known by the dealer selling the coax. You can contact the manufacturer or distributor directly to find out the actual velocity factor.

---

**E9E10 What would be the physical length of a typical coaxial transmission line that is electrically one-quarter wavelength long at 14.1 MHz? (Assume a velocity factor of 0.66.)**

A. 20 meters
B. 2.3 meters
C. 3.5 meters
D. 0.2 meters

**ANSWER C:** When you work on antenna phasing harnesses, you will need to calculate the velocity factor for a piece of coax in order to know where to place the connectors. You must be accurate down to a fraction of an inch! You can calculate the physical length of a coax cable which is electrically one-quarter wavelength long using the formula:

E $$\text{L (in feet)} = \frac{984\lambda V}{f}$$

Where: L is antenna length in **feet**
λ is wavelength in **meters**
V is **velocity factor**
f is frequency in **MHz**

For this problem, multiply 984 × 0.25, because the coax is one quarter wavelength. Then multiply the result by 0.66, the velocity factor. Remember, 0.66 is the velocity factor for coax. We are assuming that value for these problems. Divide the total by 14.1, the frequency in MHz. The length comes out to about 11.51 feet. The answer needs to be in meters, so divide 11.5 by 3 (there are about 3 feet to a meter). The result is 3.83, but the 3.5 meters in answer C is the closest. (To calculate coax in meters, simply substitute 300 for 984. This will eliminate one step in the conversion!)

**Proper coax cable length is important to maximize phased antenna performance.**

**E9E11 What is the physical length of a parallel conductor feed line that is electrically one-half wavelength long at 14.10 MHz? (Assume a velocity factor of 0.95.)**

A. 15 meters
B. 20 meters
C. 10 meters
D. 71 meters

**ANSWER C:** In this problem, the parallel conductor feed line one-half wavelength long at 14.10 MHz, with a velocity factor of .95, is calculated by multiplying 984 × 0.5 × 0.95 and dividing by 14.10, the frequency in MHz. Convert feet to meters, and you end up with 10 meters.

**E9E12 What parameter best describes the interactions at the load end of a mismatched transmission line?**

A. Characteristic impedance
B. Reflection coefficient
C. Velocity factor
D. Dielectric Constant

**ANSWER B:** A mismatched transmission line will give you REFLECTIONS.

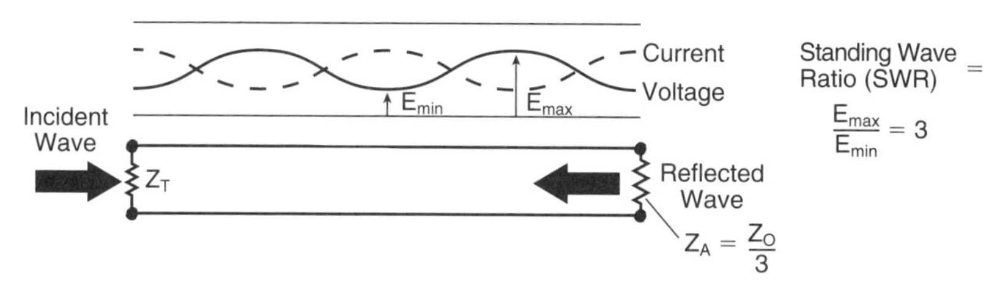


**Impedance Mismatch Causes Reflected Wave**

**E9E13 Which of the following measurements describes a mismatched transmission line?**

A. An SWR less than 1:1
B. A reflection coefficient greater than 1
C. A dielectric constant greater than 1
D. An SWR greater than 1:1

**ANSWER D:** If the SWR is greater than 1:1, this describes a mismatched transmission line.

**SWR Bridge**

**E9E14 What characteristic will 450-ohm ladder line have at 50 MHz, as compared to 0.195-inch-diameter coaxial cable (such as RG-58)?**

A. Lower loss in dB/100 feet
B. Higher SWR
C. Smaller reflection coefficient
D. Lower velocity factor

**ANSWER A:** Going to 450-ohm ladder line on the 6-meter band is one way to reduce losses over typical RG-58 coax cable and keep your expenses to a minimum. They both cost about the same. RG-58 should be relegated to test leads, never used for antenna feedline runs. The loss is simply too high. 450 ohm ladder line will have much lower loss at 50 MHz than small RG-58 coax.

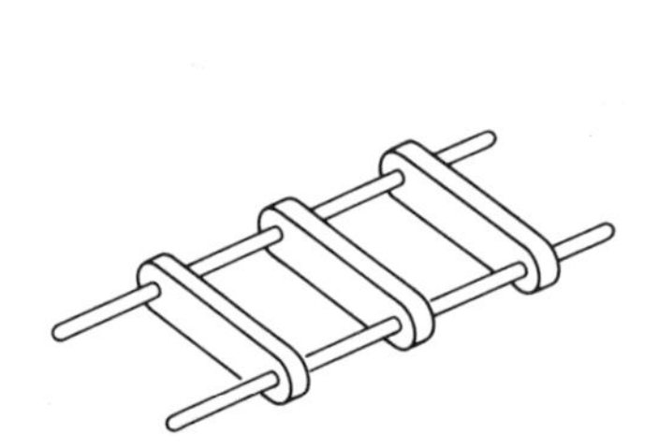

**Ladder Line**

Outer Insulation
Copper Braid Shield
Inner Conductor
Polyethylene Dielectric

**Coax**

**E9E15 What is the term for the ratio of the actual velocity at which a signal travels through a transmission line to the speed of light in a vacuum?**

A. Velocity factor
B. Characteristic impedance
C. Surge impedance
D. Standing wave ratio

**ANSWER A:** Comparing the velocity of radio signals through a piece of feed line to the speed of light in a vacuum is called the *velocity factor* of the feed line.

**E9E16 What would be the physical length of a typical coaxial transmission line that is electrically one-quarter wavelength long at 7.2 MHz? (Assume a velocity factor of 0.66.)**

A. 10 meters
B. 6.9 meters
C. 24 meters
D. 50 meters

**ANSWER B:** The coax cable is again one-quarter wavelength, so multiple 984 × 0.25, and the result by 0.66, the velocity factor. We again will assume this value. Now divide these three multiplied figures by the frequency, 7.2 MHz, and you end up with 22.55 feet of coax. Dividing by 3 to get approximate length of 7.5 meters, leads you to 6.9 meters of answer B. (See formula at E9E10.)

**E9E17 What kind of impedance does a 1/8-wavelength transmission line present to a generator when the line is shorted at the far end?**

A. A capacitive reactance
B. The same as the characteristic impedance of the line
C. An inductive reactance
D. The same as the input impedance to the final generator stage

**ANSWER C:** If a 1/8 wavelength transmission line is shorted out at the far end, it will look like an inductive reactance at the generator.

**E9E18 What kind of impedance does a 1/8-wavelength transmission line present to a generator when the line is open at the far end?**

A. The same as the characteristic impedance of the line
B. An inductive reactance
C. A capacitive reactance
D. The same as the input impedance of the final generator stage

**ANSWER C:** If that 1/8 wavelength transmission line is open at the far end, it will look like a capacitive reactance at the generator.

---

**E9E19 What kind of impedance does a 1/4-wavelength transmission line present to a generator when the line is open at the far end?**

A. A very high impedance
B. A very low impedance
C. The same as the characteristic impedance of the line
D. The same as the input impedance to the final generator stage

**ANSWER B:** If the far end of a quarter-wavelength transmission line is open, it will look like a very low impedance at the generator.

---

**E9E20 What kind of impedance does a 1/4-wavelength transmission line present to a generator when the line is shorted at the far end?**

A. A very high impedance
B. A very low impedance
C. The same as the characteristic impedance of the transmission line
D. The same as the generator output impedance

**ANSWER A:** On a quarter-wavelength transmission line, shorting it at the far end will look like a very high impedance at the generator.

---

**E9E21 What kind of impedance does a 1/2-wavelength transmission line present to a generator when the line is shorted at the far end?**

A. A very high impedance
B. A very low impedance
C. The same as the characteristic impedance of the line
D. The same as the output impedance of the generator

**ANSWER B:** If a half-wavelength transmission line is shorted at the far end, it will have a very low impedance, essentially a short.

---

**E9E22 What kind of impedance does a 1/2-wavelength transmission line present to a generator when the line is open at the far end?**

A. A very high impedance
B. A very low impedance
C. The same as the characteristic impedance of the line
D. The same as the output impedance of the generator

**ANSWER A:** If that half-wavelength transmission line is open at the far end, it will have a very high impedance, essentially an open circuit.

---

***Wow — you've made it through all 801 questions and answers! Now it's time to go back and review those questions that you aren't confident about. Remember — work your book!***

---

# 3

# Taking the Extra Class Examination

## ABOUT THIS CHAPTER

Get set for passing the Element 4 examination and gaining additional kilohertz of band privileges as an Extra Class operator! If you are upgrading from General to Extra Class, you will gain a whopping 450 kHz of *additional* frequencies on the 75/80-meter, 40-meter, 20-meter, and 15-meter worldwide service bands. If you are upgrading from a grandfathered Advanced Class license, you will gain an *additional* 175 kHz of operating area on the 75/80-meter, 40-meter, 20-meter, and 15-meter worldwide amateur service bands. Your Extra Class license gives you full privileges and full frequency coverage on every amateur service band assigned by the Federal Communications Commission.

In this chapter, we'll tell you how to prepare yourself for the exam, what to expect when you take the exam, and how you'll get your new, upgraded Extra Class license.

## EXAMINATION ADMINISTRATION

Three Extra Class accredited Volunteer Examiners are required to administer your Element 4 exam. Many of you preparing for your Extra Class examination are already part of a Volunteer Examination team. If you are, it's perfectly acceptable to allow your fellow team members to administer a computer-generated, fresh Extra Class written theory examination.

Let's take a moment to review how well the Volunteer Examination program is working. The FCC no longer conducts amateur operator examinations; all examinations are conducted by volunteer amateur operators. The examination sessions are coordinated by national or regional Volunteer Examiner Coordinators (VECs) who accredit General Class, Advanced Class and Extra Class amateur operators as Volunteer Examiners (VEs). Your Extra Class examination **must** be taken at an official examination session in front of a team of three accredited Extra Class VEs.

The Volunteer Examiners are not compensated for their time and skills, but they are permitted to charge you an examination fee for certain reimbursable expenses incurred in preparing, administering, and processing the examination. The maximum fee is adjusted annually based on inflation by the Federal Communications Commission. It is currently about $10.00.

## HOW TO FIND AN EXAM SITE

Exam sessions are held regularly at sites throughout the country to serve their local communities. The exam site could be a public library, a fire house, someone's office, in a warehouse, and maybe even in someone's private home. Each examination team may regularly post their examination locations down at the local ham radio store. They also inform their VEC when and where they regularly hold test sessions. So the easiest way for you to find an exam session that is near you and at a convenient time is to call the VEC.

A complete list of VECs is located on page 246 in the Appendix of this book. The W5YI VEC and the ARRL VEC are the two largest examination groups in the country. They test in all 50 states and in foreign countries, too. Their 3-member, accredited examination teams are just about *everywhere*. So when you call the W5YI-VEC in Texas, or the ARRL-VEC in Connecticut, be assured they probably have an examination team only a few miles from where you are reading this book right now!

***Want to find a test site fast?***
Visit the W5YI-VEC website at www.w5yi.org, or call them at 817-860-3800.

Any of the VECs listed will provide you with the phone number of a local Volunteer Examiner team leader who can tell you the schedule of upcoming exam sessions near you. Select a session you wish to attend, and make a reservation so the VEs know to expect you. Don't be a "no-show" and don't be a "surprise-show." Make a reservation. And don't hesitate to tell them how much we all appreciate their efforts in supporting ham radio testing.

Ask them ahead of time what you will need to bring to the examination session. And ask them how much the current fee is for your exam session.

## WHAT TO BRING TO THE SITE

Here is what you'll need to bring with you to the exam site for your upgrade from General or grandfathered Advanced to Extra Class:

1. The original and 2 copies of your present General or grandfathered Advanced Class license, or any CSCE issued within the last 365 days confirming examination credit. If you show up without your present license or copies of your present license, your examination or examination results could be put on a major hold.
2. A picture ID. Usually your Driver's License will suffice. Kids should bring their school ID, and very young kids should bring an adult to verify that they are who they are!
3. An examination fee of approximately $10.00 in cash, *exact change, please*. Personal checks will not be accepted, and plastic is out of the question. Ask the VE how much the fee will be when call to make your exam reservation.
4. If you have a physical disability that may require additional assistance by your examination team, let them know ahead of time so they may prepare an examination that accommodates your special requirements. This could be special reading equipment, a volunteer examiner reader, access for a wheelchair, or any other special requirements that they may be able to set up for you. But they must know ahead of time what requirements need to be met, so please contact the volunteer examiner team leader in advance.

5. You will definitely want to ***bring your calculator***. Calculators are indeed permitted and quite necessary for the Extra Class exam. Since some of the questions are from the old Advanced test, you can be sure you will have at least 8 or 10 formula problems to work on the calculator keys. The Volunteer Examination team leader may ask you to remove the battery from the calculator in order to verify memory dump. Don't be surprised if they ask you to completely clear your calculator in this fashion.
6. Bring plenty of sharp pencils and a fine-tip pen. Have backups. No red ink, please.

Try to arrive at the examination location about an hour before the exam begins. Introduce yourself to the Volunteer Examiners who will be conducting the test, and see if they need a hand in setting up the room. If it is a morning examination, Gordon West Radio School students usually bring donuts, hot coffee, tea, and hot chocolate for the Volunteer Examiners as a way of expressing their appreciation for the time and effort the examiners expend arranging and giving these tests. Start their day off with a smile and go for a dozen donuts! Believe me, they will appreciate the thought.

## RUMOR MILL

While standing in line ready to enter the examination room, rumors usually begin to fly. Someone who is ill-informed may say they are still requiring the 20-wpm code test for Extra. Nonsense!

Someone may remark that all of the test questions have been changed, and this "new" 2002 test is different than the book. Nonsense! Again, the questions in this Extra Class book are valid from July 1, 2002, through June 30, 2006.

Someone comes out of the exam room indicating the Extra Class test had numerous questions not found in his book. Again, ridiculous! This book has every single test question, exactly as they will appear on the test. Only the A, B, C, and D order of the possible answers may be scrambled.

## TAKING THE EXAMINATION

Get a good night's sleep before the exam day. Continue to study the Extra Class formulas right up to the moment you go into the examination room. Don't believe that old saying that too much study will cause you to forget the subject material. If you purchased your author's audio cassette tapes, continue to play them up to the last minute.

Listen carefully for the instructions of your Volunteer Examiners. They may be running a 5-wpm code test in another room, and will need everyone's silence in the theory exam room. Most examination sessions will conduct all levels of testing, so stay tuned in on the announcements they give on who should come forward to pick up what type of test materials.

Make no marks on the examination sheet. If you don't know an answer, slightly shade in one of the answer sheet boxes, and make yourself a big note to go back and work out the correct answer. Using this technique will keep you from accidentally skipping a hard question, and misaligning your answer boxes with the easier questions. If you finish the exam at question #49, chances are you skipped a question and all of your work will be out of synch with your answer sheet. Double check that all 50 answer boxes indeed have a mark.

When you are taking the written examination, *take your time!* Some answers start out looking correct, but end up wrong at the end. Don't speed read the exam – getting careless after all this preparation could undo all the hard work you have put in on preparing for the Extra Class exam.

If you don't know an answer, eliminate the obviously wrong answers, and then take your best guess. NEVER LEAVE AN ANSWER BLANK! Leaving a specific answer blank is counted as a wrong answer. We also find that going back and second-guessing an answer will sometimes get you into the wrong answer category. Usually your first impression is the correct one. Another trick is to read your answer first and then see if it agrees with what the question asks. Start this technique at Answer 50, and go backwards.

## AND THE RESULTS ARE...

When you are satisfied that you have passed the Extra Class exam, turn in all of your examination papers – including scratch paper and the original test paper. Thank your examiners for their courtesy, and follow their instructions while they prepare to score your paper. *Don't stand over the examiners* as they grade your test.

When you pass the test, the examiners may or may not indicate how many questions you missed. You missed only one? But which one? DON'T ASK – the examiners may tell you it is against the rules to divulge which question you got wrong, although we're not aware of any rule that states they cannot tell you what you got wrong. It's usually just a case of not enough time to spend with any one applicant pouring over their test performance. They usually will tell you your test score, and passing the exam is exactly what you want to hear!

Since we know you are going to be successful in your Extra Class written exam, we know that your examiners will issue you an official CSCE – Certificate of Successful Completion of Examination. This is an important document that is official proof that you have passed the exam. Check to see that it is signed by all three volunteer examiners. Did they get your name and call sign correct on the certificate? Write in your address if all they have is your name on the certificate, and be sure to sign your CSCE. Double-check that the date is correct. If the computer system should gobble up all of your test results, this is LEGAL PROOF that you indeed passed a particular examination element.

## COMPLETING NCVEC FORM 605

Your examiners will assist you in filling out the NCVEC Form 605. They also will have copies of the most current version of this form at the examination session. (See sample form on page 247.) *Be aware that this form is **not** FCC form 605*, which is not suitable for use when taking an amateur radio examination or for changing an address, name, routine call sign change, or renewing your license through a VEC. Note that NCVEC Form 605 asks for your Social Security Number – which is required by the *Debt Collection Improvement Act of 1996* – and your date of birth. These are kept confidential by the VEC and the FCC.

It is very important that you complete the NCVEC Form 605 legibly – preferably in black or blue ball-point pen. If the VEC and VE examiners cannot read the information you have written down, your upgrade could be delayed in the electronic processing phase.

When you are filling-out NCVEC Form 605, it is important to compare the information on your present license with what you are writing on the Form 605. Any difference in your name or address requires that you check the appropriate CHANGE box. For example, if your current license reads "Jack" as your first name and you write down "John" on Form 605, your application will automatically be rejected when it is filed electronically. Same thing for your middle initial – does it appear on your original license? And make certain that your signature matches your name. Be sure to date the application.

Only fill out the top portion (Section 1) of NCVEC Form 605 when taking an examination. Section 2 (certifying section) is completed by the Volunteer Examiners after you complete the examination.

## CHANGING CALL SIGNS

After passing the Extra Class exam, you have an opportunity to change your call sign systematically. Most examinees keep the one they have. If you check the box "CHANGE my station call sign systematically," the FCC computer will assign you the next Group "A" call sign from an alphabetized list.

The key word is *systematically*. You will not get to choose a specific call sign at an examination session. If you live in the continental United States, your assigned Group "A" systematic call sign will be in a 2-by-2 format with the first letter being "A." A 2-by-2 format has two letters followed by a numeral and two more letters, such as AK6CS. Only the AA through AL block is assigned since AM through AZ 2-by-2 letters are not internationally allocated for use in the U.S. Once all 2-by-2 Group "A"

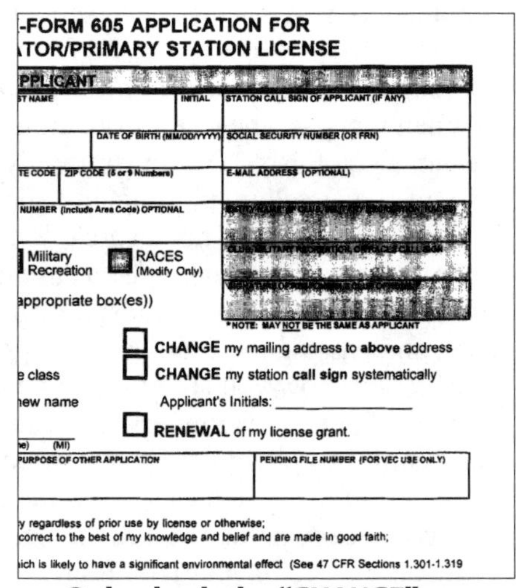
-FORM 605 APPLICATION FOR
TOR/PRIMARY STATION LICENSE

PPLICANT

T NAME | INITIAL | STATION CALL SIGN OF APPLICANT (IF ANY)

DATE OF BIRTH (MM/DD/YYYY) | SOCIAL SECURITY NUMBER (OR FRN)

TE CODE | ZIP CODE (5 or 9 Numbers) | E-MAIL ADDRESS (OPTIONAL)

NUMBER (Include Area Code) OPTIONAL

Military Recreation | RACES (Modify Only)

appropriate box(es))

*NOTE: MAY NOT BE THE SAME AS APPLICANT

CHANGE my mailing address to **above** address

e class — CHANGE my station **call sign** systematically

new name — Applicant's Initials: ______________

RENEWAL of my license grant.

e) (MI)

PURPOSE OF OTHER APPLICATION | PENDING FILE NUMBER (FOR VEC USE ONLY)

y regardless of prior use by license or otherwise;
correct to the best of my knowledge and belief and are made in good faith;

ich is likely to have a significant environmental effect (See 47 CFR Sections 1.301-1.319

**Only check the "CHANGE" my station call sign box if you want the FCC computer to give you a new call sign. If you want to choose a call sign, apply for a vanity call.**

call signs are assigned (the last will be AK1ZZ if you are in the first call district), the FCC then assigns call signs from the next lower (Group B) level to Extra Class amateurs. These will be 2-by-2 call signs beginning with K, N and W.

You must initial the request if you decide to systematically change your call sign. Once changed, it cannot easily be changed back to your previous call sign. The only way you can go back to a previously-held call sign is under the Vanity call sign program and this will cost you a small fee. So be thinking about it right now – do you want to stay with your present call sign, or do you want to change over to a new Extra Class 2-by-2 call sign?

## VANITY CALL SIGNS

The ability to select a vacant station call sign of your choice began in 1996 and it has been immensely popular, especially with Extra Class amateurs who historically like to have short call signs. While only a little more than 10% of all radio amateurs ever make it to the top Extra Class level, they hold nearly half of all Vanity call signs issued!

Extra Class radio amateurs can choose an available station call sign containing any Group, A, B, C or D Amateur format. Most try to get a 1-by-2 or a 2-by-1 format, which are only available to an Extra Class licensee. A 1-by-2 Group "A" call sign begins with either K, N or W (but not "A") followed by a numeral and two letters. For example: W5YI is a Group "A" call sign.

An example of a 2-by-1 call sign is AA1A. The two letter prefix must be from the AA to AL, KA to KZ, NA to NZ or the WA to WZ prefix block followed by a numeral and a single letter. To make matters more confusing, certain prefixes (AH, AL, KH, KL, KP, NH, NL, NP, WH, WL and WP) are reserved for amateurs who have a mailing address outside of the 48 contiguous U.S. states.

You can find out all the rules (*and there are many!*) surrounding obtaining a Vanity call sign by visiting the W5YI Group website at www.w5yi.org and clicking on the "Vanity Call Signs" button near the top of the home page.

There is an additional regulatory fee associated with a Vanity call sign of about $12.00. You can complete FCC Form 605 (*not* NCVEC Form 605) and accompanying Form 159 (Remittance Advice) and file the application yourself either using paper documents or online. Additional information on Vanity call signs is available from the FCC's Amateur Radio website located at:

http://wireless.fcc.gov/services/amateur/licensing/vanity.html

> An easier way to get a Vanity call sign is to let the W5YI Group handle everything for you. They charge a small, additional fee and do all the paperwork and filing for you. You can obtain complete details on this service by going to: www.w5yi.org/vanity.htm

As for me, I'm staying with my original-issue WB6NOA call sign. If I changed it, I would be breaking a 40+ year tradition.

## GOING FURTHER – GROL

The Extra Class license is the highest level license you can achieve in the Amateur Radio service. Did you know you can also take the commercial general radio operator (GROL) license exams, too? The commercial license allows you to work on marine and aeronautical two-way radio equipment, and the commercial GROL license often is required by two-way radio employers before being hired.

And guess what? The GROL technical element uses almost identical test questions as those you have just studied for your amateur Extra Class license. In fact, except for just a couple of numbers, the commercial radiotelephone examination may use identical math problems that you just studied.

So, why not upgrade to a professional-level FCC license? You can order the *GROL-Plus* book by calling the W5YI Group at 800-669-9594.

## BECOMING A VOLUNTEER EXAMINER

When you successfully complete your Extra Class exam, ask your 3-member accredited Volunteer Examination team for a volunteer examiner sign-up application. It only takes a couple of weeks to process, and you'll be joining a cadre of thousands of other Extra Class hams who give something back to ham radio when it comes to testing – just a few hours a month, and some paper and computer processing. It is a rewarding "extra credit" as an Extra Class operator, and you and two other accredited examiners could even give tests abroad all over the world.

***Here's how to become a volunteer examiner:***
Write the W5YI-VEC at: P. O. Box 565101, Dallas, TX 75356,
or call them at 817-860-3800 during regular business hours to obtain a VE application. You also may apply online at: www.w5yi.org.
Click on the "Become a Volunteer Examiner" button.

Accredited Volunteer Examiners also may help develop new questions when the amateur radio question pools undergo periodic updating. Your comments and suggestions will be filed with the prestigious NCVEC Question Pool Committee, and they will consider your valued contributions when each Element 2, Element 3, and Element 4 question pool is updated. Each pool gets updated once every 4 years with a public notice for radio experts to submit new or improved examination questions along with 4 suggested right and wrong answers. Join in – as an Extra Class, it's fun to see that you have contributed to the new question pool.

## SUMMARY

You made it through Extra Class! Congratulations! Allow me to send you a free certificate of achievement. Send a large envelope, self-address, with 10 first class stamps on the inside to me at Gordon West Radio School, 2414 College Drive, Costa Mesa, California 92626. It will take about 30 days to get the certificate back to you, and a hearty congratulations on making it to the top.

I hope to work you soon on the worldwide airwaves, or see you at an upcoming hamfest. Think about getting that commercial radio license, too, based on all of the study you have just completed for amateur Extra Class.

*73*
Gordon West,
WB6NOA

**I know congratulations will be in order very soon when you pass your Extra Class exam with flying colors!**

# APPENDIX

## U.S. VOLUNTEER EXAMINER COORDINATORS IN THE AMATEUR SERVICE

Anchorage Amateur Radio Club
8023 E 11th Court
Anchorage, AK 99504-2003
907-338-0662
e-mail: jwiley@alaska.net

ARRL/VEC
225 Main Street
Newington, CT 06111-1494
860-594-0300
860-594-0339 (fax)
e-mail: vec@arrl.org
Internet: www.arrl.org

Central America VEC, Inc.
1215 Dale Drive, SE
Huntsville, AL 35801-2031
256-536-3904
256-534-5557 (fax)
e-mail: dtunstil@hiwaay.net

Golden Empire Amateur Radio Society
P.O. Box 508
Chico, CA 95927-0508
530-345-3515
wa6zrt@aol.com

Greater Los Angles Amateur Radio Group
9737 Noble Avenue
North Hills, CA 91343-2403
818-892-2068
e-mail: gla.arg@gte.net
Internet: www.glaarg.org

Jefferson Amateur Radio Club
P.O. Box 24368
New Orleans, LA 70184-4368
e-mail: doug@bellsouth.net

Laurel Amateur Radio Club
P.O. Box 1259
Laurel, MD 20725-1259
301-937-0394 (6-9 PM)
e-mail: aa3of@arrl.net

The Milwaukee Radio Amateurs Club, Inc.
P.O. Box 070695
Milwaukee, WI 53207-0695
262-797-6722
e-mail: tfuszard@aero.net

MO-KAN/VEC
228 Tennessee Road
Richmond, KS 66080-9174
e-mail: wo0e@sky.net

SANDARC-VEC
P.O. Box 2446
La Mesa, CA 91943-2446
619-697-1475
e-mail: n6nyx@arrl.net

Sunnyvale VEC Amateur Radio Club, Inc.
P.O. Box 60307
Sunnyvale, CA 94088-0307
408-255-9000 (exam info 24 hours)
e-mail: VEC@amateur-radio.org
Internet: www.amateur-radio.org

W4VEC
3504 Stonehurst Place
High Point, NC 27265-2106
336-841-7576
e-mail: nq4t@aol.com
Internet: www.w4vec.com

Western Carolina Amateur Radio
Soceity/VEC, Inc.
6702 Matterhorn Court
Knoxville, TN 37918-6314
865-687-5410
e-mail: wcars@korrnet.org
Internet: www.korrnet.org/wcars

W5YI-VEC
PO Box 565101
Dallas, TX 75356-5101
817-860-3800
e-mail: w5yi-vec@w5yi.org
Internet: www.w5yi.org

NCVEC QUICK-FORM 605 APPLICATION FOR
AMATEUR OPERATOR/PRIMARY STATION LICENSE

SECTION 1 - TO BE COMPLETED BY APPLICANT

| PRINT LAST NAME | SUFFIX | FIRST NAME | INITIAL | STATION CALL SIGN OF APPLICANT (IF ANY) |
|---|---|---|---|---|
| MAILING ADDRESS (Number and Street or P.O. Box) | | DATE OF BIRTH (MM/DD/YYYY) | | SOCIAL SECURITY NUMBER (OR FRN) |
| CITY | STATE CODE | ZIP CODE (5 or 9 Numbers) | | E-MAIL ADDRESS (OPTIONAL) |
| DAYTIME TELEPHONE NUMBER (Include Area Code) OPTIONAL | | FAX NUMBER (Include Area Code) OPTIONAL | | ENTITY NAME OF CLUB, MILITARY RECREATION, RACES |

Type of Applicant: ☐ Individual ☐ Amateur Club ☐ Military Recreation ☐ RACES (Modify Only)

I HEREBY APPLY FOR (Make an X in the appropriate box(es))

*NOTE: MAY NOT BE THE SAME AS APPLICANT

☐ EXAMINATION for a **new** license grant
☐ EXAMINATION for **upgrade** of my license class
☐ CHANGE my **name** on my license to my new name

Former Name: ______________ (Last name) (Suffix) (First name) (MI)

☐ CHANGE my mailing address to **above** address
☐ CHANGE my station **call sign** systematically

Applicant's Initials: ______________

☐ RENEWAL of my license grant.

| Do you have another license application on file with the FCC which has not been acted upon? | PURPOSE OF OTHER APPLICATION | PENDING FILE NUMBER (FOR VEC USE ONLY) |
|---|---|---|

**I certify that:**

- I waive any claim to the use of any particular frequency regardless of prior use by license or otherwise;
- All statements and attachments are true, complete and correct to the best of my knowledge and belief and are made in good faith;
- I am not a representative of a foreign government;
- The construction of my station will NOT be an action which is likely to have a significant environmental effect (See 47 CFR Sections 1.301-1.319 and Section 97.13(a));
- I have read and WILL COMPLY with Section 97.13(c) of the Commission's Rules regarding RADIOFREQUENCY (RF) RADIATION SAFETY and the amateur service section of OST/OET Bulletin Number 65.

**Signature of applicant** (Do not print, type, or stamp. Must match applicant's name above.)

X ______________ Date Signed: ______________

SECTION 2 - TO BE COMPLETED BY ALL ADMINISTERING VEs

Applicant is qualified for operator license class:

☐ NO NEW LICENSE OR UPGRADE WAS EARNED
☐ TECHNICIAN Element 2
☐ GENERAL Elements 1, 2 and 3
☐ AMATEUR EXTRA Elements 1, 2, 3 and 4

| DATE OF EXAMINATION SESSION |
|---|
| EXAMINATION SESSION LOCATION |
| VEC ORGANIZATION |
| VEC RECEIPT DATE |

I CERTIFY THAT I HAVE COMPLIED WITH THE ADMINISTERING VE REQUIRMENTS IN PART 97 OF THE COMMISSION'S RULES AND WITH THE INSTRUCTIONS PROVIDED BY THE COORDINATING VEC AND THE FCC.

| 1st VEs NAME (Print First, MI, Last, Suffix) | VEs STATION CALL SIGN | VEs SIGNATURE (Must match name) | DATE SIGNED |
|---|---|---|---|
| 2nd VEs NAME (Print First, MI, Last, Suffix) | VEs STATION CALL SIGN | VEs SIGNATURE (Must match name) | DATE SIGNED |
| 3rd VEs NAME (Print First, MI, Last, Suffix) | VEs STATION CALL SIGN | VEs SIGNATURE (Must match name) | DATE SIGNED |

DO NOT SEND THIS FORM TO FCC – THIS IS NOT AN FCC FORM
THIS FORM WILL BE RETURNED TO YOU WITHOUT ACTION IF SENT TO THE FCC

NCVEC FORM 605 - OCTOBER 2001
FOR VE/VEC USE ONLY - Page 1

**NCVEC Form 605 is used to process test results, license upgrades, call sign changes, and changes of address submitted through a VEC. VECs will not accept FCC Form 605.**

FCC 605
Approved by OMB
Main Form

Quick-Form Application for Authorization in the Ship, Aircraft, Amateur, Restricted and Commercial Operator, and the General Mobile Radio Services

3060 - 0850
See instructions for public burden estimate

1) Radio Service Code:

**Application Purpose** (Select only one) ( )

| | | |
|---|---|---|
| 2) NE - New<br>MD - Modification<br>AM - Amendment | RO - Renewal Only<br>RM - Renewal/Modification<br>CA - Cancellation of License | WD - Withdrawal of Application<br>DU - Duplicate License<br>AU - Administrative Update |
| 3) If this request is for a Developmental License or STA (Special Temporary Authorization) enter the appropriate code and attach the required exhibit as described in the instructions. Otherwise enter N (Not Applicable). | | ( - ) D S N/A |
| 4) If this request is for an Amendment or Withdrawal of Application, enter the file number of the pending application currently on file with the FCC. | | File Number |
| 5) If this request is for a Modification, Renewal Only, Renewal/Modification, Cancellation of License, Duplicate License, or Administrative Update, enter the call sign of the existing FCC license. | | Call Sign |
| 6) If this request is for a New, Amendment, Renewal Only, or Renewal/Modification, enter the requested authorization expiration date (this item is optional). | | MM DD |
| 7) Does this filing request a Waiver of the Commission's rules? If 'Y', attach the required showing as described in the instructions. | | ( )Yes No |
| 8) Are attachments (other than associated schedules) being filed with this application? | | ( )Yes No |

**Applicant Information**

| | | | |
|---|---|---|---|
| 9a) Taxpayer Identification Number | | | 9b) SGIN: |
| 10) Applicant/Licensee is a(n): ( ) Individual Corporation | Unincorporated Association Limited Liability Corporation | Trust Partnership | Government Entity Joint Venture Consortium |
| 11) First Name (if individual): | MI: | Last Name: | Suffix: |
| 12) Entity Name (if other than individual): | | | |
| 13) Attention To: | | | |
| 14) P.O. Box: | And/Or | 15) Street Address: | |
| 16) City: | 17) State: | 18) Zip: | 19) Country: |
| 20) Telephone Number: | | 21) FAX: | |
| 22) E-Mail Address: | | | |

FCC 605- Main Form
July 1999 - Page 1

**FCC Form 605 is used to process routine license renewals and license changes that are submitted directly to the FCC. The FCC will not accept NCVEC Form 605.**

# SCHEMATIC SYMBOLS

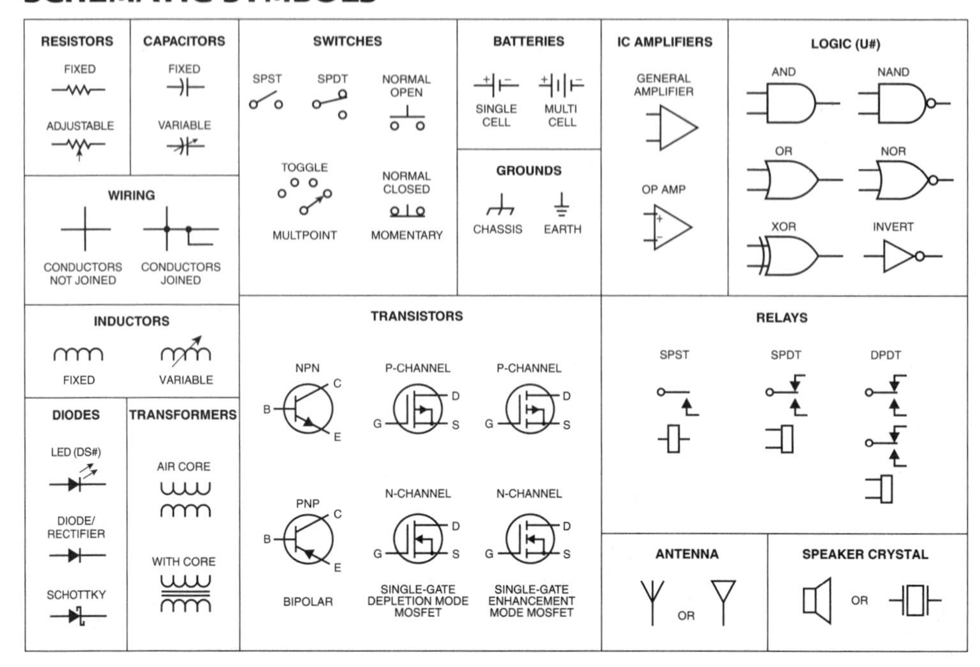


# SCIENTIFIC NOTATION

| Prefix | Symbol | | | Multiplication Factor |
|---|---|---|---|---|
| exa | E | $10^{18}$ | = | 1,000,000,000,000,000,000 |
| peta | P | $10^{15}$ | = | 1,000,000,000,000,000 |
| tera | T | $10^{12}$ | = | 1,000,000,000,000 |
| giga | G | $10^{9}$ | = | 1,000,000,000 |
| mega | M | $10^{6}$ | = | 1,000,000 |
| kilo | k | $10^{3}$ | = | 1,000 |
| hecto | h | $10^{2}$ | = | 100 |
| deca | da | $10^{1}$ | = | 10 |
| (unit) | | $10^{0}$ | = | 1 |

| Prefix | Symbol | | | Multiplication Factor |
|---|---|---|---|---|
| deci | d | $10^{-1}$ | = | 0.1 |
| centi | c | $10^{-2}$ | = | 0.01 |
| milli | m | $10^{-3}$ | = | 0.001 |
| micro | μ | $10^{-6}$ | = | 0.000001 |
| nano | n | $10^{-9}$ | = | 0.000000001 |
| pico | p | $10^{-12}$ | = | 0.000000000001 |
| femto | f | $10^{-15}$ | = | 0.000000000000001 |
| atto | a | $10^{-18}$ | = | 0.000000000000000001 |

# SUMMARY OF QUESTION POOL FORMULAS

| Question | Formula | Where: |
|---|---|---|
| E5H01-10 | **Decibels**<br>$dB = 10 \log_{10} \frac{P_1}{P_2}$ | $P_1$ = Power (usually output) in **watts**<br>$P_2$ = Power (usually input) in **watts** |
| E4B04-09 | **Counter Readout Error**<br>Readout Error = f × a | f = Frequency in **MHz** being measured<br>a = Counter accuracy in **parts per million** |
| E4C14, 27 | **Intermodulation Product**<br>$f_{IMD(2)} = 2f_1 - f_2$<br>$f_{IMD(4)} = 2f_2 - f_1$ | $f_{IMD}$ = Intermodulation product at particular f in **Hz**<br>$f_1$ = Frequency of transmitted signal in **Hz**<br>$f_2$ = Possible frequency of interfereing signal in **Hz** |
| E5A01-09, 18-25<br>E7C06 | **Resonance**<br>$X_L = X_C$<br>$2\pi f_r L = \frac{1}{2\pi f_r C}$<br>$f_r = \frac{1}{2\pi\sqrt{LC}}$ | $X_L$ = Inductive reactance in **ohms**<br>$X_C$ = Capacitive reactance in **ohms**<br>$f_r$ = Resonant frequency in **hertz**<br>L = Inductance in **henrys**<br>C = Capacitance in **farads**<br>π = 3.14 |
| E5A10 | **Phase Between V and I**<br>ELI = Voltage (E) leads current (I) in an inductive (L) circuit<br>ICE = Current (I) leads voltage (E) in a capactive (C) circuit | E = Voltage in circuit<br>I = Current in circuit<br>L = Inductance in circuit<br>C = Capacitance in circuit |
| E5A12-17<br>E5F02 | **Half-Power Bandwidth**<br>$BW_{-3dB} = \frac{f_r}{Q}$ | $BW_{-3dB}$ = Half-Power bandwidth in **kHz**<br>$f_r$ = Resonant frequency in **kHz**<br>Q = Circuit **quality factor** |
| E5B01-11 | **Time Constants**<br>$\tau = RC$<br>$\tau = \frac{L}{R}$ | τ = Time Constant in **seconds**<br>R = Resistance in **ohms**<br>C = Capacitance in **farads**<br>L = Inductance in **henries** |

**Time Constant Curve**

| | | RC | | RL | |
|---|---|---|---|---|---|
| Time Constant τ | % of Change in τ | % of Final Q or V on C When Charging | % of Initial Q or V on C When Discharging | % of Final I When Increasing | % of Initial I When Decreasing |
| 1 | 63.2 | 63.2 | 36.8 | 63.2 | 36.8 |
| 2 | 23.3 | 86.5 | 13.5 | 86.5 | 13.5 |
| 3 | 8.5 | 95.0 | 5.0 | 95.0 | 5.0 |
| 4 | 3.2 | 98.2 | 1.8 | 98.2 | 1.8 |
| 5 | 1.1 | 99.3 | 0.7 | 99.3 | 0.7 |

## SUMMARY OF QUESTION POOL FORMULAS (continued)

| Question | Formula | Where: |
|---|---|---|
| E5B06 | **Capacitors** Series: $C_T = \frac{C_1 \times C_2}{C_1 + C_2}$ Parallel: $C_T = C_1 + C_2$ **Resistors** $R_T = R_1 + R_2$; $R_T = \frac{R_1 \times R_2}{R_1 + R_2}$ | C = Capacitance in **farads**<br>R = Resistance in **ohms** |
| E5C10 | **Inductive Reactance** $X_L = 2\pi fL$ | $X_L$ = Inductive reactance in **ohms**<br>L = Inductance in **henries**<br>f = Frequency in **hertz**<br>$\pi$ = 3.14 |
| E5C11 | **Capacitive Reactance** $X_C = \frac{1}{2\pi fC}$ | $X_C$ = Capacitive reactance in **ohms**<br>C = Capacitance in **farads**<br>f = Frequency in **hertz**<br>$\pi$ = 3.14 |
| | $X_C = \frac{10^6}{2\pi fC}$<br>**Impedance in Series** $Z_T = Z_1 + Z_2$<br>**Impedance in Parallel** → $Z_T = \frac{Z_1 \times Z_2}{Z_1 + Z_2}$ | f = Frequency in **MHz**<br>C = Capacitance in **picofarads**<br>Z = Impedance in **ohms**<br>$\pi$ = 3.14<br>Special Case: When $Z_1$ = R and $Z_2 = \pm jX$<br>$Z_T = \frac{RX \angle \Theta = \pm 90°}{\sqrt{R^2 + X^2} \angle \arctan \pm X/R}$ |
| E5C11-12<br>E5D08-13 | **Rectangular Coordinates** $Z = R \pm jX$<br>**Polar Coordinates** $Z = Z \angle \pm\Theta$<br>**Conversion:**<br>$Z = R \pm jX$ to $Z \angle \pm\Theta$: $Z = \sqrt{R^2 + X^2}\ Z \angle \arctan \frac{X}{R}$<br>$Z \angle \pm\Theta$ to $Z = R \pm jX$: $R = Z \cos \Theta$; $\pm jX = Z \sin \Theta$ | Z = Impedance in **ohms**<br>R = Resistance in **ohms**<br>+jX = Inductive reactance $X_L$ in **ohms**<br>−jX = Capacitive reactance $X_C$ in **ohms**<br>$\Theta$ = Phase angle in **degrees**<br>arctan = angle whose tangent is |
| E5D01-05, 08-11 | **Phase Angle Between V and I**<br>$\tan \phi = \frac{X}{R}$ | $\phi$ = Phase angle in **degrees**<br>X = $(X_L - X_C)$ = Total reactance in **ohms**<br>R = Resistance in **ohms** |

## SUMMARY OF QUESTION POOL FORMULAS (continued)

| Question | Formula | Where: |
|---|---|---|
| E5E05-08, 16-20 | Do Addition and Subtraction in Rectangular Coordinates<br>$Z_1 = R_1 + jX_1$ $Z_2 = R_2 - jX_2$<br>**Addition**<br>$Z_T = (R_1 + jX_1) + (R_2 - jX_2)$<br>$Z_T = (R_1 + R_2) + j(X_1 - X_2)$<br>**Subtraction**<br>$Z_T = (R_1 + jX_1) + (R_2 - jX_2)$<br>$Z_T = (R_1 - R_2) + j(X_1 + X_2)$ | Do Multiplication and Division in Polar Coordinates<br>$Z_1 = Z_1 \angle \Theta_1$ $Z_2 = Z_2 \angle \Theta_2$<br>**Multiply**<br>$Z_1 Z_2 = Z_1 \times Z_2 \angle \Theta_1 + \Theta_2$<br>**Divide**<br>$\frac{Z_1}{Z_2} = \frac{Z_1 \angle \Theta_1}{Z_2 \angle \Theta_1} = \frac{Z_1}{Z_2} = \angle \Theta_1 - \Theta_2$ |
| E5G01-05, 13-16 | **Circuit Quality Factor**<br>$Q = \frac{R}{X_L}$<br>$Q = \frac{R}{2\pi f_r L}$ | $Q$ = Circuit **quality factor**<br>$X_L = 2\pi f_r L$<br>$X_L$ = Inductive reactance in **ohms**<br>$f_r$ = Resonant frequency in **Hz**<br>$L$ = Inductance in **henrys**<br>$R$ = Resistance in **ohms**<br>$2\pi = 6.28$ |
| E5G11<br>E5H13-16 | **True Power**<br>$P_T = P_A \times PF$<br>$P_A = E \times I$<br>$PF = \cos \phi$ | $P_T$ = True power in **watts**<br>$P_A$ = Apparent power in **watts**<br>$\phi$ = Phase angle in **degrees**<br>$PF$ = Power factor<br>$E$ = Applied voltage in **volts**<br>$I$ = Circuit current in **amps** |
| E6B11-16 | **Inverting IC Op-Amp Gain**<br>$G = -\frac{R_f}{R_1}$ | $G$ = Gain in **non-dimentional units**<br>$R_f$ = Feedback resistance in **ohms**<br>$R_1$ = Input resistance in **ohms** |
| E6D13 | **Turns for L Using Ferrite Toroidal Cores**<br>$N = 1000 \sqrt{\frac{L}{A_L}}$ | $N$ = Number of turns needed<br>$A_L$ = Inductance index in **mH per 1000 turns**<br>$L$ = Inductance in **millihenrys** |
| E6D14 | **Turns for L Using Powdered-Iron Toroidal Cores**<br>$N = 100 \sqrt{\frac{L}{A_L}}$ | $N$ = Number of turns needed<br>$A_L$ = Inductance index in **µH per 1000 turns**<br>$L$ = Inductance in **millihenrys** |

## SUMMARY OF QUESTION POOL FORMULAS (continued)

| Question | Formula | Where: |
|---|---|---|
| E8A08-12<br>E8D01, 06 | **Peak, Peak-to-Peak and RMS Voltages**<br>$^{**}V_P = \frac{V_{PP}}{2}$ $^{*}V_{RMS} = 0.707\ V_P$<br>$^{**}V_{PP} = 2V_P$ $^{*}V_P = 1.414\ V_{RMS}$ | $V_P$ = Peak voltage in **volts**<br>$V_{PP}$ = Peak-to-peak voltage in **volts**<br>$V_{RMS}$ = Root-mean-square voltage in **volts**<br>* Equations for sinewaves<br>** Equations for symmetrical waves |
| E8A16<br>E8D05 | **Peak Envelope Power**<br>$PEP = P_{DC} \times \text{Efficiency}$ | PEP = Peak envelope power in **watts**<br>$P_{DC}$ = Input DC power in **watts** |

| Amplifier Class | Efficiency |
|---|---|
| C | 80% |
| B | 60% |
| AB | 50% |
| A | <50% |

E8B11-12 **Modulation Index**

$$\text{Modulation Index} = \frac{\text{Deviation of FM Signal (in Hz)}}{\text{Modulating Frequency (in Hz)}}$$

E8B13 **Deviation Ratio**

$$\text{Deviation Ratio} = \frac{\text{Maximum Carrier Frequency Deviation (in kHz)}}{\text{Maximum Modulation Frequency (in kHz)}}$$

## SUMMARY OF QUESTION POOL FORMULAS (continued)

| Question | Formula | Where: |
|---|---|---|
| E8C07 | **Bandwidth for CW**<br>$BW_{CW}$ = baud rate × wpm × fading factor | $BW_{CW}$ = Necessary bandwidth in **hertz**<br>wpm = Morse code signal rate in **words per minute**<br>Fading factor = Constant of 5 for **CW** |
| E8C08-10 | **Bandwidth for Digital**<br>BW = baud rate + (1.2 × $f_s$) | BW = Necessary bandwidth in **hertz**<br>$f_s$ = Frequency shift in **hertz**<br>Baud rate = Digital signal rate in **bauds** |
| E9D03 | **Antenna Beamwidth**<br>$\text{Beamwidth} = \frac{203}{(\sqrt{10})^X}$ | Beamwidth = Antenna beamwidth in **degrees**<br>$A_G$ = Antenna gain in **dB**<br>$X = \frac{A_G}{10}$ |
| E9E10-11 | **Physical Length vs. Electrical Length**<br>$L = \frac{984 \lambda V}{f}$ | L = Physical length in **feet**<br>λ = Electrical length in **wavelengths**<br>V = **Velocity factor** of feedline<br>f = Frequency in **MHz** |

| Feedline | Velocity Factor |
|---|---|
| Coax | 0.66 |
| Parallel | 0.95 |
| Twin-Lead | 0.80 |

## ELEMENT 4 (EXTRA CLASS) QUESTION POOL SYLLABUS

The syllabus used for the development of the question pool is included here as an aid in studying the subelements and topic groups. Review the syllabus before you start your study to gain an understanding of how the question pool is used to develop the Element 4 written examination. Remember, one question will be taken from each topic group within each subelement to create your exam.

**E1 – COMMISSION'S RULES**
(7 Exam Questions – 7 Groups)

**E1A** Operating standards: frequency privileges for Extra class amateurs; emission standards; message forwarding; frequency sharing between ITU Regions; FCC modification of station license; 30-meter band sharing; stations aboard ships or aircraft; telemetry; telecommand of an amateur station; authorized telecommand transmissions

**E1B** Station restrictions: restrictions on station locations; restricted operation; teacher as control operator; station antenna structures; definition and operation of remote control and automatic control; control link

**E1C** Reciprocal operating: reciprocal operating authority; purpose of reciprocal agreement rules; alien control operator privileges; identification (Note: This includes CEPT and IARP)

**E1D** Radio Amateur Civil Emergency Service (RACES): definition; purpose; station registration; station license required; control operator requirements; control operator privileges; frequencies available; limitations on use of RACES frequencies; points of communication for RACES operation; permissible communications

**E1E** Amateur Satellite Service: definition; purpose; station license required for space station; frequencies available; telecommand operation: definition; eligibility; telecommand station (definition); space telecommand station; special provisions; telemetry: definition; special provisions; space station: definition; eligibility; special provisions; authorized frequencies (space station); notification requirements; earth operation: definition; eligibility; authorized frequencies (Earth station)

**E1F** Volunteer Examiner Coordinators (VECs): definition; VEC qualifications; VEC agreement; scheduling examinations; coordinating VEs; reimbursement for expenses; accrediting VEs; question pools; Volunteer Examiners (VEs): definition; requirements; accreditation; reimbursement for expenses; VE conduct; preparing an examination; examination elements; definition of code and written elements; preparation responsibility; examination requirements; examination credit; examination procedure; examination administration; temporary operating authority

**E1G** Certification of external RF power amplifiers and external RF power amplifier kits; Line A; National Radio Quiet Zone; business communications; definition and operation of spread spectrum; auxiliary station operation

**E2 – OPERATING PROCEDURES**
(5 Exam Questions – 5 Groups)

**E2A** Amateur Satellites: orbital mechanics; frequencies available for satellite operation; satellite hardware; satellite operations

**E2B** Television: fast scan television (FSTV) standards; slow scan television (SSTV) standards; facsimile (fax) communications

**E2C** Contest and DX operating; spread-spectrum transmissions; automatic HF forwarding; selecting your operating frequency

**E2D** Operating VHF / UHF digital modes: packet clusters; digital bulletin boards; Automatic Position Reporting System (APRS)

**E2E** Operating HF digital modes

**E3 – RADIO WAVE PROPAGATION**
(3 Exam Questions – 3 Groups)

**E3A** Earth-Moon-Earth (EME or moonbounce) communications; meteor scatter

**E3B** Transequatorial; long path; gray line

**E3C** Auroral propagation; selective fading; radio-path horizon; take-off angle over flat or sloping terrain; earth effects on propagation

**E4 – AMATEUR RADIO PRACTICES**
(5 Exam Questions – 5 Groups)

**E4A** Test equipment: spectrum analyzers (interpreting spectrum analyzer displays; transmitter output spectrum), logic probes (indications of high and low states in digital circuits; indications of pulse conditions in digital circuits)

**E4B** Frequency measurement devices (i.e., frequency counter, oscilloscope Lissajous figures, dip meter); meter performance limitations; oscilloscope performance limitations; frequency counter performance limitations

**E4C** Receiver performance characteristics (i.e., phase noise, desensitization, capture effect, intercept point, noise floor, dynamic range {blocking and IMD}, image rejection, MDS, signal-to-noise-ratio); intermodulation and cross-modulation interference

**E4D** Noise suppression: vehicular system noise; electronic motor noise; static; line noise

**E4E** Component mounting techniques (i.e., surface, dead bug (raised), circuit board; direction finding: techniques and equipment; fox hunting

**E5 – ELECTRICAL PRINCIPLES**
(9 Exam Questions – 9 Groups)
**E5A** Characteristics of resonant circuits: Series resonance (capacitor and inductor to resonate at a specific frequency); Parallel resonance (capacitor and inductor to resonate at a specific frequency); half-power bandwidth
**E5B** Exponential charge/discharge curves (time constants): definition; time constants in RL and RC circuits
**E5C** Impedance diagrams: Basic principles of Smith charts; impedance of RLC networks at specified frequencies; PC based impedance analysis (including Smith Charts)
**E5D** Phase angle between voltage and current; impedances and phase angles of series and parallel circuits;
**E5E** Algebraic operations using complex numbers: rectangular coordinates (real and imaginary parts); polar coordinates (magnitude and angle)
**E5F** Skin effect; electrostatic and electromagnetic fields
**E5G** Circuit Q; reactive power; power factor
**E5H** Effective radiated power; system gains and losses
**E5I** Photoconductive principles and effects

**E6 – CIRCUIT COMPONENTS**
(5 Exam Questions – 5 Groups)
**E6A** Semiconductor material: Germanium, Silicon, P-type, N-type; Transistor types: NPN, PNP, junction, power; field-effect transistors (FETs): enhancement mode; depletion mode; MOS; CMOS; N-channel; P-channel
**E6B** Diodes: Zener, tunnel, varactor, hot-carrier, junction, point contact, PIN and light emitting; operational amplifiers (inverting amplifiers, noninverting amplifiers, voltage gain, frequency response, FET amplifier circuits, single-stage amplifier applications); phase-locked loops
**E6C** TTL digital integrated circuits; CMOS digital integrated circuits; gates
**E6D** Vidicon and cathode-ray tube devices; charge-coupled devices (CCDs); liquid crystal displays (LCDs); toroids: permeability, core material, selecting, winding
**E6E** Quartz crystal (frequency determining properties as used in oscillators and filters); monolithic amplifiers (MMICs)

**E7 – PRACTICAL CIRCUITS**
(7 Exam Questions – 7 Groups)
**E7A** Digital logic circuits: Flip flops; Astable and monostable multivibrators; Gates (AND, NAND, OR, NOR); Positive and negative logic
**E7B** Amplifier circuits: Class A, Class AB, Class B, Class C, amplifier operating efficiency (i.e., DC input versus PEP), transmitter final amplifiers; amplifier circuits: tube, bipolar transistor, FET
**E7C** Impedance-matching networks: Pi, L, Pi-L; filter circuits: constant K, M-derived, band-stop, notch, crystal lattice, pi-section, T-section, L-section, Butterworth, Chebyshev, elliptical; filter applications (audio, IF, digital signal processing {DSP})
**E7D** Oscillators: types, applications, stability; voltage-regulator circuits: discrete, integrated and switched mode
**E7E** Modulators: reactance, phase, balanced; detectors; mixer stages; frequency synthesizers
**E7F** Digital frequency divider circuits; frequency marker generators; frequency counters
**E7G** Active audio filters: characteristics; basic circuit design; preselector applications

**E8 – SIGNALS AND EMISSIONS**
(4 Exam Questions – 4 Groups)
**E8A** AC waveforms: sine wave, square wave, sawtooth wave; AC measurements: peak, peak-to-peak and root-mean-square (RMS) value, peak-envelope-power (PEP) relative to average
**E8B** FCC emission designators versus emission types; modulation symbols and transmission characteristics; modulation methods; modulation index; deviation ratio; pulse modulation: width; position
**E8C** Digital signals: including CW; digital signal information rate vs. bandwidth; spread-spectrum communications
**E8D** Peak amplitude (positive and negative); peak-to-peak values: measurements; Electromagnetic radiation; wave polarization; signal-to-noise (S/N) ratio

**E9 – ANTENNAS**
(5 Exam Questions – 5 Groups)
**E9A** Isotropic radiators: definition; used as a standard for comparison; radiation pattern; basic antenna parameters: radiation resistance and reactance (including wire dipole, folded dipole), gain, beamwidth, efficiency
**E9B** Free-space antenna patterns: E and H plane patterns (i.e., azimuth and elevation in free-space); gain as a function of pattern; antenna design (computer modeling of antennas)
**E9C** Phased vertical antennas; radiation patterns; beverage antennas; rhombic antennas: resonant; terminated; radiation pattern; antenna patterns: elevation above real ground, ground effects as related to polarization, take-off angles as a function of height above ground
**E9D** Space and satellite communications antennas: gain; beamwidth; tracking; losses in real antennas and matching: resistivity losses, losses in resonating elements (loading coils, matching networks, etc. {i.e., mobile, trap}); SWR bandwidth; efficiency
**E9E** Matching antennas to feed lines; characteristics of open and shorted feed lines: 1/8 wavelength; 1/4 wavelength; 1/2 wavelength; feed lines: coax versus open-wire; velocity factor; electrical length; transformation characteristics of line terminated in impedance not equal to characteristic impedance; use of antenna analyzers

**AUTHORIZED FREQUENCY BANDS – AMATEUR SERVICE** (for U.S. Amateur Stations operating from ITU-Region 2–North and South America)

| **Current License Class**[1] | | Technician | Tech. w/Code | General | | Extra Class |
|---|---|---|---|---|---|---|
| **Grandfathered**[2] | Novice | Technician | Technician Plus | General | Advanced | Extra Class |
| METERS | | | | | | |
| 160 | | | | 1800-2000 kHz/All | 1800-2000 kHz/All | 1800-2000 kHz/All |
| 80 | 3675-3725 kHz/CW | | 3675-3725 kHz/CW | 3525-3750 kHz/CW<br>3850-4000 kHz/Ph | 3525-3750 kHz/CW<br>3775-4000 kHz/Ph | 3500-4000 kHz/CW<br>3750-4000 kHz/Ph |
| 40 | 7100-7150 kHz/CW | | 7100-7150 kHz/CW | 7025-7150 kHz/CW<br>7225-7300 kHz/Ph | 7025-7300 kHz/CW<br>7150-7300 kHz/Ph | 7000-7300 kHz/CW<br>7150-7300 kHz/Ph |
| 30 | | | | 10.1-10.15 MHz/CW | 10.1-10.15 MHz/CW | 10.1-10.15 MHz/CW |
| 20 | | | | 14.025-14.15 MHz/CW<br>14.225-14.35 MHz/Ph | 14.025-14.15 MHz/CW<br>14.175-14.35 MHz/Ph | 14.0-14.35 MHz/CW<br>14.15-14.35 MHz/Ph |
| 17 | | | | 18.068-18.11 MHz/CW<br>18.11-18.168 MHz/Ph | 18.068-18.11 MHz/CW<br>18.11-18.168 MHz/Ph | 18.068-18.11 MHz/CW<br>18.11-18.168 MHz/Ph |
| 15 | 21.1-21.2 MHz/CW | | 21.1-21.2 MHz/CW | 21.025-21.2 MHz/CW<br>21.3-21.45 MHz/Ph | 21.025-21.2 MHz/CW<br>21.225-21.45 MHz/Ph | 21.0-21.45 MHz/CW<br>21.2-21.45 MHz/Ph |
| 12 | | | | 24.89-24.99 MHz/CW<br>24.93-24.99 MHz/Ph | 24.89-24.99 MHz/CW<br>24.93-24.99 MHz/Ph | 24.89-24.99 MHz/CW<br>24.93-24.99 MHz/Ph |
| 10 | 28.1-28.5 MHz/CW<br>28.3-28.5 MHz/Ph | | 28.1-28.5 MHz/CW<br>28.3-28.5 MHz/Ph | 28.0-29.7 MHz/CW<br>28.3-29.7 MHz/Ph | 28.0-29.7 MHz/CW<br>28.3-29.7 MHz/Ph | 28.0-29.7 MHz/CW<br>28.3-29.7 MHz/Ph |
| 6 | | 50-54 MHz/CW<br>50.1-54 MHz/Ph | 50-54 MHz/CW<br>50.1-54 MHz/Ph | 50-54 MHz/CW<br>50.1-54 MHz/Ph | 50-54 MHz/CW<br>50.1-54 MHz/Ph | 50-54 MHz/CW<br>50.1-54 MHz/Ph |
| 2 | | 144-148 MHz/CW<br>144.1-148 MHz/All | 144-148 MHz/CW<br>144.1-148 MHz/All | 144-148 MHz/CW<br>144.1-148 MHz/Ph | 144-148 MHz/CW<br>144.1-148 MHz/All | 144-148 MHz/CW<br>144.1-148 MHz/All |
| 1.25 | 222-225 MHz/All | [3] 222-225 MHz/All | 222-225 MHz/All | 222-225 MHz/All | 222-225 MHz/All | 222-225 MHz/All |
| 0.70 | | 420-450 MHz/All | 420-450 MHz/All | 420-450 MHz/All | 420-450 MHz/All | 420-450 MHz/All |
| 0.33 | | 902-928 MHz/All | 902-928 MHz/All | 902-928 MHz/All | 902-928 MHz/All | 902-928 MHz/All |
| 0.23 | 1270-1295 MHz/All | 1240-1300 MHz/All | 1240-1300 MHz/All | 1240-1300 MHz/All | 1240-1300 MHz/All | 1240-1300 MHz/All |

[1] Effective 4-15-00 [2] Prior to 4-15-00 [3] Effective 2/1/94 219-220 MHz is authorized for point-to-point fixed digital message forwarding systems.

***Note:*** Morse code (CW, A1A) may be used on any frequency allocated to the amateur service. Telephony emission (abbreviated Ph above) authorized on certain bands as indicated. Higher class licensees may use slow-scan television and facsimile emissions on the Phone bands; radio teletype/digital on the CW bands. All amateur modes and emissions are authorized above 144.1 MHz. In actual practice, the modes/emissions used are somewhat more complicated than shown above due to the existence of various band plans and "gentlemen's agreements" concerning where certain operations should take place.

# GRID-SQUARE MAP FOR UNITED STATES

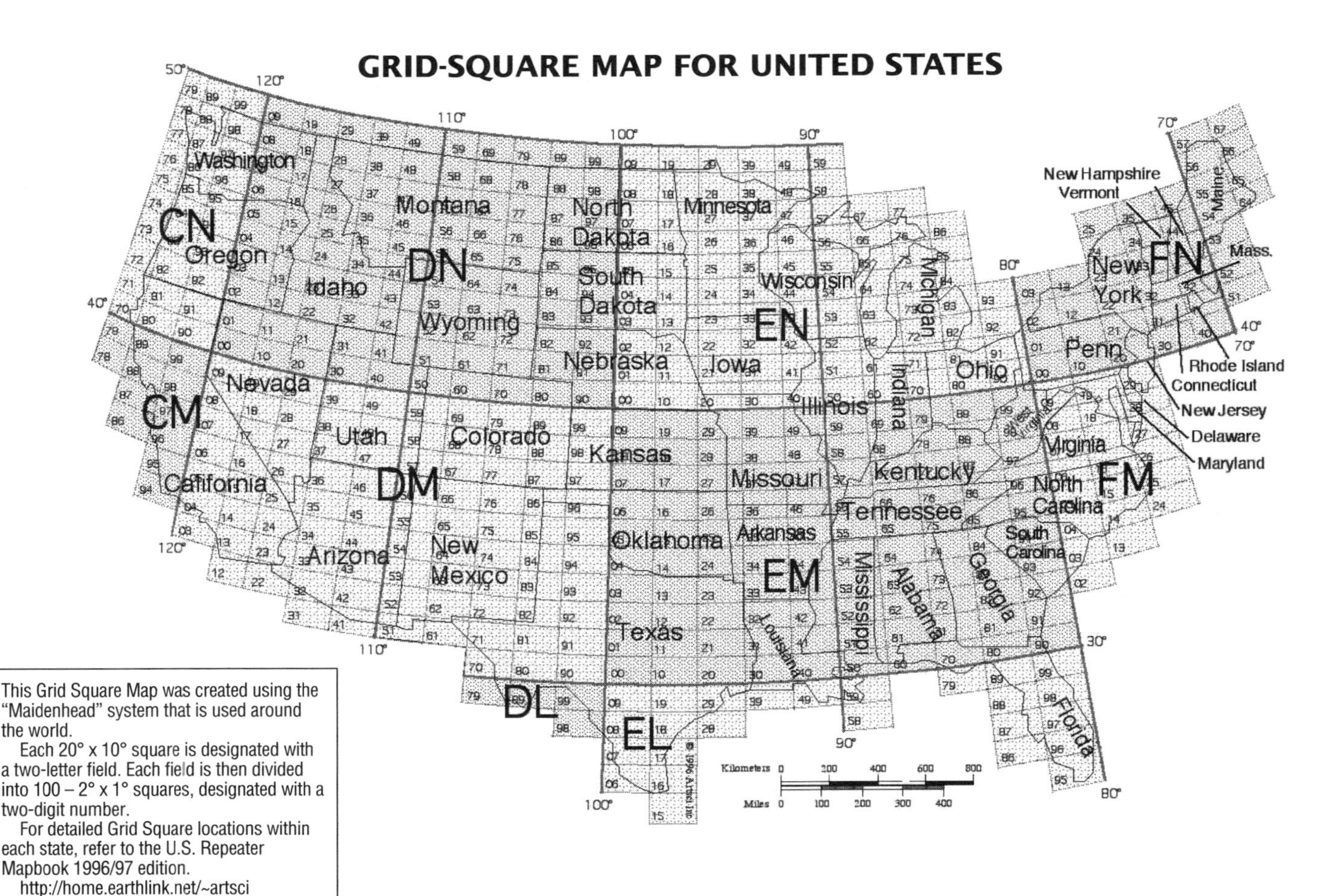


This Grid Square Map was created using the "Maidenhead" system that is used around the world.

Each 20° x 10° square is designated with a two-letter field. Each field is then divided into 100 – 2° x 1° squares, designated with a two-digit number.

For detailed Grid Square locations within each state, refer to the U.S. Repeater Mapbook 1996/97 edition.

http://home.earthlink.net/~artsci

**Reprinted by Permission of Artsci Inc.,P.O. Box 1428, Burbank, CA 91507**

## ITU Regions

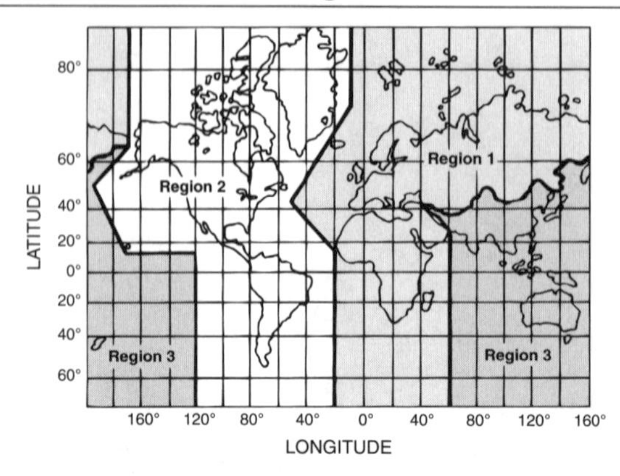


## List of Countries Permitting Third-Party Traffic

| Country | Call Sign Prefix |
|---|---|
| Antigua and Barbuda | V2 |
| Argentina | LU |
| Australia | VK |
| Austria, Vienna | 4U1VIC |
| Belize | V3 |
| Bolivia | CP |
| Bosnia-Herzegovina | T9 |
| Brazil | PY |
| Canada | VE, VO, VY |
| Chile | CE |
| Colombia | HK |
| Comoros | D6 |
| Costa Rica | TI |
| Cuba | CO |
| Dominica | J7 |
| Dominican Republic | HI |
| Ecuador | HC |
| El Salvador | YS |
| The Gambia | C5 |
| Ghana | 9G |
| Grenada | J3 |
| Guatemala | TG |
| Guyana | 8R |
| Haiti | HH |
| Honduras | HR |
| Israel | 4X |
| Jamaica | 6Y |
| Jordan | JY |
| Liberia | EL |
| Marshall Is | V6 |
| Mexico | XE |
| Micronesia | V6 |
| Nicaragua | YN |
| Panama | HP |
| Paraguay | ZP |
| Peru | OA |
| Philippines | DU |
| St. Christopher & Nevis | V4 |
| St. Lucia | J6 |
| St. Vincent & Grenadines | J8 |
| Sierra Leone | 9L |
| South Africa | ZS |
| Swaziland | 3D6 |
| Trinidad and Tobago | 9Y |
| Turkey | TA |
| United Kingdom | GB* |
| Uruguay | CX |
| Venezuela | YV |
| ITU-Geneva | 4U1ITU |
| VIC-Vienna | 4U1VIC |

## Countries Holding U.S. Reciprocal Agreements

Antigua, Barbuda
Argentina
Australia
Austria
Bahamas
Barbados
Belgium
Belize
Bolivia
Bosnia-Herzegovina
Botswana
Brazil
Canada[1]
Chile
Colombia
Costa Rica
Croatia
Cyprus
Denmark
Dominica
Dominican Rep.
Ecuador
El Salvador
Fiji
Finland
France[2]
Germany
Greece
Greenland
Grenada
Guatemala
Guyana
Haiti
Honduras
Iceland
India
Indonesia
Ireland
Israel
Italy
Jamaica
Japan
Jordan
Kiribati
Kuwait
Liberia
Luxembourg
Macedonia
Marshall Is.
Mexico
Micronesia
Monaco
Netherlands
Netherlands Ant.
New Zealand
Nicaragua
Norway
Panama
Paraguay
Papua New Guinea
Peru
Philippines
Portugal
Seychelles
Sierra Leone
Solomon Islands
South Africa
Spain
St. Lucia
St. Vincent and Grenadines
Surinam
Sweden
Switzerland
Thailand
Trinidad, Tobago
Turkey
Tuvalu
United Kingdom[3]
Uruguay
Venezuela

1. Do not need reciprocal permit
2. Includes all French Territories
3. Includes all British Territories

## POPULAR Q SIGNALS

Given below are a number of Q signals whose meanings most often need to be expressed with brevity and clarity in amateur work. (Q abbreviations take the form of questions only when each is sent followed by a question mark.)

**QRG** Will you tell me my exact frequency (or that of _____ )? Your exact frequency (or that of _____ ) is _____ kHz.
**QRH** Does my frequency vary? Your frequency varies.
**QRI** How is the tone of my transmission? The tone of your transmission is _____ (1. Good; 2. Variable; 3. Bad).
**QRJ** Are you receiving me badly? I cannot receive you. Your signals are too weak.
**QRK** What is the intelligibility of my signals (or those of _____ )? The intelligibility of your signals (or those of _____ ) is _____ (1. Bad; 2. Poor; 3. Fair; 4. Good; 5. Excellent).
**QRL** Are you busy? I am busy (or I am busy with _____ ). Please do not interfere.
**QRM** Is my transmission being interfered with? Your transmission is being interfered with _____ (1. Nil; 2. Slightly; 3. Moderately; 4. Severely; 5. Extremely).
**QRN** Are you troubled by static? I am troubled by static _____ (1-5 as under QRM).
**QRO** Shall I increase power? Increase power.
**QRP** Shall I decrease power? Decrease power.
**QRQ** Shall I send faster? Send faster ( _____ WPM).
**QRS** Shall I send more slowly? Send more slowly ( _____WPM).
**QRT** Shall I stop sending? Stop sending.
**QRU** Have you anything for me? I have nothing for you.
**QRV** Are you ready? I am ready.
**QRW** Shall I inform _____ that you are calling on _____ kHz? Please inform _____ that I am calling on _____ kHz.
**QRX** When will you call me again? I will call you again at _____ hours (on _____ kHz).
**QRY** What is my turn? Your turn is numbered _____.
**QRZ** Who is calling me? You are being called by _____ (on _____ kHz).
**QSA** What is the strength of my signals (or those of _____ )? The strength of your signals (or those of _____ ) is _____ (1. Scarcely perceptible; 2. Weak; 3. Fairly good; 4. Good; 5. Very good).
**QSB** Are my signals fading? Your signals are fading.
**QSD** Is my keying defective? Your keying is defective.
**QSG** Shall I send _____ messages at a time? Send _____ messages at a time.
**QSK** Can you hear me between your signals and if so can I break in on your transmission? I can hear you between my signals; break in on my transmission.
**QSL** Can you acknowledge receipt? I am acknowledging receipt.
**QSM** Shall I repeat the last message which I sent you, or some previous message? Repeat the last message which you sent me [or message(s) number(s) _____ ].
**QSN** Did you hear me (or _____ ) on _____ kHz? I heard you (or _____ ) on _____ kHz.
**QSO** Can you communicate with _____ direct or by relay? I can communicate with _____ direct (or by relay through _____ ).
**QSP** Will you relay to _____? I will relay to _____.
**QST** General call preceding a message addressed to all amateurs and ARRL members. This is in effect "CQ ARRL."
**QSU** Shall I send or reply on this frequency (or on _____ kHz)?
**QSW** Will you send on this frequency (or on _____ kHz)? I am going to send on this frequency (or on _____ kHz).
**QSX** Will you listen to _____ on _____ kHz? I am listening to _____ on _____ kHz.
**QSY** Shall I change to transmission on another frequency? Change to transmission on another frequency (or on _____ kHz).
**QSZ** Shall I send each word or group more than once? Send each word or group twice (or _____ times).
**QTA** Shall I cancel message number _____? Cancel message number _____.
**QTB** Do you agree with my counting of words? I do not agree with your counting of words. I will repeat the first letter or digit of each word or group.
**QTC** How many messages have you to send? I have messages for you (or for _____ ).
**QTH** What is your location? My location is _____.
**QTR** What is the correct time? The time is _____.

Source: ARRL

# Glossary

**Advanced:** An amateur operator who has passed Element 2, 3A, 3B, and 4A written theory examinations and Element 1A and 1B code tests to demonstrate Morse code proficiency to 13 wpm.

**Amateur communication:** Non-commercial radio communication by or among amateur stations solely with a personal aim and without personal or business interest.

**Amateur operator/primary station license:** An instrument of authorization issued by the Federal Communications Commission comprised of a station license, and also incorporating an operator license indicating the class of privileges.

**Amateur operator:** A person holding a valid license to operate an amateur station issued by the FCC. Amateur operators are frequently referred to as ham operators.

**Amateur Radio services:** The amateur service, the amateur-satellite service and the radio amateur civil emergency service.

**Amateur-satellite service:** A radiocommunication service using stations on Earth satellites for the same purpose as those of the amateur service.

**Amateur service:** A radiocommunication service for the purpose of self-training, intercommunication and technical investigations carried out by amateurs; that is, duly authorized persons interested in radio technique solely with a personal aim and without pecuniary interest.

**Amateur station:** A station licensed in the amateur service embracing necessary apparatus at a particular location used for amateur communication.

**AMSAT:** Radio Amateur Satellite Corporation, a non-profit scientific organization. (850 Sligo Avenue, Silver Spring, MD 20910-4703)

**Antenna gain:** The increase as a result of the physical construction of the antenna which confines the radiation to desired or useful directions. Usually specified in dB, referenced to the gain of a dipole.

**ARES:** The emergency division of the American Radio Relay League. See RACES

**ARRL:** American Radio Relay League, national organization of U.S. Amateur Radio operators. (225 Main Street, Newington, CT 06111)

**ATV:** Amateur fast-scan television.

**Audio Frequency (AF):** The range of frequencies that can be heard by the human ear, generally 20 hertz to 20 kilohertz.

**Authorized bandwidth:** The allowed frequency band, specified in kilohertz, and centered on the carrier frequency.

**Automatic control:** The use of devices and procedures for station control without the control operator being present at the control point when the station is transmitting.

**Automatic Position Reporting System (APRS):** Links a remote GPS unit to a ham transceiver which transmits the location to a base station.

**Automatic Volume Control (AVC):** A circuit that continually maintains a constant audio output volume in spite of deviations in input signal strength.

**Beam or Yagi antenna:** An antenna array that receives or transmits RF energy in a particular direction. Usually rotatable.

**Block diagram:** A simplified outline of an electronic system where circuits or components are shown as boxes.

**Broadcasting:** Information or programming transmitted by means of radio intended for the general public.

**Business communications:** Any transmission or communication the purpose of which is to facilitate the regular business or commercial affairs of any party. Business communications are prohibited in the amateur service.

**Call Book:** A published list of all licensed amateur operators available in North American and Foreign editions.

**Call sign assignment:** The FCC systematically assigns each amateur station their primary call sign.

**Carrier frequency:** The frequency of an unmodulated electromagnetic wave, usually specified in kilohertz or megahertz.

**Certificate of Successful Completion of Examination (CSCE):** A certificate verifying successful completion of a written and/or Morse code examination Element. Credit for passing is valid for 365 days.

**Coaxial cable, Coax:** A concentric, two-conductor cable in which one conductor surrounds the other, separated by an insulator.

**Control point:** Any place from which a transmitter's function may be controlled.

**Control station:** A station directly associated with the control point of a radio system, commonly used to control repeaters.

**Control operator:** An amateur operator designated by the licensee of an amateur station to be responsible for the station transmissions.

**Coordinated repeater station:** An amateur repeater station for which the transmitting and receiving frequencies have been recommended by the recognized repeater coordinator.

**Coordinated Universal Time (UTC):** Sometimes referred to as Greenwich Mean Time, UCT or Zulu time. The time at the zero-degree (0∞) Meridian which passes through Greenwich, England. A universal time among all amateur operators.

**Crystal:** A quartz or similar material which has been ground to produce natural vibrations of a specific frequency. Quartz crystals produce a high-degree of frequency stability in radio transmitters.

**CW:** Continuous wave, another term for the International Morse code.

**Dipole antenna:** The most common wire antenna. Length is equal to one-half of the wavelength.

**Direct Digital Frequency Synthesizer:** Modern frequency synthesizer employing technology to reduce phase noise to an insignificant level.

**Dummy antenna:** A device or resistor which serves as a transmitter's antenna without radiating radio waves. Generally used to tune up a radio transmitter.

**Duplex:** Transmitting on one frequency, and receiving on another, commonly used in mobile telephone use, as well as mobile business radio on UHF frequencies.

**Duplexer:** A device that allows a single antenna to be simultaneously used for both reception and transmission.

**Effective Radiated Power (ERP):** The product of the transmitter (peak envelope) power, expressed in watts, delivered to the antenna, and the relative gain of an antenna over that of a half-wave dipole antenna.

**Emergency communication:** Any amateur communication directly relating to the immediate safety of life of individuals or the immediate protection of property.

**Examination Element:** The written theory exams and Morse code test required to obtain amateur radio licenses. Technicians are required to pass written Element 2 theory; Generals are required to pass written Element 3 theory plus Element 1 Morse code test at five (5) words-per-minute; and Extras are required to pass written Element 4 theory.

**FCC Form 605:** The universal application form used to renew or modify an existing license. (See NCVEC Form 605)

**Federal Communications Commission (FCC):** A board of five Commissioners, appointed by the President, having the power to regulate wire and radio telecommunications in the United States.

**Feedline transmission line:** A system of conductors that connects an antenna to a receiver or transmitter.

**Field Day:** Annual activity sponsored by the ARRL to demonstrate emergency preparedness of amateur operators.

**Filter:** A device used to block or reduce alternating currents or signals at certain frequencies while allowing others to pass unimpeded.

**Frequency:** The number of cycles of alternating current in one second.

**Frequency coordinator:** An individual or organization recognized by amateur operators eligible to engage in repeater operation that recommends frequencies and other operating and/or technical parameters for amateur repeater operation in order to avoid or minimize potential interferences.

**Frequency Modulation (FM):** A method of varying a radio carrier wave by causing its frequency to vary in accordance with the information to be conveyed.

**Frequency privileges:** The transmitting frequency bands available to the various classes of amateur operators. The privileges are listed in Part 97.7(a) of the FCC rules.

**General:** An amateur operator who has passed the Element 3 written theory examination and Element 1 code test to demonstrate Morse code proficiency to 5-wpm.

**Ground:** A connection, accidental or intentional, between a device or circuit and the earth or some common body and the earth or some common body serving as the earth.

**Ground wave:** A radio wave that is propagated near or at the earth's surface.

**Half-Duplex:** A method of operation on a duplex (separate transmit, separate receive) frequency pair where the operator only receives when the microphone button is released. Half-duplex does not permit simultaneous talk and listen.

**Handi-Ham system:** Amateur organization dedicated to assisting handicapped amateur operators. (3915 Golden Valley Road, Golden Valley, MN 55422)

**Harmful interference:** Interference which seriously degrades, obstructs or repeatedly interrupts the operation of a radio communication service.

**Harmonic:** A radio wave that is a multiple of the fundamental frequency. The second harmonic is twice the fundamental frequency, the third harmonic, three times, etc.

**Hertz:** One complete alternating cycle per second, abbreviated Hz. Named after Heinrich R. Hertz, a German physicist. The number of hertz is the frequency of the audio or radio wave.

**High Frequency (HF):** The band of frequencies that lie between 3 and 30 megahertz. It is from these frequencies that radio waves are returned to earth from the ionosphere.

**High-Pass filter:** A device that allows passage of high frequency signals but attenuates the lower frequencies.

**Interference:** The effect that occurs when two or more radio stations are transmitting at the same time. This includes undesired noise or other radio signals on the same frequency.

**International Telecommunication Union (ITU):** World organization composed of delegates from member countries, which allocates frequency spectrum for various purposes, including amateur radio.

**Ionosphere:** Outer limits of atmosphere from which HF amateur communications signals are returned to earth.

**Jamming:** The intentional, malicious interference with another radio signal.

**Key clicks, Chirps:** Defective keying of a telegraphy signal sounding like tapping or high varying pitches.

**Lid:** Amateur slang term for poor radio operator.

**Linear amplifier:** A device that accurately reproduces a radio wave in magnified form.

**Long wire:** A horizontal wire antenna that is one wavelength or longer in length.

**Low-Pass filter:** A device that allows passage of low frequency signals but attenuates the higher frequencies.

**Machine:** A ham slang word for an automatic repeater station.

**Malicious interference:** Willful, intentional jamming of radio transmissions.

**MARS:** The Military Affiliate Radio System. An organization that coordinates the activities of amateur communications with military radio communications.

**Maximum authorized transmitting power:** Amateur stations must use no more than the maximum transmitter power necessary to carry out the desired communications.

**Maximum usable frequency (MUF):** The highest frequency that will be returned to earth from the ionosphere.

**Medium frequency (MF):** The band of frequencies that lies between 300 and 3,000 kHz (3 MHz).

**Mobile operation:** Radio communications conducted while in motion or during halts at unspecified locations.

**Mode:** Type of transmission such as voice, teletype, code, television, facsimile.

**Modulate:** To vary the amplitude, frequency or phase of a radio frequency wave in accordance with the information to be conveyed.

**Monolithic Microwave Integrated Circuit (MMIC):** A four-legged device designed for RF amplification. Well-suited for VHF, UHF and microwave applications.

**Morse code (see CW):** The International Morse code, A1A emission. Interrupted continuous wave communications conducted using a dot-dash code for letters, numbers and operating procedure signs.

**NCVEC Form 605:** The form provided by VECs that is used to apply for an amateur radio license or license upgrade.

**Noise level:** The strengths of extraneous audible sounds in a given location, usually measured in decibels.

**Novice operator:** An FCC licensed, entry-level operator in the amateur service. Novices may operate a transmitter in the following meter wavelength bands: 80, 40, 15, 10, 1.25 and 0.23.

**Ohm's law:** The basic electrical law explaining the relationship between voltage, current and resistance. The current I in a circuit is equal to the voltage E divided by the resistance R, or $I = E/R$.

**OSCAR:** Acronym for "Orbiting Satellite Carrying Amateur Radio," the name given to a series of satellites designed and built by amateur operators of several nations.

**Oscillator:** A device for generating oscillations or vibrations of an audio or radio frequency signal.

**Output power:** The radio-frequency output power of a transmitter's final radio-frequency stage as measured at the output terminal while connected to a load of the impedance recommended by the manufacturer.

**Packet radio:** A digital method of communicating computer-to-computer. A terminal-node controller makes up the packet of data and directs it to another packet station.

**Peak Envelope Power (PEP):** 1. The power during one radio frequency cycle at the crest of the modulation envelope, measured under normal operating conditions. 2. The maximum power that can be obtained from a transmitter.

**Phone patch:** Interconnection of amateur service to the public switched telephone network, and operated by the control operator of the station.

**Power supply:** A device or circuit that provides the appropriate voltage and current to another device or circuit.

**Propagation:** The travel of electromagnetic waves or sound waves through a medium.

**Q-signals:** International three-letter abbreviations beginning with the letter Q used primarily to convey information using the Morse code.

**QSL Bureau:** An office that bulk processes QSL (radio confirmation) cards for (or from) foreign amateur operators as a postage saving mechanism.

**RACES (radio amateur civil emergency service):** A radio service using amateur stations for civil defense communications during periods of local, regional, or national civil emergencies.

**Radiation:** Electromagnetic energy, such as radio waves, traveling forth into space from a transmitter.

**Radio Frequency (RF):** The range of frequencies over 20 kilohertz that can be propagated through space.

**Radio wave:** A combination of electric and magnetic fields varying at a radio frequency and traveling through space at the speed of light.

**Repeater operation:** Automatic amateur stations that retransmit the signals of other amateur stations.

**Repeater station:** An intermediate station in a system which is arranged to receive a signal from a station, amplify and retransmit the signal to another station. Usually performs this function in both directions, simultaneously.

**RST Report:** A telegraphy signal report system of Readability, Strength and Tone.

**S-meter:** A voltmeter calibrated from 0 to 9 that indicates the relative signal strength of an incoming signal at a radio receiver.

**Selectivity:** The ability of a circuit (or radio receiver) to separate the desired signal from those not wanted.

**Sensitivity:** The ability of a circuit (or radio receiver) to detect a specified input signal.

**Short circuit:** An unintended, low-resistance connection across a voltage source resulting in high current and possible damage.

**Shortwave:** The high frequencies that lie between 3 and 30 megahertz that are propagated long distances.

**Simplex:** A method of operation of a communication circuit which can receive or transmit, but not both simultaneously. Thus, system stations are operating on the same transmit and receive frequency.

**Single-Sideband (SSB):** A method of radio transmission in which the RF carrier and one of the sidebands is suppressed and all of the information is carried in the one remaining sideband.

**Skip wave, Skip zone:** A radio wave reflected back to earth. The distance between the radio transmitter and the site of a radio wave's return to earth.

**Sky wave:** A radio wave that is refracted back to earth in much the same way that a stone thrown across water skips out. Sometimes called an ionospheric wave.

**Spectrum:** A series of radiated energies arranged in order of wavelength. The radio spectrum extends from 20 kilohertz upward.

**Sporadic-E:** Radio propagation caused by refraction of signals by dense ionization at the E-layer of the ionosphere.

**Spread spectrum:** Modulation method which employs lightning-fast frequency "hopping" to allow multiple users to share a band.

**Spurious Emissions:** Unwanted radio frequency signals emitted from a transmitter that sometimes causes interference.

**Station license, location:** No transmitting station shall be operated in the amateur service without being licensed by the FCC. Each amateur station shall have one land location, the address of which appears on the station license.

**Sunspot Cycle:** An 11-year cycle of solar disturbances which greatly affects radio wave propagation.

**TAPR:** Tucson Amateur Packet Radio Corporation, a non-profit research and development organization. (8987-309 E. Tangue Verde Rd. #337, Tucson, AZ 85749-9399)

**Technician operator:** An amateur radio operator who has successfully passed Element 2. This operator has not passed Element 1, the 5-wpm code test.

**Technician with Code Credit:** An amateur operator who has passed a 5-wpm code test in addition to Technician Class requirements.

**Telecommunications:** The electrical conversion, switching, transmission and control of audio signals by wire or radio. Also includes video and data communications.

**Telegraphy:** Telegraphy is communications transmission and reception using CW International Morse Code.

**Telephony:** Telephony is communications transmission and reception in the voice mode.

**Temporary operating authority:** Authority to operate your amateur station while awaiting arrival of an upgraded license.

**Terrestrial station location:** Any location within the major portion of the Earth's atmosphere, including air, sea and land locations.

**Third-party traffic:** Amateur communication by or under the supervision of the control operator at an amateur station to another amateur station on behalf of others.

**Transceiver:** A combination radio transmitter and receiver.

**Transmatch:** An antenna tuner used to match the impedance of the transmitter output to the transmission line of an antenna.

**Transmitter:** Equipment used to generate radio waves. Most commonly, this radio carrier signal is amplitude or frequency modulated with information and radiated into space.

**Transmitter power:** The average peak envelope power (output) present at the antenna terminals of the transmitter. The term "transmitted" includes any external radio frequency power amplifier which may be used.

**Ultra High Frequency (UHF):** Ultra high frequency radio waves that are in the range of 300 to 3,000 MHz.

**Upper Sideband (USB):** The operating mode for sideband transmissions where the carrier and lower sideband are suppressed.

**Vanity call sign:** Custom-chosen call sign instead of "next-in-sequence" assignment from the FCC.

**Very High Frequency (VHF):** Very high frequency radio waves that are in the range of 30 to 300 MHz.

**Volunteer Examiner:** An amateur operator of at least a General Class level who administers or prepares amateur operator license examinations. A VE must be at least 18 years old and not related to the applicant.

**Volunteer Examiner Coordinator (VEC):** A member of an organization which has entered into an agreement with the FCC to coordinate the efforts of volunteer examiners in preparing and administering examinations for amateur operator licenses.

# Index